GENERAL ICHTHYOLOGY

THE AUTHOR

DATTA MUNSHI, Jyotiswarup (1930-), born 8 February, 1930, Ph.D Banaras Hindu University (1959), Professor and Head, Post-Graduate Department of Zoology, Bhagalpur University 1970-1992. Formerly, Lecturer in Zoology, Patna University (1952-62), Reader in Zoology, Banaras Hindu University (1962-70), Professor Datta Munshi continued to be associated with Bhagalpur University as CSIR Emeritus Scientist (1992-95) and Senior Scientist, INSA (1996-2000). Prof. Datta Munshi has been the recipient of British Council Visiting Lectureship (1971, 1977), INSA-Royal Society Exchange Programme Visiting Professorship (1985) at Bristol University, U.K., Max-Planck Institute Research Fellowship, Germany (1985), Visiting Scientist under INSA/DFG Germany Exchange Programme (1997), INSA/Polish exchange programme (1991), Rockefeller Foundation Programme (1989), Nuffield Foundation and Smithsonian Foundation Visiting Scientist (1984, 87, 89, 90, 94) at University of Notre Dame, U.S.A.

Prof. Datta Munshi's outstanding achievements, have earned him several honours such as Patna University Gold Medal (1952), S.L. Hora Gold Medal (1980), INSA, Chandrakala Hora Memorial Medal (1992), Golden Jubilee Medal of Zoological Society, Calcutta (1997), 20th Century Gold Medal of Zoological Society of India, (1998). Prof. Datta Munshi is a Fellow of Indian National Science Academy (FNA), National Academy of Sciences (FNASc) and Zoological Society of India (FZA).

Societies: Fellow of National Academy of Sciences, India; Fellow of the Indian National Science Academy (1980-); member, Ichthyological Society of Japan and Society for Experimental Biology, London; Life Member, Indian Science Congress Association.

Award and Honours: Patna University Gold Medal (1952), S.L. Hora Gold Medal (1980), Indian Society of Ichthyologists, Member, INSA Council; Bihar Gaurav Puraskar 2007, (2007).

Professor Datta Munshi has contributed significantly to the understanding of the structure, function and evolution of the air-breathing organs of teleostean fishes. His morphometric analyses of the bimodal respiratory system have indicated how changes in their relative development can be correlated with changes in the life-pattern of the fish, and seasonal characteristics of the environment.

Address: P-124/A, Usha Park, Garia, Kolkata – 700 084.

GENERAL ICHTHYOLOGY

Prof. Jyotiswarup Datta Munshi

P-124/A, Usha Park
Garia
Kolkata – 700 084

2014

Daya Publishing House®

A Division of

Astral International Pvt. Ltd.

New Delhi – 110 002

Published by : **Daya Publishing House®**
A Division of
Astral International Pvt. Ltd.
– ISO 9001:2008 Certified Company –
4760-61/23, Ansari Road, Darya Ganj
New Delhi-110 002
Ph. 011-43549197, 23278134
E-mail: info@astralint.com
Website: www.astralint.com

Laser Typesetting : **Classic Computer Services**, Delhi - 110 035

Printed at : **Replika Press Pvt. Ltd.**

PRINTED IN INDIA

Acknowledgement

I am indebted to a number of International Collaborators and National Scientists A.B. Mishra, Professor and Head of Department of Zoology, Banaras Hindu University and Research Scholars for carrying out research work with me on different aspects of structure, function, evolution, eco-physiology and phylogeny of *fishes of India*.

International Collaborators

1. Professor George Morgan Hughes, Research Unit for Comparative Animal Respiration, Bristol University, England, U.K.
2. Professor Ewald R. Weibel, Anatomical Institute, University of Berne, Switzerland.
3. Professor Kenneth R. Olson, Indiana University School of Medicine, South Bend Centre, University of Notre Dame, Notre Dame, Indiana – 46556, U.S.A.
4. Professor Hiran M Dutta, Kent State University, Ohio, U.S.A
5. Professor Peter Gehr, Anatomical Institute, University of Bern, Switzerland.
6. Professor Johannes Piiper, Max-Planck-Instut fur experimentelle Medizin, Abteilung Physiologie, Gottingen University, Germany.
7. Professor Pierre Dejours, Laboratoire d'etude des régulations physiologiques (associéa I' université Louis Pasteur) Centre national de la recherché scientifique, France.
8. Professor Giacomo Zaccone, Department of Animal Biology and Marine Ecology, Faculty of Science, University of Messina I-91866, Messina, Italy.

Research Fellows

Bali Ram Singh, Bans Narain Singh, Satyendra Prasan Singh, Ajay Kumar Mittal, Subhas Chandra Dube, Ram Kumar Singh, Jagdish Ojha, Bhupendra Narayan Pandey, Syed Aftab Khalid Nasar, Mahadeo Prasad Saha, Narendra Deo Prasad Sinha, Narendra Mishra, Abdul Hakim, Asha Lata Sinha, Ajay Kumar Patra, Niva Biswas, Devendra Prasad Choudhary, Dayanand Roy, Surendra Prasad Roy, M.A.O. Johar, Prem Kumar Verma, Shobha Sah, Arun Kumar Laal, Upendra Prasad Sharma, Shashi Kant Sinha, Onkar Nath Singh, Ragini, Amita Moitra, Gopal Krishna Kunwar, Prabhat Kumar Roy, Tapan Kumar Ghosh, Sneh Prabha Jha, Mansa Prasad Srivastava, Swapna Chowdhury, Asha Pandey, Anita Pandey, Dhrub Kumar Singh, Syed Shans Tabrez Nasar, Alakhnanda Singh, Manish Chandra Verma, Soma Adhikari, Lalan Kumar Choudhary, Suhasini Besra, Utpala Ghosh, Manoj Kumar Sinha.

I am also grateful to Mrs.Ruma Ghosh Dastidar for her help with the word processor in preparing the final manuscript.

Most of the scientific diagrams illustrated in the book have been drawn by Shri Tribhuwan Poddar.

Jyotiswarup Datta Munshi

Preface

Professor Julian Huxley, M.A., D.Sc, F.R.S. visited Patna University in 1952 and delivered a Lecture in the Senate Hall on "Evolution" which I attended. He also visited our new Department of Zoology, Patna University and we had a group photograph with him. His book on "Evolution in Action" was published in 1953 by Chatto an Hindus, London, which inspired me to write down the book on 'The Life of Air-breathing Fishes' in 2004 Narendra Publishing House in 2009.

Phyletic studies of Teleostean Fishes, with a provisional classification of Living Forms by P. Humphry Greenwood of Department of Zoology, British Museum (Natural History), Down E. Rosen, Department of Ichthyology, The American Museum of Natural History, Stanley H. Weitzman, Division of Fishes, U.S.A., Smithsonian Institution and George S. Myers, Professor of Biology Stanford University, who also visited Zoology Department of Calcutta University on February 26, 2005. I had the privilege to meet P. Humphry Greenwood in England when I visited Department of Zoology, British Museum (Natural History).

It was apparent from what was heard during the satellite Symposium held at Bhagalpur in 1974 and the others at Gottingen in Germany, Strusburg in France and Berne in 1977 under the auspices of XXVII International Congress of Physiological Science held in New Delhi and Paris in France that a tremendous amounts still remains to be done.

Professor G.M. Hughes (1976) has rightly pointed out that the application of methods and approaches developed during the last 20 years as a result of investigations by Scientists with backgrounds in mammalian physiology as well as those based on more classical aspects of Zoology has been very fruitful and it is clear that much more can be done with the fishes. Already several techniques used with water-breathing fishes have begun to be applied to air-breathing Fishes.

The particular techniques which Professor Hughes first used in comparative studies with fish was to record pressure changes within the buccal and opercular cavities (Hughes, 1960). This was applied for the first time by him on two species of Indian air-breathing fishes *Heteropneustes fossilis* and *Anabas testudineus*. It became clear that there must be a whole range of variations in the relative importance of these two pumps among air-breathing fishes.

Though morphological details are available the actual mechanics of the ventilatory movements needs further elucidation as they are quite complicated as studied by the cine-photography by Hughes (1976) and X-ray cine photography by H. Peters.

The value of morphometric analysis of the respiratory system has already been indicated in these fish and it is remarkable how changes in relative development of the respiratory organs can be correlated with changes in life habits of the fish, which in turn are related to the ecology and seasonal characteristics of the environment. But much more still remains to be done using more sophisticated analysis of these gill systems. It must not be imagined, that the full informational content of teleostean morphology has been extracted. Only the barest beginnings have so far been accomplished, even within the realm of osteology. Researches on the nervous, digestive, muscular, respiratory and vascular systems of teleosts are scattered and mostly uncoordinated and relatively few of them have been done with any specific objectives in view (Greenwood, Rosen, Weitznan and Myers, 1966). No single change in lifestyle has had more important effects on vertebrate design than the transition from life in water to life on land. This transition involved evolution of features to adapt to the new conditions including effects of gravity, a dehydrating environment and the change from aquatic to aerial respiration. Amongst the vertebrates air-breathing has evolved on several occasions. Evolution of the first bony fishes during the late Silurian and Devonian produced two new features, lobe fins and the lung. These features occurred in the fleshy finned fish (sarcoptergyii) at about the same period when the first amphibians arose from some similar line. The earliest amphibians were probably aquatic and fish like, but since they possessed lungs they could escape to the land so avoiding predators and exploit new food sources. The need to abandon the aquatic environment for the land has frequently occurred during the evolution of teleost fish, accompanied by modifications of skin, gut, opercular chamber, swimbladder etc.

A satellite symposium held at the Post-graduate Department of Zoology, Bhagalpur University, Bihar India, October 28th to November 1st, 1974 following the XXVI International Congress of Physiological Science at New Delhi. It was the second satellite symposium concerned with comparative physiology of respiration, the first having been held at Gottingen in 1971. On this occasion a topic was chosen which was particularly relevant to the local conditions in India and especially with the work done at Professor Munshi's department. Indian air-breathing fishes have long attracted the study of biologists but it is only in recent years that modern physiology and morphometric methods have been applied to them. By chosing a topic broader than air-breathing fish, an attempt was made to place these particular animals in perspective and in relation to current general problems of respiratory physiology.

Research has developed quite rapidly in this field during the past ten years and the papers presented summarized results obtained during this period.

Invitations to speakers were given so as to bring together scientists not only with specialist interest in respiratory physiology, but also biologists more concerned with the ecological conditions, morphology and ecophysiology of vertebrates which can breathe both in water and air. The meeting provided many lively discussions but perhaps the success of the symposium will be best judged by a future assessment of the stimulus that it provided not only to biologists in India but to those in other parts of the world where animals await investigation using comparable techniques.

I should like to thank all those people who helped in the conception and planning of the symposium and all who gave their time and energies most willingly at the time of the meeting at Bhagalpur. In particular, however, I am indated to Johannes Piiper and Hermann Rahn for their constant support during the planning stages. At Bhagalpur, Professor Munshi performed many miracles by completing arrangements for the first ever international symposium to be held in that city and I know that he has felt most strongly a great indebtedness to his Vice-Chancellor and the Deputy Minister of Education for their unfailing support. To all our India hosts, the visitors from overseas will feel indebted for the great efforts they made and which contributed so greatly to the success of a meeting that will always be remembered.

Animals inhabiting the water/air interface show many morphological and physiological adaptations designed to function in that unique environment, through in the present age, there are new challenges. These include climatic change and contamination of the environment through anthropogenic activities.

This volume addresses fundamental biological, morphological and functional aspects of vertebrates at the water/air interface, and then proceeds to see how these systems react and respond to change. The principal topics are :

1. Physico-chemical diversity of water.
2. Physiological adjustments required for maintaining homestasis under changing environmental conditions.
   (*a*) Respiratory responses
   (*b*) Cardio-vascular responses
   (*c*) Osmoregulation
   (*d*) Acid-base regulation.
3. Environmental pollution
4. Mechanisms of detoxiplication

This volume discusses the application of methods and approaches developed during the last twenty years in the field of eco-physiology of animals involved in water/air transitions. It is only by the adoption of such comparative approaches, one can develop new suitable techniques to understand the eco-physiology of water-breathing vertebrates which were caught in the air/water interface of inland waters. The aspects of eco-physiology will encompass the respiratory and circulatory systems, exchange and metabolism, ion and osmoregulation in animals. This volume also examines the transition from water to land environment in the context of palaleo-ecological conditions of the environment. In recent years there have been many studies on pollutants affecting our life through our freshwater systems. Most of the animals like fishes exhibit two phases in sublethal exposures to xenobiotics – the adaptive and chronic recuperative. The mechanism involved in such changes were discussed at length in the International Symposium, "Water/Air Transitions in Biology" held at Banaras Hindu University, Varanasi, India in February, 1996.

The present book on General Ichthyology of Teleostean/Air-breathing Fishes have been incorporated.

Jyotiswarup Datta Munshi

Contents

Chapter 1
Evolutionary Biology

The Process of Evolution

Science has two functions: Control and Comprehension. The comprehension may be of the universe in which we live; or of ourselves; or of the relations between ourselves and our world. Evolutionary science has only been in existence, as a special branch of scientific knowledge, for less than a century. During that time its primary contribution has been to comprehension first to that of the world around us, and then to that of our own nature. The last few decades have added an increasing comprehension of our position in the universe and our relations with it; and with this, evolutionary science is certainly destined to make an important and increasing contribution to control, its practical application in the affairs of human life is about to begin.

Evolutionary science is a discipline or subject in its own right. But it is the joint product of a number of separate branches of study and learning. Biology provides its central and largest component, but it has also received indispensable contributions from pure physics and chemistry, cosmogony and geology among the natural sciences, and among human studies from history and social science, archeology and prehistory, psychology and anthropology. As a result, the present is the first period in which we have been able to grasp that the universe is a process in time and to get a first glimpse of our true relation with it. We can see ourselves as history, and can see that history in its proper relation with the history of the universe as a whole.

All phenomena have a historical aspect. From the condensation of nebulae to the development of the infant in the womb from the formation of the earth as a planet to the making of a political decision, they are all processes in time; and they are all interrelated as partial processes within the single universal process of reality. All reality, in fact, is evolution, in the perfectly proper sense that it is a one-way process in time; unitary; continuous; irreversible; self-transforming; and generating variety and novelty during its transformations.

The inorganic sector is being dealt with extremely briefly. For further details the standard works like those of Sir James Jeans or Sir Arthur Eddington, or the more recent picture so vividly sketched by Fred Hoyle in his little book "*The Nature of the Universe*" should be studied. The chief points which have a bearing on evolution seem to be these. This sector of reality comprises all the purely physico-chemical aspects of the universe throughout the whole of space, intergalactic as well as interstellar, all the galaxies, all the stars and stellar nebulae. The diameter of that part of it visible with the new 200 inch teleoscope is nearly a thousand million light years; and there is a celestial region of unknown size beyond the range of any teleoscope that we may ever be able to construct. There are over a hundred million visible galaxies; and each of these contains anything from a hundred to ten thousands million stars. Obviously, then the inorganic sector is by far the largest in spatial extent. It is also the largest in temporal extent; astronomers put the age of our own galaxy at up to five thousand million years—probably rather less and most of them think the universe as a whole is of about the same age, though some believe it is considerably older.

But the mechanism of its transformation is of the simplest kind—physical and very occasionally chemical interaction. The degree of organization to be found in it is correspondingly simple : most of this vast sector consists of nothing but radiations, subatomic particles and atoms only here and there in it is matter able to attain the molecular level, and nowhere are its molecules at all large or complicated. Very few of them contain more than half a dozen atoms, as opposed to the many thousands of atoms in the more complex organic molecules found in living substance.

The spatial extension of the biological sector is very much restricted. Living substance could not come into being except in that small minority of stars which have produced planetary systems. Within them, it is restricted to that small minority of their planets which are of the right size and in the right stage of their history for complicated self-copying organic molecules to be produced; and in them again to an infinitesimal surface shell. The number of such potential homes of life in our own galaxy is put by a few astronomers as big as a hundred thousand, but by most at only a few thousand or even a few hundred. Whatever the truth turns out to be, the biological sector, considered spatially as the area occupied by life, cannot at the very outside constitute more than a million-million-millionth part of the extent of the visible universe and probably much less, and of course the only spot of which we have actual knowledge is our own planet, with the possibility of Mars in addition. On the earth the extension of the biological sector in time appears to be about two thousand million years.

On the other hand, the level of organization reached is almost infinitely greater than in the preceding sector. The proteins, the most essential chemical constituents of living substance, have molecules with tens or even hundreds of thousands of atoms, all arranged in patterns characteristic for each kind of protein. Each single tiny cell has a highly complex organization of its own, with a nucleus, chromosomes and genes and other cell-organs, and is built out of a number of different kinds of proteins and other types of chemical units, mostly large and complex. But that is only the beginning, for large higher mammals such as men and whales may have in their bodies over a hundred million million or even over a thousand million million cells of many different types and organized in the most elaborate patterns. As Professor J.Z. Young has set forth in his book, *Doubt and certainty in Science*, the number of cells in our "thinking parts" alone – the cerebral cortex of our brain – is about seven times the total human population of the world, and their organization is of a scarcely conceivable complexity.

Evolutionary transformation in this sector is brought about by the wholly new method of natural selection, which was not available during the thousands of millions of years before the emergence of living substance. This new method is responsible for the much higher level of organization which evolution here produces, as well as the greater variety of organization. It is also responsible for the much faster tempo of change; quite large changes in biological organization take only a few tens of millions of years; and really major ones, much more radical than any which can have occurred during the entire inorganic phase, only a hundred million or so.

At first sight the biological sector seems full of purpose. Organisms are built as if purposefully designed, and work as if in purposeful pursuit of a conscious aim.

But the truth lies in those two words "as if". As the genius of Darwin showed, the purpose is only an apparent one. However, this at least implies prospective significance. Natural selection operates in relation to the future – the future survival of the individual and the species, in the shape of actual animals and plants which are correspondingly oriented towards the future, in their structure, their mode of working, and their behaviour. A few of the later products of evolution, notably the higher mammals, do show true purpose, in the sense of the awareness of a goal. But the purpose is confined to individuals and their actions. It does not enter into the basic machinery of the evolutionary process, although it helps the realization of its results. Evolution in the biological phase is still impelled from behind; but the process is now structured so as to be directed forwards.

The human phase of evolution, Prof. Julian Huxlay (1953) have called as the psycho-social sector which is again enormously more limited in spatial extent. On this earth it is restricted to one among over a million species of organisms; elsewhere anything that could be called a psycho-social sector assuredly cannot have been attained in more than a very small fraction-perhaps a hundredth, perhaps only a ten thousand of the planets habitable by some kind of life. It is still more limited in its temporal extent; its existence on this earth, from its first dim dawn to the present, occupies only one-half of one tenth per cent of the history of life as a whole; and it has only operated at anything like full swing for perhaps a tenth of that tiny fraction of time.

Once again, a new main method of transformation has become available in this sector – the method of cumulative experience combined with conscious purpose. This has produced a new kind of result, in the shape of transmissible cultures; the main unit of evolution in the human phase is not the biological species, but the stream of culture, and genetic advance has taken a back seat as compared with changes in the transmissible techniques of cultural advance – arts and skills, moral codes and religious beliefs, and above all knowledge and ideas. It has also meant not only a more rapid tempo, but a new kind of tempo – an acceleration instead of a more or less steady average rate over long periods. In the long prologue of human evolution, each major change demanded something of the order of a hundred thousand years; immediately after the end of the Ice Age, something like a thousand years; during much of recorded history, the time-unit of major change was around a century; while recently it has been reduced to a decade or so. And again correlated with this increased tempo of change, we find an enormous increase in the variety of the results produced and in the levels of organization attained. In a way most important, purpose has now entered the process of transformation itself; both the mechanisms of psychosocial evolution and its products have a truly purposeful component, and evolution in this sector is pulled on consciously from in front as well as being impelled blindly from behind.

Then follows the purely biological problem. Description and definition are the first steps in science, the relevant features of the biological phase of reality that we have reached today. At the present time, over a million species of animals have been described, and well over a third of that number of plants, all different and distinct, while every year several thousand new species are being discovered and given names. They extend into every nook and cranny of the environment possible to life, from the polar regions to the equator, from the high mountains to the black marine abyss, from *hot springs* not much below boiling point to the oxygenless interiors of other animals. They exploit their environment in every possible way. To take only animals, there are species which feed entirely on flesh, on wood, on excrement, on nectar, on features, on the contents of others' intestines, on one particular kind of fruit or leaf. And each and every species is adapted, often in the most astonishing fashion, to its environment and its way of life. Think of the duck's webbed feet, the camel's stomach, or the luminous organs of deep-sea fish. There is no need to multiply examples; every animal and plant is from one aspect an organized bundle of adaptations of structure, physiology and behaviour; and the organization of the whole bundle is itself an adaptation.

Living things fall naturally into a number of groups, each with its own plan of structure and working with striking variations. The first grouping is into animals, plants and viruses, each characterized by a radically different chemical way of life. The first thing that strikes one about the animal group is the great variety of plan of construction and operation within it. Thus, the protozoa are all single celled, or, more accurately, are all based on the single cell as unit: the sponges are all mouthless filter-feeders: the sea-anemones and jellyfish and their relatives do possess a mouth, but have no head and are built on a radial plan without distinction of ventral or dorsal. There are the echinoderms, like starfish and sea-urchins, which have secondarily lost their bilateral symmetry for a radial construction. There are various types of worms; and the great group of mollusks, with clams, snails and cuttlefish. The two highest groups are the arthropods – insects, crustaceans and spiders – and the vertebrates, from fish to men. They both have an elaborate organization, with head, limbs, eyes, heart and brain. But while the arthropods have many limbs and have a dead horny external skeleton for their mechanical framework, the vertebrates, when they once develop limbs, have two pairs only, and their skeleton is a living tissue, of cartilage or bone, and in higher forms is entirely internal, leaving the surface of the body sensitive and free.

The plants are perhaps less varied. Certainly they never reach the complexity of some animals. They divide first into the minute bacteria; the true fungi-moulds and toadstools – that require organic compounds; and the green plants that need only simple inorganic compounds and light. Among the green plants, the algae are confined to water, and lack differentiated leaves, roots and flowers. The primitive lands forms from mosses upto ferns, have no needs and reproduce by microscopic single-celled spores; and the higher land forms, increasingly adapted to land life – including all the familiar trees and flowers are all seed plants, with seeds containing an embryo and a store of nutrient. This will at least serve as a reminder of the range of design to be found among the machines for living that we call organisms.

Organisms differ from man's machines in being able to construct themselves. In constructing itself, every organism goes through a process of individual development – what is technically called its ontogeny. In primitive forms, this may involve merely minor or negligible changes, as when a newly formed amoeba grows to double size before dividing into two. But in all higher animals and plants, ontogeny is a very elaborate process, and the developing organism passes through a whole series of transformations, surprisingly different in appearance and in mode of working. Every butterfly was once a caterpillar; every oak once an acorn; every barnacle once a tiny free-swimming crustacean. You, like me and every other human being, were once a microscopic spherical ovum, then in turn a double sheet of undifferentiated cells, an embryo with enormous outgrowths enabling you to obtain food and oxygen parasitically from your mother, a creature with an un-jointed rod what biologists call the notochord in place of jointed backbone; you once had gill-clefts like a fish, you once had a tail, and once were covered with dense hair like a monkey; you were once a helpless infant which had to learn to distinguish objects and to talk; you underwent the transformation of your body and mind that we call puberty; you learned a job. You are in fact a self-transforming process.

Ontogony is thus a pattern of processes in time, through which the inherent potentialities of the individual can be realized. Unfavourable conditions may prevent their full realization, or indeed, by killing the unfinished individual, prevent any realization at all, but in favourable conditions development proceeds freely to the bounds set by its inherent possibilities. Not only that, but in every generation ontogeny is the necessary mechanism for realizing the potentialities of heredity; any new transformation must operate through the framework of development processes which are available. Organisms tend to resemble each other more in the earlier than in the later stages of ontogeny. The vast majority of individual plants and animals, however different when adult, resemble each other at the start of their ontogeny by consisting of a single un-differented cell; all embryonic vertebrates – birds and mammals, fish and reptiles – look remarkably alike when in the notochord phase of their development; and anyone but a specialist would be hard put to it to tell rabbits from men, giraffes, or whales even in much later embryonic stages.

Sometimes the revelation is spectacular, Sea-squirts (Ascidians) are fixed sessile filter-feeders, with two siphons for taking in an expelling water. They lack nerve-cord and sense-organs, and were for a time classed with clams and similar mollusks. Then, in the 1870's, the Russian Zoologist Kovalevsky discovered their larva-a little free-swimming creature with tail and notochord-rod (precursor of the backbone), a nerve-cord along the back, primitive hollow brain-vesicle, eye-spots, gill-clefts; its ground-plan is revealed as similar to that of an embryonic vertebrate, but with no resemblance to the ground-plan of any other group.

Striking evolutionary effects can be expected when the general tempo of differentiation is speeded up or slowed down, or one phase of development prolonged into another, as Dr. de Beer has shown in his book *Embryos and Ancestors*. Thus the slowing down of the thyroid gland's growth, and the delaying of the moment when it liberates its secretion, has given us the Mexican axolotl. This is a salamander tadpole which never metamorphoses into a land animal, but becomes sexually mature in the gill-breathing larval form. A simple meal of thyroid will reconvert it into a land salamander.

It has been suggested with some plausibility that insects were derived from myriapods by a somewhat similar process. The newly hatched myriapod has only three pairs of legs instead of a whole series all down the body. If this stage were prolonged until sexual maturity, the adult "thousand-leg" phase would drop out of existence, and you would have a creature very like a primitive insect.

Chapter 2

Origin of Fishes

Origin of Chordates and Fishes-Protostomia-Deuterostomia-Spiral Cleavage-Origin of Coelom-Evolutionary relationship larvae of a Deuterostoma, mouth, larvae and the origin of chordates and vertebrates-Phylogenetic significance of Echinoderm larvae-Evolution of different groups of Echinoderms and their relationship with Hemichordates-Protochordate-Ascidian tadpole-Evolution of Protovertebrates from Ascidian stage of Urochordate-Amphioxus Larvae-Palaeontological Evidences-Physiological Evidence-Life History of Lamprey (Cyclostomes)-Urinogenital Tubules of Vertebrates-Osmoregulation in Cyclostomes-Development of Glomerular Kidney-Mechanism of Osmoregulation in Lampreys-A Comparative Accounts of Water and Salt Regulation in the vertebrates-Freshwater fishes-Marine fishes-Osmoregulation in Osteichthys and Chondrichthys-Osmoregulation in Freshwater and Marine teleosts-Branchial glands and Osmoregulation – Characteristics of Chloride cells-Effects of Marine Habitat upon the Kidney-Osmoregulation in Chondrichthys-Salient Points-Paraneurone cell system of skin and gills for fishes.

Introduction

Fishes may have been in existence since the beginning of the Ordovician period about 400 million years ago. It is difficult to obtain direct proof of their origin and nature of the environment in which they flourished. The problem of freshwater or marine origin of fishes still remains in the realm of probability, and has not yet been settled. Different interpretations have been given of the palaentological facts, but strong inferences have been drawn from morphological and physiological studies of the present day vertebrates in favour of their freshwater origin.

Romer (1946) studied the North American fossil vertebrates in relation to the sediments in which they were found and argued in favour of freshwater origin of vertebrates. Watson (1925, 1934) regarded the freshwater or marine origin of fishes still as a debatable question and has shown that both fresh and sea waters had representatives of the ostracoderm group.

1. Origin of Chordate and Fishes

Grarstang (1894) was the first zoologist to trace the origin and evolution of vertebrates from early youthful stages of invertebrates (Figures 2.1–2.3).

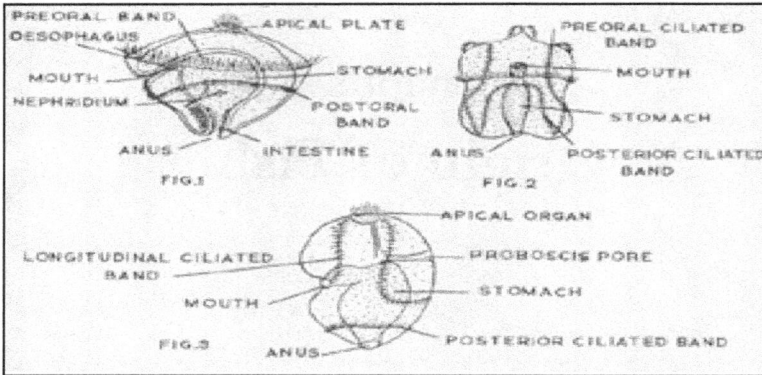

Figure 2.1: Trochosphore larva of Annelida
Figure 2.2: Auricularia larva of Echinodermata
Figure 2.3: Ternaria Larva of Balanoglossus

He focused our attention on the structure of the larvae of the echinoderms and suggested that the vertebrates have evolved through neoteny and paedomorphosis by retaining certain features.

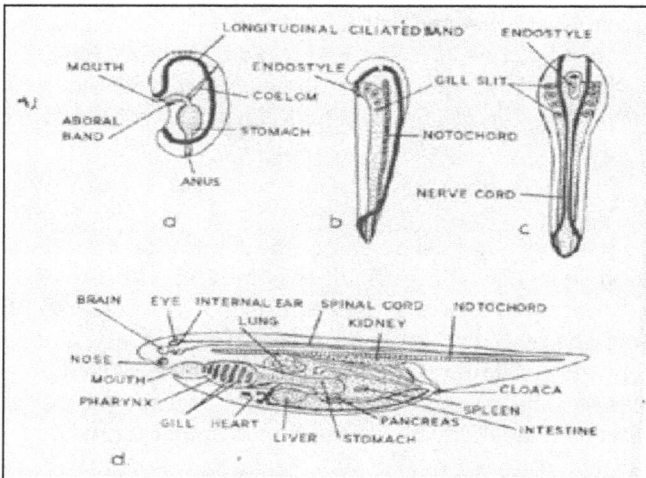

Figure 2.4(a-d): Diagram depicting Garstang's auricularia theory which speculates the steps by which protochordate and then a protovertebrate animal might have been derived from an echinoderm larva a) Lateral view of auricularia b) Protochordate – side view c) Protochordate dorsal view d) Sagital section of a protovertebrate.

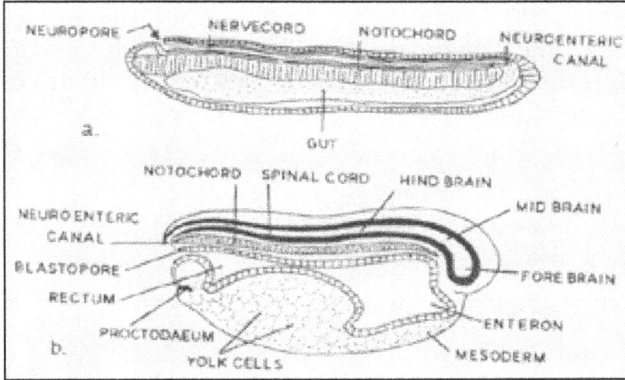

Figure 2.5: (a) Young amphooxus shortly after hatching showing neuropore – nerve cord, neuroenteric canal, notochord and gut; (b) Early tadpole of frog showing development of notochord, nervous system and its connection with the enteric system through neuro-enteric canal.

Gastang's auricularia theory which speculates the steps by which protochordate and then a protovertebrate animal might have been derived from an echinoderm larva.

Perhaps in the beginning the tube use to carry and convey the food particles into the alimentary system. This explain the formation and existence of the neuroenteric canal in the developmental stages of *Amphioxus*. As the larva grows and body becomes elongated it required support. The notochord developed as a row of vacuolated cells filled up with fluid from the dorsal side of the alimentary canal to form a sort of axial hydroskeleton for support. Further, when the body mass increased the animal had to change from its ciliary mode of feeding to the pharyngeal one. The adoral band of cilia of *auricularia* which probably served to carry food particles (Plankton) into the mouth, was eventually incorporated into the floor of the pharynx to form the endostyle. The pharynx got perforated and developed into a large food (Plankton) trapping device. The pharyngeal slits were lined by ciliated cells and eventually developed into gills for gaseous exchange. Thus all the three main characteristic features of the chordate, *viz*. (*i*) the dorso-tubular nervous system, (*ii*) the notochord and (*iii*) the gill slits developed.

Affinities between Two Main Groups of Triploblastic Animals

In general, two main streams of triploblastic animals have been recognized, *viz*.:

☆ *Protostomia*: Annelid superphylum comprising the annelids, molluscs and arthropods;

☆ *Deuterostomia*: Echinoderm superphylum comprising the echinoderms, the protochordates, higher chordates and vertebrates.

Protostomia – First Mouth

Coelomates with a true coelom, having spiral cleavage of the egg, with the blastopore forming mouth, or mouth and anus, and having a trochophore as a larva.

Deuterostomia – Second Mouth

Coelomates having a true coelom, with radial cleavage of the egg, the blastopore becoming the anus, the coelom being formed by enterocoel(=the process of making an enterocoel) and a dipleura larva.

Based mainly on details from embryology, these two groups differ from each other in the following basic characters:

1. *Cleavage*: It is radial in echinoderms and chordates but spiral in annelid group of animals and molluscs;

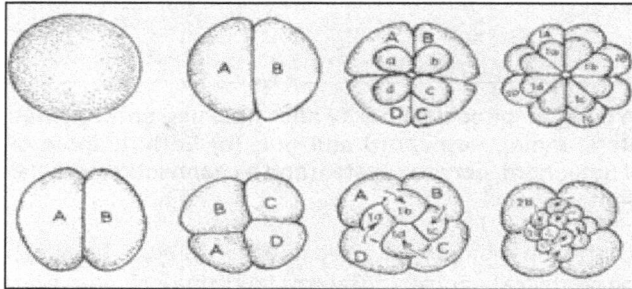

Figure 2.6: (a) Radial cleavage of the egg, characteristics of echinoderm and chordates; (b) Special cleavage of the egg, characteristic of flatwork and annelids.

2. *Origin of Coelom:* In the annelid group of animals the coelom develops as a schizocoele by splitting of the mesoderm, whereas in echinoderms and chordates the coelom develops as enterocoelic pouches from the alimentary tract. In the vertebrates, the enterocoelic origin of body cavity is, however, not clearly traceable.

3. *Larval forms*: Trochosphere type of larvae having ciliated bands around their bodies are found in annelids. In echinoderms, on the other hand, the auricularia or bipinnaria have longitudinal and adoral bands of cilia. The tornaria of *Balangoglossus* resembles the larvae of echinoderms.

4. *Potency of eggs*: The annelids have mosaic type of eggs in which if a part of the blastula is removed, the adult develops without certain part. In contrast, the ehinoderms and chordates have equipotential eggs, where if some cells are removed at an early stage, it can compensate the loss and develop into a complete adult.

5. *Nervous system*: In echinoderms and chordates the central nervous system develops from the ectoderm by invagination, whereas in annelids it develops by delamination.

6. *Muscle biochemistry*: Energy-rich phosphate compounds are, as a rule, linked with muscle activity in animals. Fish muscles have creatine as the phosphate material in place of arginine found in invertebrates. Some echinoderms and hemichordates have both creatine and arginine.

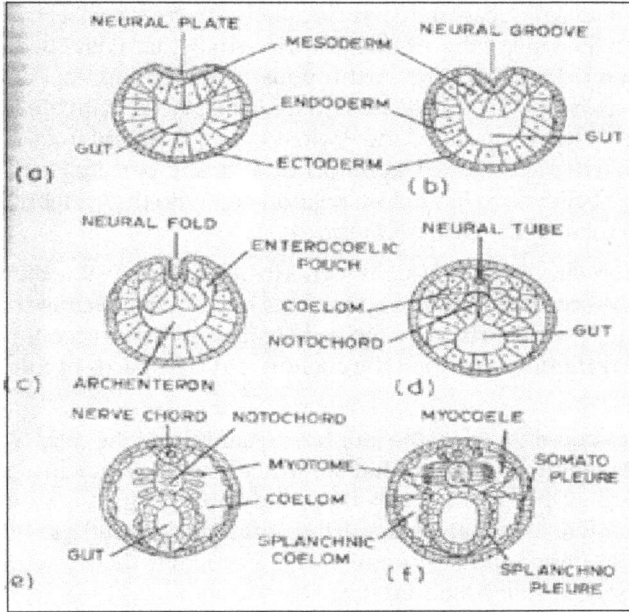

Figure 2.7: (a-f) Enterocoelic mode of formation of coelom in echinoderm and chordates. Also showing the formation of neural tube, development of notochord and myotome.

Consideration of these facts indicate a close relationship between the two groups – the echinoderms and the chordates (Datta Munshi and Srivastava, 1988, 2002).

2. Significance of the Coelom

This can be defined as a cavity that arises within the endomesoderm, covered on its outer surface by the somatic mesoderm and on its inner surface by the splanchnic mesoderm. It contains a fluid, the coelomic fluid, and it is lined by an epithelium, the coelomic epithelium or peritoneum.

It is usually referred to as the secondary body cavity, because it is preceded in development by the primary body cavity or blasotocoel.

2.1 The question of the evolutionary origin of the coelom has been much debated, but there is a lack of definite information, and conclusions can only be reached by inference. Four main therories have been proposed; i) that it originates from outgrowths of the alimentary canal (enterocoel theory), ii) from the enlarged cavities of nephridia (nephrocoel theory), iii) from splits in the mesoderm (schizocoel theory), or iv) from the enlarged cavities of gonads (gonocoel theory).

It was first proposed by Lankester in 1874 at a time when the relationship of the coelom with nephridia and coelomoducts had still to be clarified; it has since found few supporters. The schizocoel and enterocoel theories have been prologued on the basis of two methods of development of the coelom, which may arise either by splitting of the mesoderm, in which case the cavity is called a (i) schizocoel, or by the evagination of pouches from the wall of the archenteron, in which case it is called an *enterocoel*. A schizocoel is found in the Annelids, Arthropods, Molluscs and Pogonophora – groups that are believed to be closely related and that can be placed, with the Platyhelminthes and Nemertines, in a group called the Protostomia. (ii) An enterocoel is found in the Echinodermata, Hemichordata, Cephalochordata, and perhaps in the Urochordata – groups that are believed to have close relationship with the vertebrates and that can be placed, with them in the *Deuterostomia*.

The *schizocoel theory* is at first sight seems to be more plausible than the preceding theories, yet it has received little support. This is probably because of its essentially indefinite character, and particularly because of its inability to account for the clearcut morphological difference between the coelom and the spaces of the blood vascular system.

The most favoured theory of the four has certainly been the *gonocoel theory*. Among the points in its favour is the fact that some acelomate nemertines pass through a stage in which they possess a series of gonads with large and sometimes empty cavities, it is possible to visualize the further enlargement of these into the spacious coelomic cavities of annelids. Moreover, this particular theory would account for gonads commonly developing from the coelomic wall.

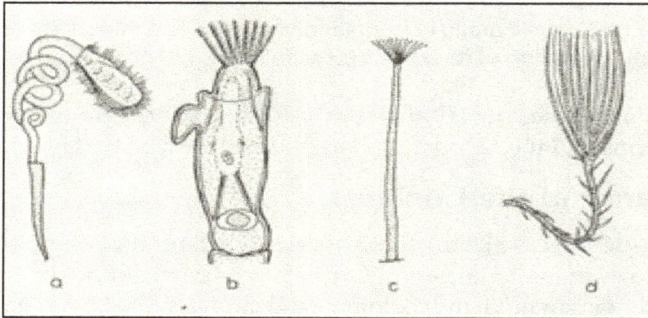

Figure 2.8: (a) Lingula (Brachinopoda); (b) Bugula (Bryozoa); (c) Phoronis (Phormida); (d) Metacrinus (Echinoderms).

Coelom with its contained fluid, provides the structural basis for a hydrostatic skeleton more highly organized than that of Coelenterates and Platyhelminths. In this respect it probably played a major part in ensuring the successful survival of the lower coelomates during the period when calcium and phosphorus metabolism had not been elaborated to a point at which rigid and jointed skeletons could be constructed. The movements of annelid worms, echinoderms, and many of the smaller groups, are based upon hydrostatic principles in essentially the same way as are the movements of coelenterates.

3. Evolution of Vertebrates

The palaeozoic ancestors of Echinoderms – the Cystoids, Blastoids, Crinoids and their surviving related animals like Polyzoa, Ectoprocta and Phoronids have the following three main features.

1. Ciliary mode of feeding by lophophore – like structures, bearing tentacles;
2. Sedentary, often stalked;
3. A larval stage in the life cycle – with their characteristic longitudinal bands.

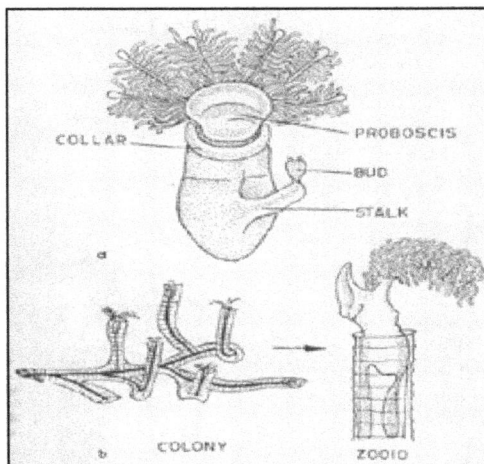

Figure 2.9: Living genera of Pterobranchia: (a) Cephalodiscus (Pterobranchia); (b) Rhabdopleura.

1. The Pterobranchs (*Hemichordata*), represented by the living genera *Cephalodiscus, Rhabdopleura* and *Atubaria* possess both lophophore and gill slits. This is a stage where the pharyngeal mechanism got gradually substituted for the ciliary mode of feeding by tentacle bearing lophophore. The adoral band of cilia of the *auricularia* which served the function of carrying food into the mouth became incorporated into the pharynx to form the endostyle. The pharynx became perforated so that a large volume of water may filter through it. Atrium developed to protect the perforated pharynx.

2. In the second stage of the evolutionary process the tunicates (Ascidians), in which the branchial pharyngeal feeding is fully replaced by the tentacle-feeding of sessile adults.

   If the adult tunicate has evolved from a modified lophophore-feeding animals, then how the ascidian tadpole has developed from the auricularia type of larva? The Garstang's auricularia theory again provides the answer. Garstang suggested that the fish-like form arose by development of muscle cells along the sides of the elongated body of the ascidian tadpole. We may regard the adult tunicate organization as being directly derived from that of a sessile lophophore-feeding creature, and the larval organization from an echinoderm-like larva.

3. At the last stage of vertebrate evolution, the protovertebrates were obliged to get rid of the tunicate stage of sedentary life from their life history. It is

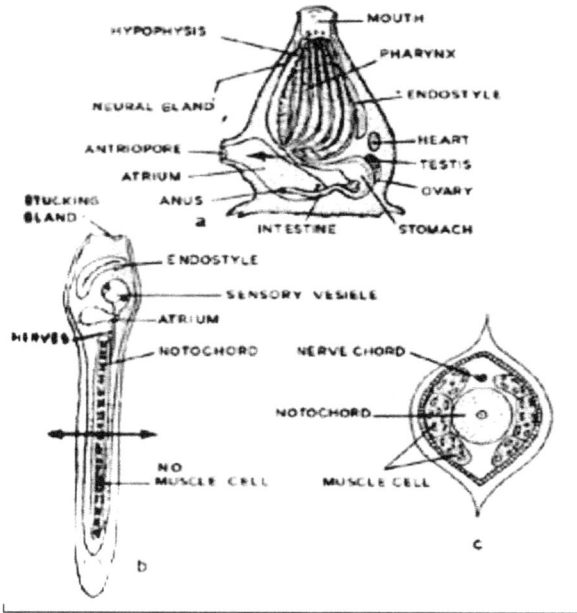

Figure 2.10: (a) Adult stage of tunicate; (b) Ascidian tadpole larva; (c) Cross section of ascidian tadpole larva.

reasonable to consider that the elimination of sedentary stage might have been accomplished by *paedomorphosis* in their course of evolution, so that the larval forms need not undergo the process of retrogressive metamorphosis. The gonads matured, while they still retained in the larval body structure. Precocious development of gonads is well known in certain special cases like *axolotl* larva of the amphibians when they were fed with iodized food. There are strong grounds for supposing that similar processes were responsible for the origin of several of the main animal groups like Insecta from Myriapoda, and Mollusca from certain annelids.

The Subphylum Tunica (= Tunicata Urochordata), has three main orders – 1. Acidiacea (Ex-*Herdmania*) 2. Thaliacea (Ex-*Doliolum, Salpa, Pyrosoma*) 3. Larvacea (Ex-*Oikopleura, Appendicularia*. The Larvacea group (=*Appendicularia*) has evolved by the same process of paedomorphosis.

It is logical to think that at sometime during the Ordovician period tadpole larva of ascidians became neotenous and matured sexually as a free swimming organism. In the ascidian tadpole of modern tunicates three rows of muscle cells numbering eighteen on each side of the notochord develop for locomotion. The muscle cells contain cross striated myofibrils. It is quite likely that in the neotenous forms the muscles became more organized and developed to facilitate locomotion against water currents. These changes took place for the exploitation of the rich pasture of oceanic surface waters and the shallow continental seas. The thaliacians (*Doliolum, Salpa,*

Pyrosoma) and the Larvacea (*Appendicularia, Oikopleura*) group of animals are the direct but modified descendants of this original neotenous form (Figure 2.11a-c).

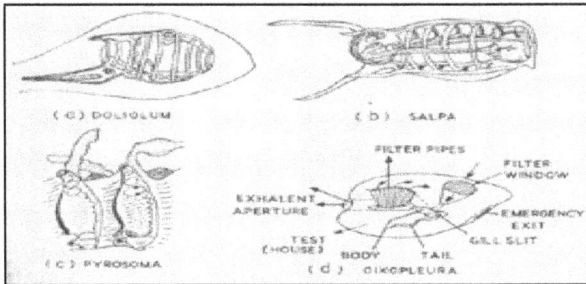

Figure 2.11: (a) *Doliolum*; **(b)** *Salpa*; **(c)** *Pyrostoma*; **(d)** *Oikopleura*

Acquisition of Locomotory and Sense-organs

At an early period of pelagic evolutionary phase, some ascidian tadpole-like neotaneous forms exploited the rich detritus containing organic matter descending from the river systems. They entered the estuaries using their (*i*) well developed body myotomes as locomotory power and (*ii*) sense organs as navigational equipment. The segmentation of the body was called forth and the streamlined body was achieved in relation to the flowing current of water, as they ascended the rivers. At first the rivers were invaded by only partly grown or adult organisms solely for the purpose of feeding. They returned to the sea to breed, where they produced pelagic eggs.

Amphioxus is a representative of such early chordates that ascended the rivers to feed but returned to sea to breed.

Within the rivers systems the chordates (neotenous ascidian type of life forms) evolved directly to relatively simple unarmoured type of Ostracoderms. Such an Ostracoderm was the vertebrate proto-type, which gave rise to the heavily armoured and specialized Ostracoderms and then to the true jaw bearing group of fishes known as the Placoderms.

4. Evolutionary Relationship of Deuterostoma and the Origin of Chordates and Vertebrates

Professor Barrington has discussed in length the evolutionary relationship of Deuterostoma larvae like Echinodern larvae, and tornaria larva and ascidian tadpole larvae of tunicates and the origin of chordates and vertebrates. The fundamental characteristics of deuterostome larvae are to be seen in the larval forms produced by the Phylum Echinodermata. The range of modification found in these larvae, their drastic metamorphosis, and the possible bearing of their own organization upon the problem of the origin of vertebrates have long been of interest of evolutionary biologists. The characteristic features of echinoderm development are as follows: (*i*) the process of gastrulation, (*ii*) the appearance of mesenchyme and an enterocoel formation of the *dipleurula* stage, (*iii*) at this stage the mouth is ventral and the anus is derived from the blastopore, (*iv*) the digestive canal differentiates into an oesophagus, stomach, and intestine, (*v*) the general surface ciliation of the embryo becomes reduced to a ciliary band which runs round the margin of a sadle-shaped depression that includes the mouth but excludes the anus, (*vi*) the region of the body anterior to the mouth is the pre-oral lobe, at the apex of which an apical sensory plate and tuft of cilia may be formed.

Externally, the *dipleurula* stage is distinguished from the trochophore larva of annelid, by the circumoral course of the ciliary band. Internally, it is distinguishable by the paired coelomic enterocoelic sacs and by the absence of protonephridia.

Pluteus Larva

In the Echinoidea and the Ophiuroidea the dipleurula stage develops into the pluteus larva. They have developed long arms for better locomotion and of suspension in the water. The long outgrowths of long paired arms carry cilia. The *echinopluteus* differ from the *ophiopluteus* in having pre-oral arms and they also differ, in general appearance, the arms being more widely opened out in the *ophiopluteus*.

These larvae may live for weeks or months as plankton before undergoing their metamorphosis which may be completed within an hour.

4.1 Metamorphosis of Echinoderm Larva

A new mouth appears on the left side of the larva, while the anus sifts to the right, with the gut forming a loop between them. The continued development of the coelomic sacs gives rise to the left and right body cavities. The water-vascular system is derived from the left hydrocoel, the right one disappears. Soon the hydrocoel begins to grow around the gut to form the water-vascular ring, and outgrowths from it establish its pentamerous symmetry. The pre-oral lobe region corresponds to the stalk of the crinoid larva. The larval arms are absorbed and their skeletal supports are lost. With the hydrocoel and primary podia already established, the young urchin, probably less than 1mm in diameter, is immediately capable of independent movement.

The metamorphosis of the ophiopluteus is similar in general principle to that of the echinopluteus. The radially symmetrical organization of the adult laid down within the larva until this sinks to the bottom under the influence of the increasing weight of the developing skeletal material. The arms and their skeleton, together with the anterior end of the larva, are then either absorbed or discarded.

The characteristic larva of the Asteroidea is the bipinnaria. The ciliated band of the dipleurula stage extends into anterior and posterior folds and then subdivides to form two ventral loops—a smaller pre-oral one and a larger circumoral one. The locomotion and support of the growing organism are aided by further extension of the ciliated band. The arms have no skeletal supports, and so are much more flexible in use.

The metamorphosis of the asteroid larva is often, but not always, proceded by temporary fixation. In preparation for this they develop, after some weeks of plantonic life, three brachioler arms, lying anterior to the pre-oral loop. These arms differ from the others in possessing extensions of the coelom, and in bearing adhesive cells at their tips. An adhesive glandular area or sucker develops between their bases. The larva is now known as a *brachiolaria larva*. It attaches to the substratum, first by the brachiolar arms, and then by the sucker, and begins metamorphosis. The future starfish develop in the posterior region of the brachiolaria. Its formation follows the same general course as that of the young sea-urchin, the left side of the larva becoming the oral part of the disc and the right side the aboral part. The metamorphosis may be completed within one day from fixation. The juvenile starfish is now able to pull itself from the remains of the larva.

A larva closely resembling the early bipinnaria is found in many of the Holothuria, which hatch at around the third day of development as an auricularia larva (*auricular* means a little ear). This has a circumoral ciliated band which is thrown into anterior and posterior folds, breaks up into sections that become rearranged and extended to form three to five transverse ciliated bands. This is the doliolaria stage, so called because of its obvious resemblance to the characteristic larva of crinoids.

The development of crinoids is little known. There is no bilaterally symmetrical larvae. *Antedon* lead to the hatching of a *vitallaria* (*doliolaria*), with four or five ciliated bands. The anterior or apical end bears an apical sensory plate with a tuft of cilia. At the opposite end is the blastopore. The *vitellaria* swims for a limited period, probably for not more than a few days, and then becomes attached by an adhesive pit. Later the internal organs, which have been differentiating progressively throughout this period, rotate through about 90°. The original anterior end of the larva becomes the attachment stalk, while the expansion of the free oral region makes the larva look like a stalked metazoa. This is the *cystidean stage* which resembles one of the extinct cystids. At first it cannot feed, because the stomodaeum remains as a vestibule which within a few days opens up and feeding then begins. After remaining in this form for some 6 weeks the larva develops arms, and it is now referred to as the *pentacrinoid larva*.

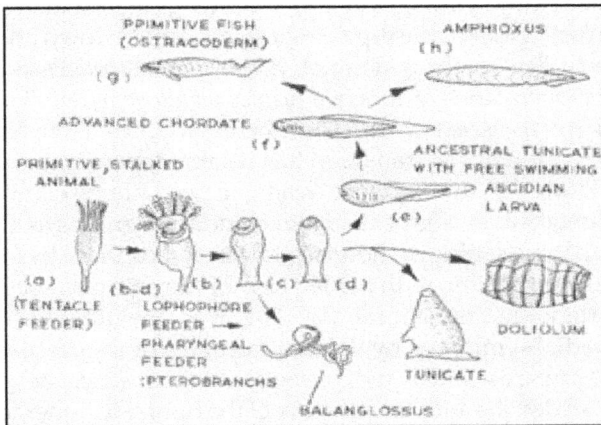

Figure 2.12: (a-g): Diagram showing the possible steps of evolution of fishes from echinoderm through tunicates.

GK. *Deuteros*; Second; *Stoma* (mouth). Appl. Coelomates having a true coelom, with radial cleavage of the egg, the blastopore becoming the anus, the coelom being formed by enterocoel and a dipleurula larva.

Finally, after some further months of sessile life, it breaks away from the stalk and begins the free-swimming life characteristic of *Antedon* and of most living crinoids.

5. Phylogenetic Significance of Echinoderm Larvae

The phylogenetic significance and the interrelationships of these several types of echinoderm larvae have been postulated. The development within each group has been subjected to much secondary modification, leading to abbreviation of the life-cycle which exemplify the convergence and divergence of different adult stages of echinodermata. Abbreviation of the larval life history, and even direct development have been observed in echinoderms.

Professor Fell has argued that remarkable plasticity in echinoderm development makes it impossible to attach phylogenetic significance to their larvae. Certainly one cannot possibly justify interpreting them as a strict recapitulation, in the Haeckelian sense of phylogenetic history. Larvae in general undergo independent adaptation to their particular modes of life. Their primary function is of ensuring development and dispersal of the species. They cannot, therefore, be precise guides to the organization of adult ancestors.

Most of the *echinoderm* larvae have certain common features of organization, *viz.* (*i*) bilateral symmetry, (*ii*) an apical plate and tuft, (*iii*) an anteroventral mouth, a posterior anus from the blastopore, (*iv*) a tripartite and paired entercoel, and (*v*) a ciliated band curving to run around the mouth and in front of the anus. These we may regard as primary features, which give the foundation of the structure of the Deuterostomia. *They also suggest the possibility that the dipleurula stage may represent a common ancestor of Echinoidea, Holothuroidea, and the Asteroidea.*

The drastic character of echinoderm metamorphosis reflect a time in their past history when a bilaterally symmetrical and free-living stage, which may or may not have had the characteristics of the dipleurula stage, settled down and in due course of time developed a pentamerous symmetry. *To regard the behaviour of contemporary larvae as an exact recapitulation of ancestral history would be wrong.* Events that must have been spread over a vast extant of time are here compressed into hours or minutes; moreover, the execution of the transition has been independently modified in the various groups. However, the evidence available need not be completely ignored. Viewing the metamorphic events of all of the echinoderm groups as a whole, we may reasonably infer that the setting of the ancestral stock took place by the anterior end; this became a stalk of fixation, with the left side of the organism becoming the oral surface and the right side the aboral one. This was presumably followed by the development of radial symmetry, by the rotation of the oral surface upwards and of the aboral one downwards, and by the consequent asymmetrical development of the coelome of the two sides, the anterior structure of the right side undergoing regression.

The resemblance between the auricularia larva of the Holothuria and the early bipinnaria of the Asteroidea might be taken as evidence of a close relationship between these two groups, while a similarly close relationship between the Echinoidea and the Ophiuroidea would be implied by their common possession of a pluteus. We might thus conclude that the ophiuroids, despite their resemblances to starfish in the adult stage, are actually more closely related to the sea-urchins.

6. Evolution of different Groups of Echinoderms and their Relationship with Hemichordates

The evidence from palaeontology shows that asteroids and ophiuroids must be closely related, with common ancestors in the early Palaeozoic, and that the echinoids already constituted an entirely separate line of evolution at that time. It follows that the organization of echinoderm larvae can be of guide to the phylogenetic history of the several subphyla. We must regard their special features as independent responses to the demands of planktonic life, influenced, by the common genetic potentialities of the phylum as a whole.

Tornaria larva : The larval history of the other deuterostomes is no less complex than that of the echinoderms. Some members of the phylum Hemichordata have an indirect development with the production of a characteristic *tornaria larva*. This larva resembles in its mode of origin and in its organization that differ with the *dipleurula*.The larval organization is fundamental to the echinoderms, and is an indication of a common inheritance. *It is therefore reasonable to regard the tornaria as indicating a close relationship between hemichordates and the echinoderms.* The development of the temporary post-end tail during the metamorphosis of the *Saccoglossus horsti* is an illuminating feature. This is probably a vestige of the attachment stalk of the sessile hemichordates, seen today in *Rhabdopleura* and *Cephalodiscus*. In other respects the metamorphosis of the larva involves little more than simple growth and transformation, without the drastic reorganization that occurs in echinoderms. *We may safely conclude that the Hemichordates were derived from the same type of sessile, microphagous ancestor as were the echinoderms.* We may further infer that they have diverged much less from the ancestral type of the echinoderms. The Enteropneusta have evolved from the Pterobranchia by assuming the limited degree of independent movement involved in their burrowing life. The temporary development of the tail recalls the appearance of the stalk in asteroid metamorphosis.

In some enteropneusts development is direct. As with echinoderms, this is best regarded as a result of secondary loss of the larva, the more so in that an intermediate stage of this is seen in *Saccoglossus horsti*; this has a free-swimming stage of short duration, which does not develop into a tornaria. Undoubtedly the differences in modes of development are to be associated with the yolk content of the egg as in echinoderms. The tornaria is planktotrophic, whereas the larva of *S. horsti* is lacithotrophic, relying entirely upon the yolk laid down in its egg for the whole of its free swimming activity, which lasts for only 1 or 2 days, at the end of which time the organism is ready to settle and metamorphose.

7. Protochordate

The larvae of the Urochordata and the Cephalochordata do not show obvious relationship with any of those that we have so far considered. However, these groups show some developmental characters that justify their inclusion in the Deuterostomia. In *amphioxus*, where these characters are easiest to appreciate, we find total non-spiral cleavage, gastrulation by invagination, the origin of the anus from the blastopore, and an enterocoelic coelom with indications of a tripartite differentiation. The larva, however, are of more advanced organization than those of echinoderms and enteropneusts, possessing distinctively chordate features as, in the case of the urochordates, which provide the main reason for including the adults within the phylum Chordata.

8. Ascidian Tadpole

Biologists became very much concerned with the two main larval types, - the ascidian tadpole of tunicates and the larva of amphioxus. The larval stages of the pelagic urochordates (Thaliacea and Larvacea) are considered to have been modified or lost in correlation with the Pelagic life of thse animals. The development, dispersal, and habitat selection are the main functional working principles of the ascidian tadpole, but it is the last of the three that predominates. The larva possesses the embryonic rudiment of the adult alimentary canal, but it does not itself feed, for its free-swimming life is restricted to a matter of hours or perhaps to a few days at the most. Its chordata features are particularly well seen in the dorsal nervous system, and also in the locomotor mechanism, which is, however, confined to the tails. This mechanism, like that of the vertebrates, does not employ the principle of the hydrostatic skeleton, but relies instead upon the association of lateral muscle bands with an axial support, the notochord.

Closely associated functionally with locomotion are the ocelli and statocyst in the cerebral vesicle, at the anterior end of the central nervous system. *Ascidian tadpoles* are at first positively phototropic, and swim to the surface of the water. This behaviour pattern marks the distributive phase of their activity. Soon however, they become negatively phototropic and positively geotropic, moving downards into shaded rock crevices and overhanging surfaces that are suitable for the life of the adults.

The significance of the tadpole in the ascidian life history is apparent for habitat selection, but its phylogenetic relationships are puzzling. The most illuminating interpretation is due to Professor Garstang. He suggested that the tadpole could have been evolved from a larva like the auricularia of echinoderms as a consequence of the further elaboration of locomotor adaptation and this could have led to the approximation of the loops of the ciliary band in the mid-dorsal line, where, with their associated nerve fibres, they could have given rise to the neural folds. The adoral ciliated band, situated inside the mouth and responsible for maintaining a feeding current, might have been the forerunner of the endostyle. Certainly this possibility can not be ignored. We can visualize the ascidian tadpole evolving as a special product of the urochordate line without weakening the arguments for associating the group with the Deuterostomia.

9. Evolution of Protovertebrates from Ascidian Stage of Urochordata

The larval history of urochordates is of outstanding importance in connection with the second main line of Garstang's theory, in which he attempted to trace the line of evolution from the protochordates to the vertebrates. This, indeed, is a crucial evolutionary issue. The deuterostomes stem from a sessile and microphagus ancestory, a supposition that leaves us with difficult of seeing how vertebrates organization could have evolved from such a background. Garstang's answer was analogous to his explanation of the torsion of gastropods as resulting from the persistence of a larval adaptation in the adult. He suggested that the fundamentals of chordate organization, *the nervous system and notochord were initially established in ancient larval forms like ascidian tadpoles.* In due course of time their metamorphosis was eliminated, so that their larval organization became that of sexually mature adults. This process, is well known in axolotl larvae of urodela (Amphibia) when they were fed with iodized food. We call this phenonmena as neoteny in which there is retention of larval characters even when there is sexual *maturity. In this instance it is visualized as leading to a new type of adult organization,* one containing the immense potentialities that flowered in the evolution and diversification of vertebrate animals.

The developmental and structural organization of larvacea gives the clue of possibility of evolution of protovertebrates *from neotenous ascidian tadpole like forms.* It is like that the larvae that have not become specialized in this way, but are rather tending to prolong their independent life. It seems, therefore, that the origin of vertebrates must be sought in some remote and primitive deuterostome, urochordate-like in fundamental organization, but perhaps not very far from the level of organization of hemichordates, with a generalized type of tailed larva. Hypothetical though the *argument may seem, the contention that new forms of adult organization may arise through neoteny is supported by protochordate life histories. It is generally agreed that the Larvacea are neotenous urochordates, preserving their tail into the adult stage, but even more illuminating are conditions within the Cephalochordata.*

These ascidian tadpole like animals equipped with navigational sensory vesicle (with otolite and ocellus) entered the estuaries of rivers, running against current of water and developed the characteristic shape of fishes. But due to the osmotic changes of body fluids could not advance much and started dying. In order to protect themselves these spindle shaped fish like animals secreted thick scale like armours, to prevent movement of water molecules into the body.

These armoured fish like animals became very heavy. In course of time of evolution these ectodermal shields were converted into thin scales (Flat, small plate – like structure dermal or epidermal, bony or chitinuous outgrowth as in present day fishes.

The respiratory and excretory organs are in the form of atrium (L. atrium, chamber) surrounding branchial chamber which is pulsatile). An endostyle is present in the pharynx to collect minute plankton, the pharynx has many gill slits. Many muscle cells are present in the tail region having nerve supply.

10. *Amphioxus* Larvae

Amphioxus has a metamerically segmented planktotrophic larva, highly specialized in its asymmetry. The mouth is relatively enormous, and lies on the left side of the body, probably as an adaptation to the intake of food by a cilliary feeding mechanism. Clearly this larva is much more highly organized for independent life than is the ascidian tadpole; yet some affinity with the latter is seen in the tendency of young *Amphioxus* larvae to attach themselces temporarily to the bottom of aquaria by the secretion of the club-shaped gland. Moreover, older larvae become temporarily attached by three anterior adhesive papillae, which strikingly recall the attachment papillae of the ascidian tadpole.

11. Evolution of Vertebrates from Nemertines

This view leaves the general concept of the Deuterostomia (=Deuterostome) Coelomates having a true coelom, with radial cleavage of the egg, the *blastopore becoming the anus*, the coelom being formed by enterocoely, and a dipleurula larva undisturbed, but another view, deriving from an idea first suggested by the Dutch embryologist Hubrecht in 1883 and recently elaborated by Professor Willmer is more radical. According to this, the protochordates could have been derived from nemertines, the proboscis being converted into a notochord and the pharynx into a filter-feeding mechanism, with, of course, many other concommitant changes. Professor Willmer argues further that the process may have developed along more than one line. *Amphioxus* and the vertebrates, for example, might have started from different nemertine ancestors and have followed parallel course of evolution.

Professor Barrington concludes that these unconventional approaches to a perpetully fascinating problem merit more attention than can be given to them here, if only because of their salutary warning that larval life histories may not be the only repositories (place where valuables are deposited for safe keeping stores much like vaults of Banks of the phylogenetic record).

12. Palaeontological Evidences

The evidences concerning the nature of the habitat of the early vertebrates have been analysed on the basis of available fossils. The main conclusions are as follows:

1. That Ostracoderms, - the earliest vertebrates, were entirely inhabitants of freshwater in Ordovician time.

2. That the Ostracoderms remained essentially as freshwater forms to the end of their history in the Devonian, although there was some tendency towards a marine life in the Class Heterostraci (= Pteraspida) of Superclass Agnatha.

3. That the Placoderms were of freshwater origin; the Antiarchs (Subclass Antiarchi) remained almost entirely in freshwaters. The Arthrodiras (Subclass Arthrodira) in the lower Devonian migrated into the seas, where the greater part of their short but remarkable evolutionary history took place. All these fishes were jaw bearing animals belonging to the superclass-Gnathostomata.

4. That the group of cartilaginous fishes – Chondrichthyes (= Elasmobranchii) also originated in freshwater; where the oldest known species *Cladoselache* flourished from the Devonian to the Permian. A strong trend for sea water habitat began at about the beginning of Devonian time and the sea remained its main centre of evolution (Jurassic or Recent)

5. That the bony fishes, the Palaeoniscids had their origin in freshwaters in the late Devonian period, though the fossils of *Cheirolepis* have been encountered in the early Devonian. The higher bony fishes, the

Actinopterygians, however, showed an early tendency towards the seas in the late Devonian, and flourished during the period Jurassic to Recent.

The lung fishes (Sarcopterygii = Subclass Crossopterygii comprising both the Crossopterygii and Dipnoi remained in freshwaters from the very beginning of the Devonian but gave rise to a group which migrated to the seas in the late Devonian.

On the basis of Palaeontological evidences available it is now more or less agreed that it was in the freshwaters of the Palaeozoic continents that the first chordates and from them the Ostracoderms have evolved.

13. Physiological Evidences

The theory of the freshwater origin of fishes is supported by the physiological evidence obtained from the study of structure and function of glomerular kidney of the vertebrates. The evolution of the kidney is essentially related with the evolution of regulation of water balance of the body. Most marine invertebrates, polychaetes, echinoderms, crustacea and molluscs are generally in osmotic equilibrium with the sea water and they have, therefore, to face no problem of water regulation. It may be assumed that the renal organs of the protochordate ancestor of the vertebrates were wholly concerned with the excretion of nitrogenous wastes and were not concerned at all with the excretion of water. As in most invertebrates the excretory organs consisted of a coelomic membrane together with open tubules with ciliated funnels which connected the body cavity or coelom with the exterior. With this type of equipment the first chordates ventured to enter the freshwaters of the Palaeozoic continents. They had immediately to encounter the problem of osmotic regulation. In entering into freshwater via the estuaries they were entering a medium with a very low osmotic pressure as compared with that of the salt water of their original home. The fossil remains of the first vertebrates, the Ostracoderms were found in abundance in freshwater sediments of the Silurian and Devonian time. These early vertebrates developed impregnable armour of bony plates and scutes or scales to protect themselves from the inflow of freshwater into their bodies. The heavy armour was also believed to protect them from the attacks of Eurypterids. The seas of the early Cambrian-Ordovician period had probably only one half or less of the present day salinity of the sea. Even then the immediate problem that the protovertebrates had to face in this new freshwater habitat was the regulation of water balance of the body. The salts and proteins of the body fluid would have drawn water from outside by osmotic pressure and the organism would suffer from oedema due to hydration. It has been surmised that the water proof armour of the Ostracoderms was evolved as a first step towards arresting the infiltration of water.

But with the insulation of the body by the water proof armour, the segmental openings of the coelomic excretory tubules had to be closed and these tubules though still arranged segmentally pierced the armoured skin only at the posterior end of the body. This is how the three sets of kidney-tubules, the pro- meso-, and meta-nephros have evolved in fishes with their common archinephric ducts.

Even after a water proof covering had been evolved, still the danger of excessive hydration through the mouth, gills and the alimentary canal remained. In order to solve this problem of water regulation, devices of the glomerular kidney have evolved. In many fishes and amphibia, the mesonephric tubules still retain an early connection with the body cavity as is found in the developmental stages of fishes. The evolution of the renal glomerulus as a filtering device was a very important step in the evolution of the kidney. Freshwater taken into the body would be successfully filtered out through these glomeruli under cardiac pressure. But this high pressure filtration system of the glomerulus allows not only water to be pumped out of the body, but also most of the useful constituents of the plasma, glucose, phosphates, chloride etc. The excretion of these would mean a serious loss to the organism. With the evolution of the glomeruli it became necessary to modify the kidney tubules so that they could reabsorb these valuable constituents from the filtrate. Thus the kidney tubules were made capable of reabsorbing large quantities of glucose and valuable salts and this produces urine that is hypotonic to the blood.

Because of the many structural resemblances of kidney it is almost universally held that all vertebrates evolved from a common stock. The physico-chemical evidence from the regulation of water through the glomeruli in response to freshwater environment supports the palaeonotological evidence that fishes originated in freshwater or in estuaries rather than in the sea.

14. Life History of Lampreys (Cyclostomes)

The only living Agnatha of the present day are the lampreys and hag fishes. Two distinct phases have been recognized in the life history of lampreys (i) the ammocoete larva living for about five years in the muds of freshwater rivers feeding on micro-organisms using endostyle and muscular contraction of the pharynx, and (ii) the adults living in sea water as parasite on fishes for about two and half years.

After metamorphosis the young ones migrate to the sea. These animals show double migrations and as such are called anadromous. It will be interesting to study their osmoregulatory mechanism.

15. Urinogenital Tubules of Vertebrates

In the invertebrates the genital ducts having coelomostomes carry the genital products from the coelom to the exterior. In all coelomate animals the gonads develop from the walls of the coelom and it is believed that the coelom represents an enlargement of a gonadal sac. As such the primary function of urinogenital tubules of vertebrates seems to be to serve the purpose of carrying genital products to the exterior. The conversion of these tubules to excretory purpose and water regulation may have been due to their evolution in freshwater system.

16. Osmoregulation in Cyclostomes

The blood of lampreys contains a higher concentration of salts than the surrounding water, when they migrate into the rivers, but into a lower concentration of salts when they are in the sea.

When the animals come to freshwater they face the problem of hydration as there is a tendency to flow in water into the body in response to existing osmotic gradient. This water must be removed without losing salts.

17. Development of Glomerular Kidney

The nephrotome region between the scleromyotome and lateral plate mesoderm gives rise to the kidney. The tissue differentiation during development takes place from in front backwards. A series of segmental funnels open into a common archinephric duct. Usually four such funnels open into the pericardium in a new hatched ammocoete larva. Associated with each such ciliated funnel develops a network of blood vessels in the form of glomerulus. Due to blood pressure water is filtered out from the glomeruli into the coelom's fluid, which is then removed by the funnels. The tubules become longer and coiled in course of time of development of the larvae and become specialized for reabsorption of salts and other essential substances.

As the ammocoete grows the anterior tubles along with other funnels constituting the pronephros kidney gradually disappear to form a mass of lymphoid tissue. The mesonephros develops as a much larger fold hanging into the coelom. In contrast to pronephros here the tubules do not open directly into the coelom but into the minute coelomic space of the Bowman's capsule which lodges the glomerulus. This structural formation is known as the Malpighian corpuscle. This mesonephros kidney is obviously more efficient than the open funnels of the pronephros for filtration of excess water from the blood in freshwater habitat. There is a gap of several segments between the two segments of the kidney in which no tubules appear. No explanations are available for this characteristics discontinuity which is common to all vertebrates.

18. Mechanism of Osmoregulation in Lampreys

"The mechanism of osmoregulation is remarkably similar to that of teleostean fishes, considering some 500 million years of independent evolution" (Young, 1981).

Freshwater Adaptation

In freshwater the excess of water absorbed through the skin and gills is removed by the pronephric/mesonephric glomerular kidney in the form of hypotonic urine. The essential salts are reabsorbed by tall columnar cells provided with microvilli. These fishes can also take up sodium and chloride from the environment by ionocytes present on the gills. Localisation of ^{3}HCL in ammocoete gills have been demonstrated by autoradiography.

Sea Water Adaptation

In the sea the same ionocytes of the gills excrete ions as in marine teleosts. Water is swallowed and the monovalent ions are absorbed in the gastrointestinal tract and then the excess of ions are removed by the *chloride cells*. Divalent ions are removed along with the faeces. During the anadromous spawning migration the alimentary canal degenerates and as such if the animal is returned back to sea it will not be able to osmoregulate by drinking sea water and will die.

19. A Comparative Account of Water and Salt Regulation in the Vertebrates

The osmotic pressures of the body fluids of animals are customarily measured by the depression of the freezing point they cause. This is expressed by the symbol Δ, the internal medium being represented by Δ_i and the external medium as Δ_o

20. Freshwater Fishes

Fishes first evolved in freshwater and all present day seawater species have a freshwater ancestry. For fish living in freshwater the external medium has a lower osmotic pressure than their internal fluids ($\Delta_i = 0.57$, $\Delta_o = 0.03$), so water will tend to diffuse in across permeable surfaces (mainly the gills and gut) and elimination of excess water will tend to carry out valuable salts in the dilute urine. The osmotic and ionic problems of freshwater fish are thus to eliminate excess water and to conserve salts. The excess water is filtered out by the glomeruli of the kidneys, and freshwater fish have large Malpighian bodies to deal with the passage of considerable quantities of water. Ionic balance is maintained by replenishing salts by way of the diet, reabsorbing salts in the kidney and absorbing salts through special cells the chloride cells on the gills (Marshall and Hughes, 1980).

21. Marine Fishes

1. *Teleosts*: Seawater has an osmotic pressure greater than that of the blood of marine teleosts, whose body fluids ($\Delta_i = 0.6$) tend to have a similar composition to those of their freshwater relatives. The osmotic pressure of the sea corresponds to $\Delta_o = 1.7$. Marine teleosts thus tend to lose water through their gills, gut or any permeable surfaces, and their osmotic and ionic need is to dilute the internal medium. In order to do this the fish swallow large quantities of salt water, absorbing both the water and the salts from the gut. They conserve the water, producing only small quantities of urine, and eliminate excess salts via special secretory cells on the gills as well as in the urine. Some of the excretion of nitrogenous waste takes place through the gut lining. Because of the small amount of water eliminated the capsule and glomeruli of the kidneys are very small or even absent.

2. Elasmobranchs that live in the sea are unusual in using the retention of urea in the body as a means of raising the internal osmotic pressure. In these fishes the $\Delta_i = 1.8$ and that of seawater $\Delta_0 = 1.7$. For this reason the marine elasmobranch has no tendency to lose water to the medium and its kidney has a large Malphigian capsule and well developed tubule for selective reabsorption of salts. The young elasmobranch is provided with a supply of urea from its mother. Freshwater elasmobranchs descended from marine ancestors retain this characteristic of urea in the bloodstream. And it is interesting that modern diponoans and coelacanths, although more closely related to teleosts than elasmobranchs, also show the same osmoregulatory adaptation (Marshall and Hughes, 1980).

Amphibians

The amphibians evolved from a group of freshwater fishes in the Devonian (a geological period that commences some 300 million years ago) and they have many of the same problems of osmoregulation and ionic maintenance. The amphibians live largely in freshwater and have a body fluid which is hypertonic to the surrounding medium (for *Rana*, the common frog, $\Delta_i = 0.56$). The skin is partly permeable to water as is the gut lining and thus amphibians absorb large quantities of water from their freshwater environment. Although the kidney has a large glomerulus and capsule as well as long tubule for salt reabsorption, some leaching out of salts occurs. The urine is copious and hypotonic to the body fluids ($\Delta_i = 0.17$). The loss of salt is made up by taking in salts in the diet as well as by selective absorption through the skin.

22. Osmoregulation in Osteichthys and Chondrichthys

In freshwater teleosts the internal osmotic pressure is greater than the external, and water, therefore, passes into the fish, dilutes the blood, but is filtered off by the glomeruli of Kidneys and the salts are reabsorbed by the kidney tubules from the glomerular filtrate. This helps the fish to maintain a high osmotic pressure aided by active salt absorption *by special chloride cells in the gills*. A freshwater teleost, therefore, is osmotically independent like other freshwater organisms (Figure 2.1). In the case of marine teleosts, however, the danger is not that of death by flooding but of death from desiccation. It must obtain water somehow from the surrounding sea which it does by swallowing salts and water, are therefore, both absorbed by the gut of the fish but the kidney of the marine teleosts has not evolved the power of secreting a hypertonic urine. *The excess salts are excreted not by the kidneys but by special cells called the "Chloride Secretory Cells" situated in the gills.* Water is thus a valuable necessity for the marine teleosts but the urine is really isotonic with the blood. The possession of glomeruli would be a disadvantage and, therefore, a marine teleost has closed them down, so to speak, or reduced them in number. The marine elasmobranchs, on the other hand, have devised a new method. They have two percent or more of urea in their blood and thus maintain a higher osmotic pressure than that of sea water. Their kidneys still retain glomeruli but they conserve their blood-urea by having a special urea absorbing segment in their kidney tubules. The osmotic gradient driving waters in fish is small. Water enters fairly slowly and a small quantity of hypotonic urine is formed. The freshwater elasmobranchs still maintain a mark of their marine ancestry, although they have become so accustomed to uraemia that the heart of a freshwater elasmobranch is unable to beat in the absence of urea.

23. Osmoregulation in Freshwater and Marine Teleosts

It has been claimed that the bulk of the sodium, potassium and chloride absorbed in the gastrointestinal tract of marine teleost is excreted by the 'Chloride Secreting Cell' in the gills (Smith, 1930; Keys, 1931, 1933; Keys and Willmer, 1932), but Bevelander (1935, 1946) believed that the supposed excretory cells are nothing but intra-epithelial mucous glands and that the general respiratory epithelium might be a site of chloride excretion.

Liu (1942) attempted to acclimatize the freshwater air-breathing fish, *Macropodus opercularis*, to different concentrations of salt solution. He stated that even an exclusively freshwater fish can tolerate a salt solution nearly as saline as sea-water by virtue of the enormous development of latent 'Chloride Secreting Cells'. From this experiment he concluded that freshwater teleosts possess 'Chloride Secreting Cells' in the gills in a dormant condition.

Copeland (1948) noted cytological changes in the chloride cells of *Fundulus heteroclitus* during adaptation to varying degree of salinity, and by using the Leschke test demonstrated the presence of a copius amount of chloride in the secretory cells of fishes adapted to the salt water condition but only a limited amount of it at the free ends of the cells in fishes adapted to freshwater life. Getman (1950) working on *Anguilla rostrata* came to a similar conclusion.

Vickers (1961) found that when the gills of *Lebistes reticulatus* (a freshwater teleost) were subjected to hypertonic-salt solutions of varying concentrations, its mucous cells became functionally transformed into chloride cells.

24. Branchial Glands and Osmoregulation

The structure of the gills of *Labeo rohita, Hilsa ilisha, Rita rita* and *Ophiocephalus (= Channa) striatus* has been described in detail by Munshi (1960).

Catla catla : In this species only two kinds of specialised cells are present, the mucous glands and the acidophil mast cells.

Mucous glands are present in large number in the epithelium covering the gill arches and the primary and the secondary gill lamellae. In a horizontal section of the epithelium of the head of the gill arch, the mucous glands are found to be unicellular and rounded, and to posses flattened nuclei owing to the presence of the secretory substance. The glands are scattered on the general surface of the primary gill lamellae and in the interlamellar space close to the base of the secondary lamellae. A few of them may occur on the secondary lamellae as well.

It seems that the mucous glands present on the primary gill lamellae and those lying between the bases of the secondary lamellae are concerned with the excretion of chlorides also. These cells react matachromatically with thionin and give a positive reaction with PAS.

Heteropneustes fossilis : In this species large acidophil gland cells and mucous glands are present. The large acidophil-gland cells are present in the epithelium covering the surface of the gill-arches.

The mucous glands are abundant in the epithelium covering the head of the gill arches and occur in smaller numbers on the primary gill lamellae. However, these glands are absent from the secondary lamellae.

Only a few epithelial cells of the primary-gill lamellae respond to the chloride test.

Channa punctata: In this species only two kinds of specialized cells, namely mucous glands and acidophil Mast cells are present in the gills.

Mucuous glands are present in the epithelium covering the heads of the gill arches and the primary and the secondary gill lamellae. Each mucous gland is unicellular, flask-shaped structure, opening to the exterior through a small pore. These glands are also found on the primary lamellae, at the bases of the secondary lamellae and may occur even on the secondary lamellae. They give a strong reaction with PAS, showing thereby that the secretary substance is a mucopolysaccharide.

Some of the epithelial cells of the primary and the secondary lamellae are prominent in form and react sharply to the $AgNO_3/HNO_3$ test. Sometime they appear to be vasicular, with a spout-like opening. It seems probable that some, if not all, of the mucous glands are capable of acting as chloride-excreting cells.

Mastacembalus armatnus: Mucous glands are present in the epithelium covering the heads of the gill-arches and the primary and the secondary gill lamellae. They are of the usual goblet type.

25. Characteristics of Chloride Cells

It may be useful to list here the characters of the 'Chloride cells'

1. The cells are ovoid, or sometimes columnar;
2. They are large;
3. Their cytoplasm is finely granular;
4. There is a marked affinity for eosin;
5. The nucleus is nearly spherical, and frequently eccentric;
6. The cells are present in the epithelium of the secondary and the primary lamellae. They are closely packed together in the epithelium of the filament, between the bases of the respiratory lamellae. They may reach the sub-epithelium and connective tissue layer;
7. They are close to the vascular layer. The corelation between the number of chloride cells and the degree of vascularity is very well marked;
8. They respond to the chloride test;
9. They are secretory; and
10. They are rich in mitochondria with reticulam endoplasmic: Diagrammatic view of epithelium of gill showing chloride cells, mucous cells, basement membrane and microridge epithelial cells.

In a comparative study of the branchial epithelium of fishes, Bevelander (1936) found several types of 'intra-epithelial gland cells' which, according to him, belong to unicellular, multicellular, and transitional types. Only three types of specialised cells have been found in the gills, - the mucous glands of goblet type, the mast cells, and the large eosinophilic, multi-nucleate gland cells. Branchial glands, possibly multicellular in composition, are represented by the large eosinophil glands that are bi- or tri-nucleate and are found chiefly in the siluroid fishes. The function of these hypertrophic-multicellular glands is not clear.

Mucous glands of the ordinary goblet type are present in large numbers in the gill epithelium of the freshwater fishes examined, except in *Trichogaster fasciatus*. They generally occur on the head of the gill-arches and may even extend on to the surface of the primary gill lamellae. In certain cases, such as *Catla catla, Labeo rohita, Channa punctata, C. striatus* and *Mastacembalus armatus*, they extend into the epithelial covering of the secondary gill lamellae. In the gill rakers of _Labeo rohita_ a modified type of these glands occurs, which may be looked upon as having been derived from the usual kind of mucous glands. The mucous glands of the primary lamellae of _Catla catla_ respond to the 'Chloride test', and this indicates that these glands, besides discharging mucous, they also play some part in the process of elimination of the chlorides. In _Channa punctata, Clarias batrachus_ and _Heteropneustes fossilis_ only a few of the mucous glands react positively to a limited extent. In _Hilsa ilisha, Rita rita, Channa striata_ and _Mastacembalus armatus_ only a network of silver is formed on the surface of the primary and secondary lamellae. Shelbourne (1957 a, b) also obtained a similar kind of silver network in the integument of the marine plaice larva, which he took to mean that the excretion of chlorides also takes place through the general integumental surface. In the fishes reported here, the silver network is evidently caused by the presence of chlorides in the intercellular cementing substance. The 'opennings' of the mucous glands can be seen distinctly in the interstices of the network. Occasionally, a little of the secretory matter containing chloride is detectable in the small pores in the silver network.

According to the theory of osmoregulation, extrarenal excretion of chloride is not to be expected in purely freshwater teleosts like _Catla catla, Channa punctata, C. striata, Clarias batrachus_ and _Heteropneustes fossilis_. The occurrence of 'Chloride cells' in the gills of the freshwater fishes is therefore paradoxical. Copeland (1948 a,b) explains this away by saying that the cell reverses its polarity and serves as a physiological mechanism to absorb chloride ions from freshwater and Krogh (1937) thinks that a mechanism for the absorption of chlorides from the surrounding medium exists in freshwater fishes. From the fact that the chloride cells are abundantly present in some of the freshwater teleosts, it is difficult to escape the conclusion that extra-renal excretion (_i.e._, excretion of chloride by the gills) occurs in freshwater teleosts also.

Smith (1931) and Bevelander (1935) have questioned the nature of the cells that perform this 'electrolyte excretion' in the gills of fishes and think that the whole of the respiratory epithelium of the gill lamellae might be involved in this work. But histochemical tests have disclosed the presence of only a limited number of epithelial cells concerned in this 'electrolyte excretion'. Histological studies of freshwater fishes by Munshi (1964) have shown that mucous gland cells excreting chlorides occur in the gill epithelium of some species of freshwater fishes.

The hypertonicity of the blood and of other body fluids of different species of freshwater fishes varies within a certain range, so that chloride cells are called into play according to the needs of the fish. It seems probable that the silver network is due to the presence of chlorides in the intercellular-cementing substance. Recent studies using SEM and TEM have confirmed the presence of mitochondria rich ionocytes/chloride cells in the gills of freshwater fishes (Munshi and Hughes, 1986).

H.W. Smith (1931) was aware of chloride excretion in freshwater fishes, and remarked that 'it is a very primitive process being common to fresh and salt water teleosts'.

26. Effects of Marine Habitat upon the Kidney

It has been suggested that the renal glomeruli represent a primitive vertebrate character, and that the absence or degeneration of glomeruli in certain marine fishes is a specialisation associated with the decreased excretion of water, relative to the freshwater ancestral stock (Smith, 1930). With the secondary assumption of a marine habitat, where the osmotic gradient is reversed and water excretion is greatly reduced the teleost is at a disadvantage if glomerular activity is maintained at its full level, and there is consequently a need for reducing this activity to a minimum. Studies of glomerular activity in the marine teleost indicate that in the normal fish the rate of glomerular filtration is of the order of magnitude of 14 cc per kg per day in contrast to 200-400 cc per kg per day in the freshwater teleost.

Freshwater fish show typically numerous and well developed glomeruli, and marine forms in general show a reduction in both number and size. Among the more highly specialised marine fishes many are known to have no glomeruli, viz., *Opsanus, Lophius, Syngnathus, Hippocampus*.

27. Osmoregulation in Chondrichthys

The elasmobranch or cartilaginous fishes are unique in normally possessing in the blood, body fluids and tissues, large quantities of urea in amounts exceeding 2.5 percent (NH_2–CO–NH_2). Since in other animals urea is a waste product formed by the degradation of protein nitrogen, and as such is excreted from the body as rapidly as it is formed, this phenomenon of retention by the elasmobranchs is of considerable interest.

The term 'uraemia' has long been used to denote a condition of elevated blood urea in man or other mammals, due to renal insufficiency. It is now accepted that the pathological consequences of this condition are not due to the urea itself. It has beeen established that the retention of urea in the elasmobranchs is a consequence of a normal physiological process.

Salient Points

1. The high urea content (2.0-2.5 per cent) that characterises the blood, body fluids and tissues of the elasmobranchs owes its origin to the relative impermeability of the gills and integument to this substance and to the circumstance that the urea is actively conserved by the elasmobranch kidney.

2. In consequence of this physiological uraemia, the elasmobranch is osmotically superior to its environment, even in sea-water, and is able to absorb at least a minimum quantity of water for the formation of urine that is isotonic or hypotonic to the blood, in accordance with the osmotic limitations of the fish kidney.

3. The uremic state tends to develop and is regulated more or less automatically. Urea is constantly being formed by the ordinary metabolic combusion of protein, and the accumulated urea in the blood raises the osmotic pressure of the latter to a point where water is again available by

direct absorption. Water plethora (as in freshwater) leads to diuresis and increased urea excretion, which in turn lowers the osmotic pressure of the blood and in some measure, at least, reduces the rate of water absorption.

4. Trimethylamine oxide ($N-CH_3)_3-O$) which imparts about one-quarter as much osmotic pressure to the blood as does urea, is also conserved by the elasmobranch. The fact that this substance is present in the urine in lower concentration in the blood suggests that, like urea, it is actively reabsorbed from the glomerular filtrate.

5. This physiological uremia is apparantly an archaic biochemical habit acquired early in elasmobranch evolution, since it is shared by the divergent orders of the subclass. Presumably it is a secondary mode of osmotic regulation superimposed upon the more primitive one of branchial regulation, as observed in the teleostomes.

6. The cleidoic egg, unique (among the fishes) in the elasmobranchii, and the viviparous mode of reproduction, are viewed as adaptations to urea retention, protecting the embryo against the loss of urea during its early development.

7. Urea retention enables the elasmobranchs to maintain a constant rate of urine formation (water excretion) in contrast to the marine teleosts, a fact that pehaps explains why the former do not show the glomerular degeneration or the aglomerular development as has been observed in the marine teleosts.

28. Paraneurone Cell System of Skin and Gills of Fishes

Various kinds of paraneuronal cells are the sources of neurologically active substances, typical of the endocrine cells belonging to the diffuse neuroendocrine cells system scattered throughout the animal body.

In fishes the paraneuronal cells of the skin and the gills are likely to control the complex epithelial functioning by a paracrine mode of action. The uptake of ions, cell homeostasis and the recepto-secretory functions of the epidermal and gill paraneurons are now matters of pioneering studies showing that the skin and the gill are neuroreceptor and endocrine organs.

Certain non-paraneuronal cells in fish skin and gill epithelia also produce neuroactive substances. The sacciform gland cells produce bomesin/gastrin-releasing peptide, caerulein and serotonin and the club cells in the skin are a storehouse of bioactive compounds. Serotonin is also ectopically produced by fish photophores. Similar bioactive substances occur in frog cutaneous exocrine glands and in the Leydig cells of urodelan larvae which contain both serotonin and B-endorphin substances.

Closed and open types of neuroendocrine cells were recently shown by immunohistochemistry in fish gill and regarded as paracrine cells.

Sites for the production of nitric oxide (NO) in the neuroendocrine cell system of the gill epithelium and the gas bladder of the catfish *Pangasinus sutchi* have been demonstrated. 'NO pathway' is now regarded as an efficient regulatory system implicated in a multiplicity of physiological actions and recognized for the first time in fish epithelia.

Although fish studies have reported very little information about 'paraneurons' as a concept, all the above skin and gill cells showing a paracrine appearance await investigation. It is assumed that they have a regulatory function.

Historical Background

The term 'paraneuron' is currently used to define a large group of cells previously considered as endocrine and sensory cells because they share biological features with neurons. Paraneurons are recepto-secretory cells possessing a receptive site and a secretory site on their plasma membrane (Fujita, 1994). They release peptidic, aminic and other signal substances that are common to neuronal secretions.

The functions of all the epithelial paraneuronal cells, described in this chapter, are associated with the production of bioactive substances. These cells synthesize and store these substances in the cytoplasmic granules. Ultrastructural and immunohistochemical markers render possible easy recognition of some cells types, but the physiological role of the paraneurons varies between different species. Due to their 'biphasic' (sensory and endo or paracrine) function and subsequent polarization, it should be appropriate to include some paraneurons (like those of the lung, air-bladder and gill), in the group of specialised 'sensoneuroendocrine cells'. Information about specific signals (like hypoxia) are providing by the release of peptides and amines, and the receptors of these substances, together with their binding affinities, include the modulation of the responses of the above cells by central nervous mechanisms.

Paraneurons in the Skin Epithelium

Merkel Corpuscle

Merkel corpuscles are specialized epithelial secretory cells present in the Stratified squamous epithelia of most vertebrate classes. They remain isolated or dispersed in or near the basal layer of epidermis and oral mucosa. Merkel cells and apposed nerve endings are considered to be mechanoreceptive.

Fish Merkel corpuscle comparatively received little attention in comparison with those of higher vertebrates. The distribution and the morphology of these cells in the lower vertebrate skin were reviewed by Whitear (1989). Whitear (1989) describe Merkel cells in 12 species of teleost fishes and their distribution over the body from barbels to fins, also in the oral epithelium. These have dense core vesicles and granules of varied size. These vesicles are often closely adjacent to the plasma membrane, both facing the nerve and elsewhere, suggesting a tactile receptor activity of the Merkel cells.

Merkel cells are still not identified in the cartilaginous fishes, nor in the myxinoids. In lampreys Merkel cells occur at various levels in the epidermis of adult and larval stages and their structure is comparable to that reported from the gnathostome classes. In the sea lamprey, *Petromyzon marinus* numerous cells immunoreactive for both leu-5-enkephalin and met-5-enkephalin, resembling Merkel cells, are found in the epithelium of oral papillae.

Antibodies to neuron-specific enolase, serotonin, enkephalin and vasocative intestinal polypeptide were used to study the distribution of the Merkel cells in the skin of the conger eel, *Conger conger,* and the Indian catfish, *Heteropneustes fossilis*. The demonstration of serotonin was reported for the first time in the Merkel cell of the conger eel by Zaccone (1986). Merkel cells are present in the head skin of the catfish *H. fossilis* and the lip skin of conger eel.

Most of the immunohistochemical results obtained in mammals indicated that the Merkel cells are a soure of numerous bioactive substances. Some of the above substances are found in the secretary granules of Merkel cells.

The immunohistochemical and electron-microscopic characterization of the Merkel cells in fish skin lead to the conclusion that these cells closely resemble those in higher vertebrates, though a possible neuroendocrine function has not been proposed.

Recent studies demonstrated that these cells have a multifunctional role in the skin of humans. Merkel cells may play a role in touch perception and/or nerve growth may be induced by nerve growth factor produced by these cells. Besides the neurohormonal factors, Merkel cells contain neuro-peptides, such as substance that may induce the proliferation of mesenchymal cells and so the hair follicle differention of memsenchymal cells. Some investigators have also postulated that the Merkel cells serve to stimulate the local proliferation and differentiation of keratinocytes as well as to maintain normal epidermal structure. Studies are needed to clarify the paracrine functions of the Merkel Cells of fishes.

Chapter 3
Evolution of Air-breathing Habit in Vertebrates

The oxygen level of the atmosphere further went down from Devonian to Carboniferous and drastically low in the Permian and Triassic periods. The last placoderms were seen giving rise to the actinopterygians in the Permian. They were the primitive palaeoniscids and the holosteans which had both gills and lungs. In the Triassic about 190 milion years ago, warm conditions returned, without seasonal variation. The bony fishes (Actinopterygii) underwent adaptive radiation in the waters and the reptiles on the land during this period. Warmer conditions persisted throughout the Jurassic period.

In the Cretaceous, about 120 million years ago, cold conditions returned with seasonal variations. Mountain building and glaciation occurred in some parts of the earth. The temperature was considerably low. Atmospheric oxygen level went up. The hot conditions set in again for the main part of the tertiary which begin 70 million years ago, gradually diminishing towards its close when the Himalayas and the Alps were formed.

It is relevant to quote from Professor Birbal Sahni's Presidential address at the Indian Science Congress held at Madras in 1940, where he has described the dawn of the 'Tertiary era – the era when the modern air-breathing teleosts were evolving with adaptive radiation.

"Competent authorities place the dawn of the Tertiary era between sixty and seventy million years ago. It is the birth of a new era in a very real sense. Stupendous forces, surging in the womb of the earth, had already caused gigantic rifts in the crust, and these rifts are gaping out into oceans. From smaller fissures in the crusts, molten rock is now pouring forth in repeated floods of lava which will cover millions of square miles of land and sea. Vast areas are being convered into desert by showers of volcanic ash. A new type of landscape develops, with high volcanic plateau (like the Deccan plateau) as a dominant feature. The face of the earth is rapidly changing. She puts on 3 more modern garb of vegetation. The land, lakes and rivers become peopled by creatures more familiar to us. Still there is no sign of man. But the stage is being set for his arrival. For this critical period foreshadows the birth, out of the sea, of the mightiest mountains of the world (Himalayas and Alps), and the heavy bosom of the earth; somewhere to the north of India, is to be the cradle of man."

The volcanic activity developed in big scale consuming oxygen at high rate consequently the atmospheric oxygen level went down drastically (0.1 per cent PAL). Then about one million years ago, came the great Ice-Age, with its four episodes of glaciation and three interglacial periods. Mammals continued evolving during this period, towards the end of which man appeared.

The most important early changes in the environment as far as the vertebrates were concerned were the drying up of the lagoons and estuaries in the Devonian and, the variation of temperature and ambient oxygen during the Mesozoic era, and Tertiary and Quaternary periods of Cenozoic (Munshi, 1990).

Atmospheric Gases

Life, it seems, has evolved in a chemically reducing atmosphere of methane, ammonia and hydrogen (Tappan, 1968). The production of oxygen in the environment was the result of the evolution of blue-green algae capable of photosynthetic activity. The oxygen level remained low for a considerable period because any oxygen produced was rapidly utilized in oxidation of the substrates and flowing molten lava. In the Pre-Cambrian period, oxygen levels in the atmosphere were probably around 3 per cent of the atmosphere as compared to the present 16 per cent (Odum, 1971). The oxygen in the Tertiary and Quaternary periods was about 6 per cent of the Present Atmospheric level (PAL).

The gas analysis of thermal springs of Bihar reveals the presence of rare gases like helium and argon along with oxygen, nitrogen and carbon dioxide. At the time of fault formations in the earth's crust probably the atmospheric gases got entrapped deep inside the lithosphere. The composition of the thermal spring gases indicates the nature of the atmosphere during the Tertiary and Quaternary periods (Table 3.1; Munshi *et al.*, 1984)

The Evolution of Air-Breathing Habit

Air-breathing teleosts of the modern time inhabit both freshwater and intertidal environments. The crossopterygians are generally regarded as being a freshwater group since the mid-Devonian. As such salinity does not seem to be an important factor limiting the evolution of air-breathing vertebrates (Thomson, 1969; Packard, 1974). Aquatic hypoxia in freshwaters is probably the more important selective force in the evolution of air-breathing vertebrates. There have been oscillations in atmospheric oxygen levels over more recent times in the Quaternary period and these changes would have been duplicated in the water. Under such conditions the air-breathing habit in fishes will definitely have selective advantage (Dejours, 1988).

Table 3.1: Gaseous Composition of the Hot Springs in India

Sl.No.	Source	Temperature °C		He per cent	Ar per cent	O_2 per cent	N_2 per cent	CO_2 per cent	H_2 per cent	CH_4 per cent
		Water	Air							
1.	Surajkund	85.0	25.0	1.4	1.6	1.8	95.0	0.6	Nil	Nil
2.	Bhimbandh	63.0	23.5	55×10^{-4}	1.2	6.0	65.2	27.6	Nil	Nil
3.	Tantloi	61.0	23.0	0.6	0.9	6.0	91.5	Nil	Nil	1.0
4.	Rishikund	46.0	24.0	31×10^{-4}	1.3	5.2	73.1	20.4	Nil	Nil

We envisage that the following sequence of events occurred in the evolution of air-breathing in vertebrates. In conditions of aquatic hypoxia there is a clear selective advantage to those animals with some means of utilizing oxygen from the atmosphere. The first step in this direction was the habit of taking air into the pharynx when at the surface in the air-water interface. The fish ventilates its air-breathing pharyngeal or supra-branchial organs by using the buccal pump, which is used for gill ventilation. Thus the first step was probably the evolution of an air-breathing organ ventilated by a buccal pump. Aspiratory modes of ventilation, which have developed in a variety of groups of air-breathing animals allow greater control over tidal volumes and separation of ventilatory mechanisms of air-breathing organs from those associated with feeding. Aspiration has evolved in aquatic as well as terrestrial vertebrates as a mode of ventilating their air-breathing organs (Liem, 1988).

Evolution of Accessory Respiratory Organs in Teleosts

During the early Devonian, the atmospheric oxygen was more or less like the present period which started declining in the late Devonian with drying up of lagoons, swamps and freshwater bodies. These adverse ecological conditions led to the development of lungs from pharyngeal/branchial pouches in the Dipnoi, crossopterygians and Amphibia. The extinct fish *Cheirolepis* which was the ancestor of the group Palaeoniscoidea, acquired lungs besides the gills. The palaeoniscids later on gave rise to the Chondrostei group of fishes represented by the modern forms *Polypterus* and *Polyodon*, and the Holostei represented by *Amia* and *lepisosteus*. The teleostean fishes became the masters of fresh as well as sea water during the Tertiary and Quaternary periods of the Cenozoic era. The lungs got transformed into the specialized hydrostatic organ, the swimbladder (Fange, 1976). During the Tertiary and Quaternary periods the oxygen level of atmosphere fell down considerably affecting the oxygen content of the water. The gills were unable to cope up with the depletion of oxygen in water especially in the freshwaters of rivers and swamps. The lungs were no longer available for aerial respiration as they got modified into the specialized organ of swimbladder. As such the advanced group of teleostean fishes represented by the modern forms of *Anabas, Trichogaster, Channa (=Ophiocephalus), Clarias, Heteropneustes, Monopterus (Amphipnous)* had to develop some other types of accessory respiratory organs. These air-breathing organs are essentially modification of the gills (Munshi, 1980, 1985; Munshi and Hughes, 1972). The swimbladder in these fishes is either absent or very much reduced. The linings of the accessory respiratory organ are essentially mucoid in nature, while the lungs have a surfactant lining of phospholipids (Pattle, 1976; Hughes and Weibel, 1976).

Bimodal Gas Exchange

In air-breathing fishes the accessory respiratory structures can easily cope with serving the oxygen demand, but are unable to function efficiently in the elimination of carbondioxide (Hughes and Singh, 1970a,b, 1971; Singh and Hughes, 1971, 1973). In general, the gills or skin are used for CO_2 elimination and ion, water and pH regulation (Johansen, 1970). As such at initial stages of the evolution of air-breathing, bimodal breathing is required where the air-breathing organs are used for oxygen uptake, and the gill respiratory surface is used for CO_2 elimination and pH regulation (Singh, 1976). Transitional forms went through stages of bimodal gas exchange with a large exchange ratio initially in the gill-skin system and a low ratio initially in the accessory respiratory organ or primitive lung.

In reptiles, birds and mammals the lungs play an important role in acid-base regulation by controlling CO_2 levels in the blood via ventilation. This enables the animal to become less dependent on water for ion and H^+ regulation and then they could eliminate the need for bimodal breathing through gills/skin and accesory respiratory organs (Randall *et al.*, 1981). The moistening of the skin and the respiratory surface by gill ventilation is no longer required and the cooling effects of water on the surface, due to its high specific and evaporative heat have disappeared. Then heat loss could be controlled and body temperature could be regulated at a level higher than the ambient temperature. The evolutionary changes in the vertebrate circulatory system reflect this sequence of events of aquatic forms evolving from uni-modal water breathing into bimodally breathing amphibians and then to unimodal air-breathing animals (Munshi *et al.*, 1986, 1990; Munshi and Hughes, 1987; Olson *et al.*, 1986).

Macro- and Micro-circulation of the Accessory Respiratory Organs

Plastic vascular corrosion replica were prepared of gills and air-breathing organs and examined under light and scanning electron microscopes.

Macro-circulation

Unlike obligate water-breathing teleosts, the bulbus arteriosus in the facultative air-breathing fish, *Heteropneustes fossilis* gives rise to three separate ventral aortae, with an aberrant small vessel arising from the first ventral aorta that represents Ist aortic arches of mandibular segment (Olson *et al.*, 1990). In the obligate air-breathing fish, *Anabas testudineus* and *Channa punctata* the bulbus arteriosus gives rise to a very short ventral aorta which divides into two distinct, - dorsal and ventral branches. In the mud-eel, *Monopterus cuchia* unlike other fish the single long ventral aorta gives rise to six aortic arches, supplying blood to mandibular, hyoid and four branchial arches. A seventh aortic arch (5th branchial) is also present which supplies blood to the hypopharyngeal region. The relationship between the respiratory (gills, air-breathing organs), and the systemic circulation in *H. fossilis, A. testudineus, C. punctata* and *M. cuchia* have been worked out. In *H. fossilis* the discovery of an alternative delivery route of oxygenated blood from the efferent air-sac artery to the dorsal aorta is interesting from physiological point of view. In the obligate air-breathing fish, *Anabas* and *Channa* there seems to be some mechanism in the heart to separate oxygenated blood of the respiratory organs from the venous blood of the systemic circulation.

Micro-circulation

The micro-circulation of gills and accessory respiratory organs of *H. fossilis* and *M. cuchia* have been worked out. Air-sac lamellae of *Heteropneustes* are less than 1/3 the size of gill lamellae, with their afferent and efferent supply. The outer marginal channels and the next ones carry most of the RBCs, increasing their haematocrit values 62-68 per cent from normal value of 37 per cent (Hughes, *et al.*, 1992). Seventy five per cent of the lamellar sinus is discontinues and contain plasma and WBCs (Munshi *et al.*, 1986).

In *M. cuchia* the respiratory islets of buccopharynx and air-sacs consist of clusters of blood capillaries. Each vascular unit (rosette) within the respiratory islet (RI) consists of a somewhat centrally placed afferent arteriole that gives rise to several spiral capillaries that radiate outward from the artery. The capillaries are drained by a venous plexus that runs circumferentially around the capillaries. There is evidence that these respiratory islets have been designed from lamellae like structures (Munshi *et al.*, 1990)

The spiral nature of the islet capillaries is an unique feature of islet circulation of *M. cuchia*. These capillaries seem to be homologous to the outer marginal channels of gill lamellae. Several functions for the spiral vessels and their related endothelial cell valves have been suggested (i) to increase resistance, (ii) to increase residence time of red cells – thereby maxitimize gas equilibrium; (iii) the valves force the red cells to the epithelial surface of the capillary wall and minimize diffusion distance between the air and red cells.

Salient Features of Vascular System of Air-Breathing Teleosts

(i)   Secondary blood capillaries have developed on arterial supply vessels of air-breathing organs, brain and retina.

(ii) Evidence of plasma skimming at micro-circulation level led to higher haematocrit value at gas exchange surface.

(iii) Shunt systems have developed at different levels of macro- and micro-circulatory systems of air-breathing fishes.

(iv) The air-breathing respiratory sacs and their vascular supply of *Heteropneustes fossilis* seem to be most efficient so far as oxygen delivery to systemic circulation is concerned.

(v)   In *Anabas, Channa* and *Monopterus* the arteriovenous pathway existing between the efferent filementar arteries and venous system of gills could have been utilized in draining blood from RI to veins.

(vi) The cardiac output in the air-breathing fishes seem to have been put differentially into several ventral aortae according to the needs of the fish in relation to the hypoxic-hypercarbic conditions of the environment.

(vii) There is clear division of drainage of blood of the cephalic region of *Anabas* and *Channa* into the two systemic circulation of coeliaco-mesenteric (CM) and dorsal aorta (DA).

(viii) There may be some mechanism in the heart of air-breathing fishes to prevent mixing of oxygenated and deoxygenated blood.

Capillary Loading (V_c/S_a) of Respiratory Membrane

Capillary loading, which gives information about the volumes of blood relative to the surface available for gas exchange, is larger in fish than mammals. In *Monopterus*, it is about 2.5 times greater than an average for mammals of the same body weight. In *Dipnoi, Lepidosiren* has larger red cells and capillary volume (Hughes and Weibel, 1976). The large capillary loading of fish lungs and air sacs are partly due to gas exchange being restricted to one side of the Capillary, but in some cases (Hughes, 1978) this is compensated by added curvature of the air/blood barrier. The *Monopterus* respiratory papillae with their curved surface are good examples of this type of adaptation.

Angiotensin-Converting Enzyme and Hormone Metabolism

Recent studies (Olson *et. al*, 1987) show that there is considerable angiotensin-converting enzyme (ACE) activity in the respiratory organs of most air-breathing fish. In all but two of the species examined, ACE activity in the accessory respiratory (air-breathing) organs was as high as, or higher than, that found in the gill. In nearly all of these fish, the vasculature circuit is modified from the typical aquatic breathing teleost to include the accessory respiratory organs and compensate for a diminished role of the gills in gas exchange. As the vascular surface of the gills becomes attenuated it is also able to metabolize circulating hormones. Clearly, the presence of ACE in the accessory respiratory organs ensures maximal exposure of this enzyme to the plasma. It is possible that the evolutionary transition from aquatic to air-breathing not only encompassed by variety of physiological and anatomical adjustments in gas exchange and osmoregulatory tissues but that the variety of biochemical (*i.e.* nonrespiratory) lung functions were originally developed in the fish gill and were subsequently shifted to the pulmonary tissues. Interestingly, in this regard, the gill and lung show the same preference for norepinephrine over epinephrine during both accumulation from the plasma and metabolism (Nekvasil and Olson, 1986).

Angiotensin converting enzyme was measured in tissue homogenates from the African lungfish and six species of air-breathing teleosts (*Heteropneustes fossilis, Clarias spp. Anabas testudineus, Notopterus chitala* and *Monopterus cuchia*) using a standard spectrophotometric assay. In most species, the highest levels of ACE activity were found in the respiratory organs (gills and/or accessory respiratory organs). ACE was also found in heart and kidney tissues from most species and occasionally in liver. Converting enzyme was not found in skin or skeletal muscle from any species and only in blood from *H. fossilis* and brain from *C. batrachus*, Captopril, a potent inhibitor of mammalian ACE, inhibited enzymatic activity from all tissues except *C. gachua* heart and liver and *A. testudineus* heart. As fish make the transition from aquatic to aerial respiration the gill microcirculation is usually reduced in size and the accessory respiratory organs become elaborated and occupy a more central position in the vascular tree. The presence of ACE in accessory respiratory organs of air-breathing fish appears to greatly enhance the metabolic efficiency of this enzyme on circulating substrates (Olson *et al.*, 1987).

Structural Evolution of Vascular Papillae

The vascular papillae are specialized blood capillaries connecting the branchial arterial system with those of veins of the jugular system for gas exchange. In *Monopterus, Channa* and *Anabas* oxygenated blood from the capillaries of the air-breathing organs returns to the anterior cardinal vein or its branches. In *A. testudineus* the respiratory islets are composed of a series of parallel blood capillaries, each capillary being made up of a single row of endothelial cells. Their prominent cell bodies project into the capillary lumen. The endothelial cells have tongue like processes which may act as minute valves controlling the flow of blood (Hughes and Munshi, 1975a, in Munshi *et al.*, 1986). These parallel capillaries are separated from each other, not by pillar cells but by epithelial cells. In *M. cuchia* these capillaries have become more complex due to their spiral disposition giving rise to the vascular papillae in series. As each turn of the spiral the epithelial cells along with the basement membrane get tucked into the bases of the curvature of the capillary forming the vascular papillae. At these points the endothelial cells develop enormously to take up the role of valve-like structures much like these of *Anabas* and *Channa*. These endothelial valves are metabolically active with mitochondria and vesicles. The structural similarities of the vascular papillae and their similar disposition in the buccopharynx both in *Channa* sp. and *M. cuchia* are very much apparent (Hughes and Munshi, 1986). At this same time while *Anabas* and *Channa* have similar gill structure with their characteristic shunt vessels, *M. cuchia* has lost functional gills. The structural similarities found in the microcirculatory system of air-breathing organs in the three species of fishes studied suggests a morphological series, but whether it represents an evolutionary sequence is conjectural but at least represents one possibility.

The *Anabas* type may represent one of the first steps in the evolution of air-breathing organs in the Percomorphi group of fishes, where the arteriovenous connections are straight endothelial tubes situated just below the epithelium. In the second stage, they became arranged in a wave like fashion as in *Channa* sp. in the third stage, as in *M. cuchia*, the capillaries take the forms of spirals. This spiral type of capillary also exists in the gill system of *M. cuchia*. The functional significance of these wave or spiral like arrangements of capillaries seem to be that more blood could be accommodated in a small space of respiratory islets. Moreover, the system develops an efficient mechanism to bring every individual RBC into close contact with the respiratory surface for gas exchange. It also provides enough resistance to increase their residence time and ensure proper oxygenation (Munshi *et al.*, 1989).

Homology of Lungs, Swimbladder and Acessory Respiratory Organs of Animals

It is generally considered that the lungs of Dipnoi and of primitive actinopterigians acipenser fish of group Chondrostei (*Polypterus, Polyodon*) and Holostei (Amia, Lepisosteus) which posses lungs are homologous with the swimbladder of teleosts. The lungs of these primitive fishes resemble those of Dipnoi, Crossopterygii and tetrapods (Amphibia, Reptiles, Aves and Mammalia including man) in having "lamellated osmiophilic bodies (LOB) in their epithelial cells (Hughes, 1970, 1973; Pattle 1976, Gonialowska-Witalinska, 1971, 1979, 1980, 1982, 1984, 1986). These structures (LOB) secrete di-pulmonary lecithins – which is a sort of "lung surfactant" forming a lipid lining of lung alveoli. The lipids contain 74 per cent phospholipids, 8 per cent cholesterol, 10 per cent triglycerids and 8 per cent fatty acids and 1 per cent protein (Pattle, 1976). The main function of this surfactant material is so reduce surface tension (Weibel, 1964). The great interest taken in the surfactant arises mainly from its connection with the respiratory distress syndrome of the newborn human. These lamellated osmiophilic bodies are absent in the accessory respiratory organs of *Anabas, Channa, Amphipnous, Clarias* and *Heteropneustes*. The lining of these organs is essentially mucoid in nature (Munshi *et al.,* 1989) These studies rule out their homology with lungs. They have developed from gills andserve the function of gas exchange from air (Hughes and Munshi, 1979, Munshi, 1985).

Chapter 4

Morphometrics of the Respiratory Organs of the Indian Green Snake-Headed Fish, *Channa punctata*

Introduction

Channa (=Ophicephalus) punctata (=punctatus) (Menon, 1974; 90-91) is a very common air-breathing fish of Indian swamps and ponds. Gunther (1880) reported that "this fish is able to survive droughts, living in semi-fluid mud or resting in a torpid state beneath the hard-baked crust at the bottom of tanks from which every drop of water has disappeared". Boake (1866), Day (1868), Dobson (1874) and Ghosh (1934) experimented with this species and the time taken to die when denied access to air *i.e.* "drowning" varied from 1 hour 30 min (Day) to 16 hours (Ghosh). Day (1976) observed *C. punctatus* in swamps, ponds and ditches of stagnant muddy waters of the Indo-Gangetic plain and even from hilly tracts. According to Hora (1935). "The snake-headed fishes are found in the freshwaters of eastern and south-eastern Asia but are more abundant in confined waters specially marshes with thick growth of vegetation. Some live in burrows along the edges of ponds, rivers and brooks. In an aquarium they lie quietly at the bottom, rising occasionally to breathe air."

These fishes are capable of living a normal life even in waters having very little dissolved oxygen. They can live out of water for long periods (about 24 hours) if the skin remains moist. They have functional gills with accessory respiratory organs for aerial respiration (Munshi, 1962; Saxena, 1967; Munshi, 1976). The accessory respiratory organs of *C. punctata* comprise the suprabranchial chamber, the hyomandibular partition, and the dendritic plate. The suprabranchial chamber is developed dorsal to the gill arches above the pharynx and lateral to the prootic bone of the auditory capsule on either side of the postero-lateral region of the skull. The hyomandibular partition is a vertical shelf-like outgrowth projecting into the chamber and dividing the chamber into a small anterior and a large posterior chamber. The anterior and the posterior chambers both have an opening on either side of the hyomandibular partition. These inhalant apertures are closed by two knob-like projections in the roof of the buccopharynnx. The posterior chamber is contractile due to the presence of the constrictor supra-branchialis muscle. The posterior chamber opens into the opercular chamber and the opening is covered by the dendritic plate, which acts as a valve. This opening probably serves for the exit of air from the suprabranchial chamber.

The lining epithelium of the suprabranchial chamber, the hyomandibular partition, the labyrinthine organ and the inner-surface of the dendritic plate is highly vascularized surface at which gaseous exchanges occur.

Several studies have been made of the gill dimensions of teleostean fishes (Hughes, 1966; Hughes and Morgan, 1973) of which some have paid particular attention to the relationship with body weight.

Morphometric studies of the lung of mammals and other tetrapods have been carried out by Tenney and Remmers (1966), Tenney and Tenney (1970), Weibel (1963, 1969), Hutchison, Whitford and Kohi (1968) and others, but comparable studies on air-breathing fishes are somewhat limited. Recently, Rahn, Howell, Gans and Tenney (1971) made morphometric measurements of the lungs of *Lepisosteus asseus*; Hughes, Dube and Munshi (1973) and Hughes, Singh *et al.* (1974a,b) have established a definite relationship between bimodal gas exchange organs and body weight in some air-breathing fishes of India. These studies have shown variations between species in relation to their air-breathing habits.

Other works available on Indian fishes are from Dubale (1951), Saxena (1958, 1959, 1962 and 1963) and Ojha and Munshi (1974).

The present work is an attempt to study the morphometrics of the respiratory organs of *Channa punctata* in relation to body weight and their gas exchange function.

Materials and Methods

Live specimen of *Channa punctata* were collected from ponds of Bhagalpur and nearby places and were maintained in aquaria for 4-6 days. During this period the fishes were treated with potassium permanganate ($KMnO_4$) and methylene blue, so that any wounds on the gills could heal.

Fresh body weights were determined and the fishes divided into 11 weight groups, each having a range of 10g, the total weight range of the samples being 1-110g. The fishes were killed, opercula removed, and fixed in 5 per cent formalin. After 2-4 days fixation, all four gills of one side together with the lining of the suprabranchial chamber were dissected out carefully with the hyomandibular partition and dendritic plate and were washed thoroughly in running tap water and slowly processed through the different grades of alcohol and were preserved in 70 per cent ethanol.

For morphometric studies, the gill arches were stained overnight with borax carmine and measurements made under a dissecting binocular microscope. Each gill arch was divided into sections of 20 filaments and an average filament length determined for each section by measuring the length of the middle pair of filaments, and the total filament length calculated for each section. To measure the number of secondary lamellae/mm, 6-10 counts were made along the length of each filament of the middle pair of filaments and an average value obtained. This number of secondary lamellae/mm was doubled because they are present on both sides of the filament. This figure was multiplied by the total length of the filaments of that section to give the total number of secondary lamellae for that section. The average bilateral surface area of a secondary lamella was obtained graphically. Hand cut sections from the tip, middle and base of each of the middle pair of filaments were mounted in glycerine and viewed under a microscope. Camera lucida drawings of the shape were traced on graph paper and an average unilateral surface area calculated from the drawings of secondary lamellae from different regions.

From this average bilateral area the total area of the section was calculated. The numbers of secondary lamellae of all the sections were summed and divided by the total filament length to obtain a weight value for secondary lamellae/mm. Similarly the summed total surface area of all the sections divided by the total number of secondary lamellae gave a weight value for bilateral surface area. The product of these two weight values multiplied by total length of the filaments of the gill arch gave the total surface area of the gill arch. The total surface area of all four arches was summed and doubled to obtain the total gill surface area of the fish.

Morphometrics of the Suprabranchial Chamber

Surface area of the suprabranchial chamber was determined by tracing the shape of the membrane pieces on graph paper. Volume was determined by filling the suprabranchial chambers with coloured water from a micro burette in smaller fishes and through a 50 ml burette for larger fishes. The volume was verified by filling the suprabranchial chamber with fine sand (BDH).

The general equation $Y=aW^b$ or $\log Y = \log a + b \log W$ was used to show the various allometric relationships, where Y is the parameter analysed.

W is the weight, a and b are respectively the intercept and the slope of the regression line.

The data obtained from the measurements were analysed by linear logarithmic transformation, using the least square repression method. A Facit hand calculator and HP 9810A computer were used for the computations.

Drowning Experiment

The term "drowing' is used in the same context as used by Hora (1935) *i.e.* when an air-breathing fish is prevented from coming to the surface for air and ultimately dies. The experiment was started on 10.8.1974 at 9.30 a.m., the ambient temperature being 31°C. Eleven laboratory adapted fishes, 12,14,17,20,33,44,63,100,101 and 108g were used. These fishes were divided into two groups and placed in separate aquaria containing seven and four fish. The wirenet lid of each aquarium was tied tightly so that the fish could not escape. Both aquaria were suberged under water in a large reservoir through which a continuous flow of fresh tap water passed at a rate of about 2 litres/min. This flow was sufficient to maintain the oxygen content at 5.5-6.4 mg/l. The temperature ranged from 30 to 32°C. The oxygen content of the reservoir and the ambient temperature were noted every 2-3 hours.

Diffusing Capacity of the Respiratory Organs

The diffusing capacity of the different gills and the suprabranchial chamber were calculated using a modified Fick equation (Hughes, 1972).

The value for the permeation coefficient used in the calculations was (0.00015 ml/min/µm/Cm^2/mmHg) as obtained by Krogh (1919) for frog connective tissue. Very few data are available on the permeation coefficient for different tissues, and for fishes no such measurements are available.

Results

All dimensions of the gills appear to increase with an increase in body weight. The dimensions in any particular specimen vary from the 1st arch and the smallest in the 4th (Figure 4.1). Individual filaments of the anterior hemibranchs were smaller

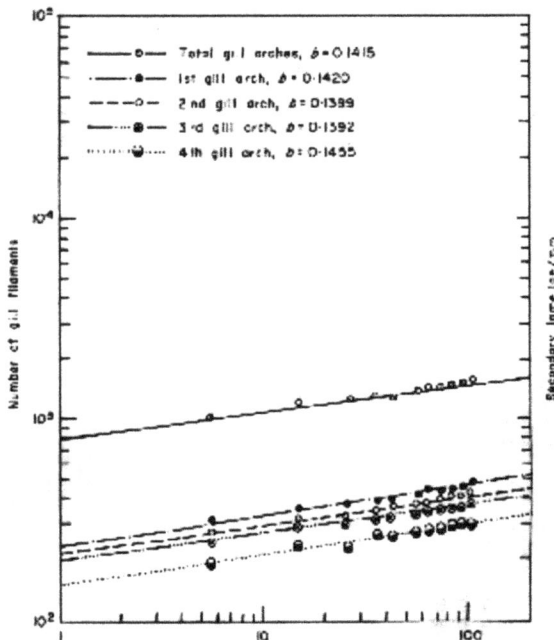

Figure 4.1: Plots on logarithmic coordinates of the number of gill filaments for each of the gill arches and their total number, against body weight.

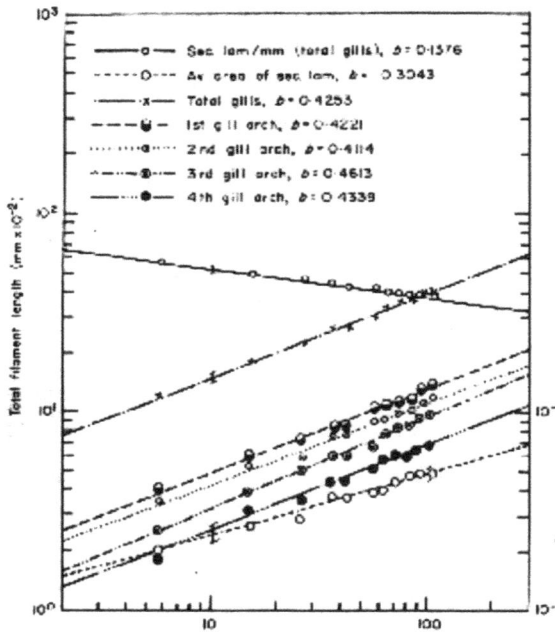

Figure 4.2: Bi-logarithmic plots of the secondary lamellae/mm bilateral surface area of an average secondary lamella, and total filament length against body weight (Brackets show 95 per cent confidence limits for fish of 10 and 100 g). Separate plots are also given for the filament length for each of the four gill arches.

than those of the corresponding posterior hemibranch, though they are more or less equal in size in the middle section of each arch.

The size of the secondary lamellae appears to be directly proportional to filament length, but the frequency of the secondary lamellae is inversely proportional to the length of the gill filaments. The data obtained for 35 fishes are summarized in (Tables 4.1–4.6).

Regression analysis was carried out and the values for the intercept a, the regression coefficient b (slope) and the correlation coefficient r are given in Tables 4.7 and 4.8. Separate analyses were made for each arch and the corresponding values when all measurements for the gill arches were taken together. Relationship based on the allemetric equations are given in Tables 4.7 and 4.8. From these relationships dimensions for 1,10 and 100 g fish were calculated (Table 4.9) together with 95 per cent confidence limits and standard deviations for the slope and intercept values (Sb, Sa).

Relationships between Body Weight and Total Number and Average Length of the Filaments

The relationships between both measurements and body weight showed significant correlation (Tables 4.7 and 4.8). Differences in respect of parameters found for the four different gill arches are also summarised in these Tables.

Table 4.1: *Channa punctata* (Bloch) mean values for the body weight and other component parameters of the first gill arch in each of the 11 weight groups.

Sl.No.	Body Wt. (g)	Total Gill Filaments (Numbers)	Average Filament Length (mm)	Total Filament Length (mm)	Secondary Lamellae/ mm	Total Secondary Lamellae	Average Surface of a Secondary Lamellae (mm²)	Total Gill Area (mm²)	Gill Area/g Body Weight (mm²)
1	2	3	4	5	6	7	8	9	10
1.	5.67	310.67	1.3036	404.989	54.574	22102.0	0.024590	543.49	95.853
2.	15.00	347.33	1.7061	392.569	48.304	28624.0	0.027501	787.18	52.479
3.	25.67	475.33	1.9028	714.175	47.051	33602.0	0.029438	989.19	38535
4.	36.40	391.30	2.1319	834.015	43.840	36563.0	0.032730	1196.70	32.877
5.	43.33	382.00	2.1986	839.870	41.473	34832.0	0.037483	1305.52	30.130
6.	57.33	424.00	2.3958	1015.83	41.377	42032.0	0.042846	1800.88	31.413
7.	64.00	426.67	2.5345	1081.38	39.688	42918.0	0.045253	1942.16	30.346
8.	74.00	414.00	2.6992	1117.46	38.200	42687.0	0.047530	2028.91	27.418
9.	84.00	436.67	2.6175	1142.91	37.684	43070.0	0.052780	2273.22	27.154
10.	93.67	464.00	2.8557	1325.06	37.808	50098.0	0.052637	2637.14	28.154
11.	105.67	490.67	3.0035	1475.73	36.258	53434.0	0.059812	3195.91	30.244
Average	54.98	416.776	2.3045	958.3623	42.4052	39087.0	0.041145	1700.03	38.592

Table 4.2: *Channa punctata* (Bloch) mean values for the body weight and other component parameters of the second gill arch in each of the 11 weight groups.

Sl.No.	Body Wt. (g)	Total Gill Filaments (Numbers)	Average Filament Length (mm)	Total Filament Length (mm)	Secondary Lamellae/ mm	Total Secondary Lamellae	Average Surface of a Secondary Lamellae (mm²)	Total Gill Area (mm²)	Gill Area/g Body Weight (mm²)
1	2	3	4	5	6	7	8	9	10
1.	5.67	275.33	1.2589	346.62	54.869	19018.7	0.022840	434.39	76.611
2.	15.00	333.33	1.5866	528.85	49.277	26060.1	0.028193	734.71	48.981
3.	25.67	332.00	1.7865	593.11	45.809	27169.5	0.028270	768.08	29.921
4.	36.40	357.60	2.0990	750.59	44.523	33418.0	0.031789	1062.36	29.186
5.	43.33	366.67	2.0920	767.08	42.663	32726.0	0.035615	1168.97	26.978
6.	57.33	373.33	2.3241	867.65	41.017	36688.0	0.040066	1425.89	24.872
7.	64.00	384.00	2.3708	910.40	40.061	36471.0	0.039658	1446.45	22.601
8.	74.00	400.00	2.4495	979.80	38.335	37561.0	0.043647	1639.43	22.154
9.	84.00	408.00	2.5088	1025.57	38.354	39258.0	0.050134	1968.15	23.430
10.	93.67	409.33	2.6500	1084.75	38.971	42273.0	0.053771	2273.01	24.266
11.	105.67	426.67	2.7992	1194.35	37.989	45373.0	0.052435	2379.35	22.517
Average	54.98	369.66	2.175	822.432	42.897	34137.85	0.038774	1390.98	31.974

Table 4.3: *Channa punctata* (Bloch) mean values for the body weight and other component parameters of the third gill arch in each of the 11 weight groups.

Sl.No.	Body Wt. (g)	Total Gill Filaments (Numbers)	Average Filament Length (mm)	Total Filament Length (mm)	Secondary Lamellae/ mm	Total Secondary Lamellae	Average Surface of a Secondary Lamellae (mm^2)	Total Gill Area (mm^2)	Gill Area/g Body Weight (mm^2)
1	2	3	4	5	6	7	8	9	10
1.	5.67	249.33	1.0208	254.52	54.135	13778.7	0.019630	270.48	47.703
2.	15.00	289.33	1.3978	404.43	49.676	20090.5	0.024327	488.74	32.583
3.	25.67	310.67	1.6518	513.16	46.709	23969.0	0.026740	640.93	24.968
4.	36.40	352.80	1.8703	622.45	43.516	27086.5	0.033453	906.12	24.894
5.	43.33	329.33	1.8970	624.74	42.645	26641.6	0.035549	947.08	21.857
6.	57.33	326.67	2.0402	666.47	39.134	26081.5	0.035737	932.15	16.259
7.	64.00	352.00	2.2517	792.68	40.200	31862.9	0.035622	1134.96	17.734
8.	74.00	358.67	2.3785	854.47	38.436	32842.4	0.040606	1333.60	18.022
9	84.00	368.00	2.3426	862.08	38.939	33568.2	0.046565	1576.53	18.768
10.	93.67	369.33	2.3690	874.94	39.360	34437.1	0.045868	1579.39	16.863
11.	105.67	381.33	2.5727	981.04	37.226	36520.5	0.041279	1507.57	14.267
Average	54.98	339.133	1.9811	677.3555	42.724	27898.081	0.035034	1028.868	23.083

Table 4.4: *Channa punctata* (Bloch) mean values for the body weight and other component parameters of the fourth gill arch in each of the 11 weight groups.

Sl.No.	Body Wt.(g)	Total Gill Filaments (Numbers)	Average Filament Length (mm)	Total Filament Length (mm)	Secondary Lamellae/ mm	Total Secondary Lamellae	Average Surface of a Secondary Lamellae (mm^2)	Total Gill Area (mm^2)	Gill Area/g Body Weight (mm^2)
1	2	3	4	5	6	7	8	9	10
1.	5.67	194.00	0.9542	185.11	57.560	10654.9	0.017369	185.06	32.639
2.	15.00	236.67	1.3639	322.79	49.340	15926.7	0.024575	391.40	26.093
3.	25.67	230.67	1.5778	363.95	47.493	17285.0	0.027754	479.72	18.688
4.	36.40	265.20	1.6705	443.00	43.625	19326.0	0.030637	592.10	16.266
5.	43.33	258.00	1.7163	442.80	43.051	19063.0	0.036809	701.68	16.194
6.	57.33	268.00	1.9083	511.45	40.639	20784.0	0.030894	642.10	11.200
7.	64.00	282.67	2.0475	578.77	40.453	23413.0	0.033421	782.49	12.236
8.	74.00	280.00	2.1657	606.39	37.965	23021.5	0.044297	1019.78	13.781
9.	84.00	289.33	2.0536	594.16	38.030	22596.0	0.041389	935.44	11.136
10.	93.67	302.67	2.1163	640.47	38.696	24784.0	0.037712	934.64	9.978
11.	105.67	297.33	2.3304	692.89	36.634	25383.0	0.041279	1051.75	9.953
Average	54.98	264.049	1.8095	489.251	43.062	20203.372	0.033299	701.469	16.196

Table 4.5: *Channa punctata* (Bloch) mean values for the body weight and other component parameters of the fifth gill arch in each of the 11 weight groups

Sl.No.	Body Wt. (g)	Total Gill Filaments (Numbers)	Average Filament Length (mm)	Total Filament Length (mm)	Secondary Lamellae/ mm	Total Secondary Lamellae	Average Surface of a Secondary Lamellae (mm^2)	Total Gill Area (mm^2)	Gill Area/g Body Weight (mm^2)
1	2	3	4	5	6	7	8	9	10
1.	5.67	1032.00	1.13963	1176.1016	55.8524	65688.0	0.021741	1428.12	251.874
2.	15.00	1200.00	1.51357	1816.2880	49.7723	90401.0	0.026570	2401.95	160.130
3.	25.67	1248.67	1.72970	2159.8245	47.2421	102035.0	0.028210	2878.40	112.131
4.	36.40	1346.80	1.94244	2616.0782	44.4935	116398.0	0.032280	3757.34	103.224
5.	43.33	1336.00	1.97598	2639.9137	42.9031	113260.0	0.036410	4123.81	95.238
6.	57.33	1392.00	2.16711	3016.6171	41.6086	125517.0	0.038864	4878.10	85.098
7.	64.00	1445.33	2.30117	3325.9576	40.4913	134672.0	0.039402	5306.36	82.912
8.	74.00	1452.67	2.41937	3514.5462	38.7293	136119.0	0.044094	6002.02	81.108
9.	84.00	1502.00	2.38057	3575.6161	38.7901	138698.0	0.043690	6753.21	80.295
10.	93.67	1545.33	2.49770	3859.7707	39.2746	151591.0	0.048976	7424.93	79.267
11.	105.67	1596.00	2.67643	4271.5823	37.6234	160711.0	0.050616	8134.56	76.981
Average	54.98	1372.44	2.0676	2906.5724	43.34	121372.0	0.037781	4826.3	109.844

Table 4.6: *Channa punctata* (Bloch) mean values for the body weight and
other component parameters of the sixth gill arch in each of the 11 weight groups

Sl.No.	Body Wt.(g)	Total Gill Filaments (Numbers)	Average Filament Length (mm)	Total Filament Length (mm)	Secondary Lamellae/ mm	Total Secondary Lamellae	Average Surface of a Secondary Lamellae (mm^2)	Total Gill Area (mm^2)	Gill Area/g Body Weight (mm^2)
1	2	3	4	5	6	7	8	9	10
1.	5.67	487.34	85.95	1428.1	1915.4	337.81	0.10766	0.01899	4524.0
2.	15.00	1170.90	78.06	2402.0	3572.9	238.19	0.24733	0.01649	4734.0
3.	25.67	1647.70	64.19	2878.4	4526.1	176.32	0.39153	0.01525	4208.6
4.	36.40	1911.20	52.51	3757.3	5668.5	155.73	0.52777	0.01450	3623.0
5.	43.33	2126.40	49.07	4123.8	6250.2	144.25	0.61251	0.01414	3471.6
6.	57.33	2671.00	42.73	4873.1	7549.1	131.68	0.77822	0.01340	3432.0
7.	64.00	2639.00	41.24	5306.4	7945.6	124.15	0.85500	0.01336	3086.8
8.	74.00	3225.60	43.59	6002.0	9227.6	124.70	0.96802	0.01308	3332.0
9.	84.00	3428.70	41.46	6753.2	10236.0	121.86	0.07880	0.01284	3228.3
10.	93.67	3704.70	39.55	7424.9	11130.0	118.82	0.18430	0.01262	3128.2
11.	105.67	4113.00	43.91	8134.6	12248.0	115.91	0.32270	0.01242	3136.5
Average	54.98	2470.89	52.93	4826.3	7297.22	162.61	0.73308	0.014281	3627.73

Table 4.7: Table summarizing relationships between body weight (W) and other component parameters (y) of the different gill arches as based on equation Y=aW^b, where a is the value for 1 g fish (log Y intercept) and b is the slope value of the regression line. The correlation coefficient r have also been shown.

Body Weight vs. Gill Dimension Parameter	First Gill Arch			Second Gill Arch			Third Gill Arch			Fourth Gill Arch		
	Intercept (a)	Slope (b)	Correlation Coefficient (r)	Intercept (a)	Slope (b)	Correlation Coefficient (r)	Intercept (a)	Slope (b)	Correlation Coefficient (r)	Intercept (a)	Slope (b)	Correlation Coefficient (r)
Gill filament nos.	236.604	0.1420 $y=236.6\,W^{0.1420}$	0.9653	217.428	0.1399 $y=217.43\,W^{0.1399}$	0.9865	196.624	0.1392 $y=196.62\,W^{0.1392}$	0.9882	151.983	0.1455 $Y=151.98\,W^{0.1455}$	0.9766
Average filament length (mm)	0.7870	0.2799 $y=0.7870\,W^{0.2799}$	0.9973	0.7677	0.2715 $y=0.7677\,W^{0.2715}$	0.9981	0.5974	0.3117 $y=0.5974\,W^{0.3117}$	0.9981	0.5981	0.2883 $Y=0.5981\,W^{0.2883}$	0.9939
Total filament length (mm)	185.83	0.4221 $y=185.83\,W^{0.4221}$	0.9924	166.817	0.4114 $y=166.82\,W^{0.4114}$	0.9981	112.60	0.4613 $y=112.60\,W^{0.4613}$	0.9903	90.86	0.4339 $Y=90.86\,W^{0.4339}$	0.9957
No. of secondary lamellae/mm (both sides)	71.293	-0.1408 $y=71.293\,W^{0.1408}$	0.9884	70.271	-0.1336 $y=70.27\,W^{0.1336}$	0.9711	70.320	-0.1354 $y=70.32\,W^{0.1354}$	0.9876	75.785	-0.1534 $Y=75.79\,W^{-0.1534}$	0.9933
Total No. of secondary lamellae	13273.2	0.2811 $y=13273.2\,W^{0.2811}$	0.9824	11699.9	0.2785 $y=11699.9\,W^{0.2785}$	0.9906	8196.31	0.3182 $y=8196.3\,W^{0.3182}$	0.9862	6879.61	0.2807 $Y=6879.61\,W^{0.2807}$	0.9872
Bilateral surface area of an average lamella (mm²)	0.01212	0.3156 $y=1.6196\,W^{0.3156}$	0.9479	0.01201	0.3028 $y=0.01201\,W^{0.3028}$	0.9465	0.01115	0.2973	0.9721	0.01072	0.2943 $Y=0.01072\,W^{0.2943}$	0.9513
Gill area (cm²)	1.6196	0.5953 $y=1.6196\,W^{0.5953}$	0.9512	1.4263	0.5781 $y=1.4263\,W^{0.5781}$	0.9580	0.9212		0.9792	0.7454	0.5722 $Y=0.7454\,W^{0.5722}$	0.9691
Gill area/g (cm²/g)	1.6302	0.4065 $y=1.6303\,W^{0.4065}$	0.9507	1.4360	0.4237 $y=0.4360\,W^{0.4237}$	0.9633	0.9270		0.9763	0.7505	-0.4297 $Y=0.7505\,W^{-0.4297}$	0.9747

Table 4.8: Summary table showing intercept (a), slope (b) and relationship equations of the different dimensions of the respiratory organs as based on equation Y=aW^b. Correlation coefficient have also been shown.

Body weight (W) Vs. Dimensional parameters (Y)	Intercept(a)	Slope(b)	Equation	Correlation Coefficient
(a) All gill arches				
Total number	802.37	0.1415	$Y=802.37W^{0.1415}$	0.9905
Average length (mm)	0.6908	0.2856	$Y=0.6908W^{0.2856}$	0.9995
Total length (mm)	557.99	0.4253	$Y=577.99W^{0.4253}$	0.9935
Secondary lamellae:				
Number/mm	72.0552	-0.1376	$Y=72.0552W^{-0.1376}$	0.9845
Total number	40023.2	0.2889	$Y=40025.2W^{0.2889}$	0.9940
Average bilateral surface/lamella (mm²)	0.0117	0.3043	$Y=0.0117W^{0.3043}$	0.9470
Gill area:				
Total area (cm²)	4.7039	0.5919	$Y=4.7039W^{0.5919}$	0.9837
Area/g wt.(cm²/g)	4.7037	-0.4081	$Y=4.7037W^{-0.4081}$	0.9663
(b) Suprabranchial chamber Surface area:				
Total Surface area (cm²)	1.5908	0.6957	$Y=1.5908W^{0.6957}$	0.9908
Total surface area/g wt. (cm²/g)	1.5334	-0.2937	$Y=1.5334W^{-0.2937}$	0.9214
Volume:				
Volume (ml)	00244	0.8550	$Y=0.0244W^{0.8550}$	0.9942
Volume/g (ml)	0.0245	-0.1458	$Y=0.0245W^{-0.1458}$	0.9991
Total respiratory area (cm²) (Gill+suprabranchial chamber)	6.2976	0.6225	$Y=6.298W^{0.6225}$	0.9929
Total respiratory surface area/g wt. (cm²/g)	6.2974	-0.3775	$Y=6.2974W^{-0.3775}$	0.9810

Table 4.9: Computed respiratory dimension values for 1, 10 amd 100 g fishes using logarithmic transformations based on equation log Y =log a + b log W are presented with 95 per cent confidence limits, slope (b) and intercept (a). The standard deviations for (a) and (b) are also shown.

Dimensional Parameters	1 g Fish		10 g Fish		100 g Fish		Slope (b)		Intercept (a)	
	Value	95 per cent Conf. Limits	Value	95 per cent Conf. Limits	Value	95 per cent Conf. Limits	Value (b)	Deviation (sb)	Value (b)	Deviation (sb)
Total gill filaments	802.372	792.10 812.03	1111.5	1102.5 1120.2	1539.5	1515.2 1563.8	0.1415	0.0062	802.37	1.009
Total filaments length	557.99	505.00 616.44	1485.56	1422.45 1551.46	3955.04	3833.04 4080.92	0.4253	0.0115	557.99	1.0450
Secondary lamellae/ mm (both sides)	72.06	68.51 75.78	52.48	51.34 53.64	38.23	37.63 38.84	-0.1376	0.0059	72.06	1.0225
Bilateral surface area of an average lamella (mm²)	0.0117	0.00950 0.01441	0.02357	0.02153 0.02582	0.0475	0.04449 0.05072	0.3043	0.0240	0.0117	0.0197
Total gill area (cm²)	4.7039	3.7725 5.8653	18.3807	16.6957 20.2356	71.8229	67.0072 76.9846	0.5919	0.0254	4.7037	1.1025
Surface area of suprabranchial chamber (cm²)	1.5908	1.3109 1.9305	7.8938	7.2554 8.5884	39.1705	36.9570 41.6291	0.6957	0.0223	1.5908	1.0893
Volume of the suprabranchial chamber (ml)	0.0244	0.0230 0.0259	0.1749	0.1687 0.1815	1.2522	1.1654 1.3455	0.8550	0.0312	0.0244	0.0028
Total respiratory surface (cm²)	6.2976	5.4090 7.3322	26.4054	24.7121 28.2148	110.7166	105.5445 116.1421	0.6225	0.0175	6.2976	1.0096

Table 4.10: *Channa punctata* (Bloch)

Diffusing capacity (D_t) for the tissue barrier of the respiratory organs as based upon morphometric findings: Corresponding data for Anbas, Saccobranchus and Amphipnous are also given for comparison

Fish Species	Respiratory Surface for 1 g Fish (mm²)	Thickness of Tissue Barrier (µm)	Area/g wt. 100 g Fish (mm²)	Diffusing Capacity 100 g fish(ml/ min/mmHg/kg	Reference
Channa punctata					
(i) First gill arch	161.96	2.0333	25.1189	0.0185	Present authors
(ii) Second gill arch	142.63	2.0333	20.4489	0.0151	Present authors
(iii) Third gill arch	92.13	2.0333	15.5431	0.0115	Present authors
(iv) Fourth gill arch	74.54	2.0333	10.3925	0.0077	Present authors
(v) Total gills	470.39	2.0333	71.8229	0.0530	Present authors
(vi) Suprabranchial Chamber	159.08	0.7800	39.1705	0.0753	Present authors
Anabas testudineus					
(i) Total gills	278.00	10.0000	47.2000	0.0071	Hughes, Dube and Munshi (1973)
(ii) Suprabranchial chamber	55.40	0.2100	7.6500	0.0539	Hughes, Dube and Munshi (1973)
(iii) Labyrinthine organs	80.70	0.2100	32.0000	0.2286	Hughes, Dube and Munshi (1973)
Saccobranchus fossilis					
(i) Total gills	186.10	3.5800	37.7000	0.0242	Hughes *et al.* (1974)
(ii) Air sac	145.90	1.6000	30.7000	0.0288	Hughes *et al.* (1974)
(iii) Skin	851.10	98.000	200.0000	0.0031	Hughes *et al.* (1974)
Amphipnous cuchia (i) Air-sac	12.90	0.44	4.8400	0.0165	Hughes *et al.* (1974a)

Table 4.11: Slope 9 lb) for the total secondary lamellae and bilateral surface area of an average size secondary lamellae and their sum in different fish species for comparison with the slope (b) for the total gill area in relation to body weight.

Fish Species	Slope (b) Total Sec. Lamellae 1	Slope (b) Bilateral Surface of a Sec Lamellae 2	Sum of Columns 1 and 2 3	Slope (b) Total Gill Area 4	Reference
Channa punctata	0.2889	0.3043	0.5932	0.5919	Present authors
Anabas testudineus	0.1770	0.4360	0.6030	0.6150	Hughes, Dube and Munshi (1973)
Saccobranchus fossilis	0.3360	0.4070	0.7430	0.7450	Hughes et al. (1974)
Opsanus tau	0.4200	0.3600	0.7800	0.7700	Hughes and Grey (1972)
Coryphaena hippurus	0.3900	0.3270	0.7170	0.7100	Hughes (1970a)
Macrognathus aculeatum	0.4042	0.3469	0.7511	0.7330	Ojha and Munshi

Table 4.12: Computed respiratory dimension values for 1, 10 and 100 g fishes using logarithmic transformations based on equation log
Y =log a + b log W with 95 per cent confidence limits, slope (b) and intercept (a).
The standard deviations for (a) and (b) are also shown.

Fish Species	Slope (b) Total Filament Length	Slope (b) Sec Lam./ mm	Slope (b) Bilateral Surface of an Average Sec. Lam.	Sum of Columns 1, 2, and 3	Slope (b) Total Gill Area	Reference
Coryphaema hipparus	0.4310	-0.0360	0.3270	0.7230	0.7130	Hughes (1970a)
Secomber sccombras	0.4110	-0.0234	0.5560	0.9904	0.9970	
Seyllorhinus canicula	0.3510	-0.0710	0.6840	0.9640	0.9610	Hughes (1970b)
Tinca tinca	0.3190	-0.0160	0.1860	0.5210	0.5230	
Opsumas tou	0.4850	-0.0750	0.3720	0.7820	0.7900	Hughes and Grey (1972)
Macrognathus aculeatum	0.4670	-0.0690	0.3470	0.7450	0.7330	Ojha and Munshi (1974)
Saccobranchus fossilis	0.4350	-0950	0.4080	0.7480	0.7460	Hughes et al. (1974a)
Anabas testudineus	0.3350	-0.1250	0.4260	0.6090	0.6150	Hughes, Dube and Munshi (1973)
Channa punctata	0.4253	-0.1376	0.3043	0.5920	0.5919	Present authors

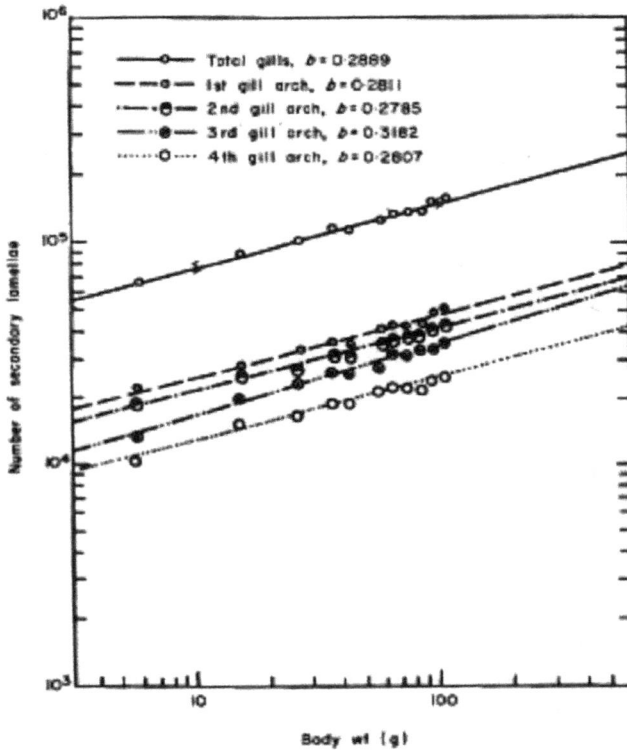

Figure 4.3: *Channa punctata*, Bi-logarithmic plots of the total number of secondary lamellae for the whole fish against body weight. (95 per cent confidence limits are indicated for fish of 10 and 100 g.) Numbers of secondary lamellae for each of the constituent arches are also plotted.

Relationship between Body Weight and Total Filament Length

The two acts of measurements of the last section multiplied together gave values for the total filament length. Results for all four gill arches showed correlation coefficients exceeding 0-99. Relationships between body weight and total filament length for all the gills are given in Table 4.7. Log-log plots gave straight lines with slopes ranging from 0.411 to 0.461 (Table 4.7 and Figure 4.2). The weight specific total filament lengths for fish of 1, 100 and 55g (average weight of 35 specimens used ranging from 4 to 109g) were 557.99, 39.55 and 52.89 mm respectively (Table 4.9).

Relationship between Body Weight and Secondary Lamellae/mm on both Sides of a Filament

Correlation coefficients for the gill arches separately, as well as when the results were combined. In all cases the number of secondary lamellae/mm decreased with body weight, the slopes ranging from -0.1336 to -0.1534 (Table 4.7, Figure 4.2).

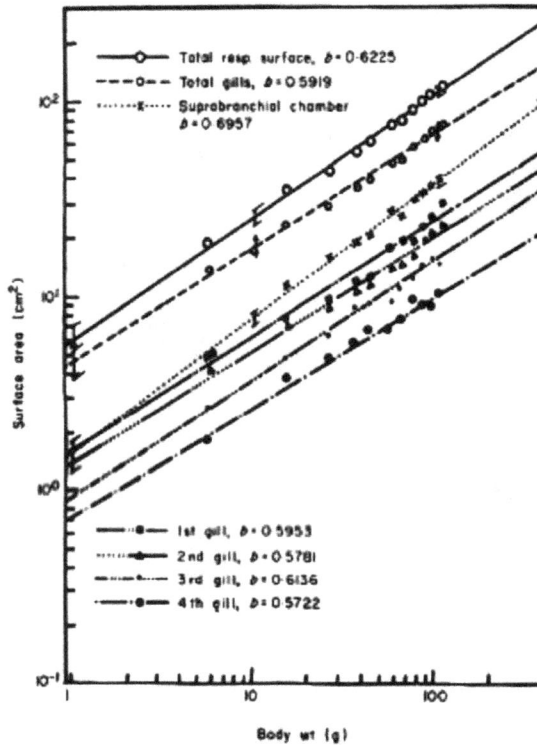

Figure 4.4: Areas of the suprabranchial chambers, total gill surface and their combined areas are plotted against body weight on logarithmic coordinates. Surface areas of the individual gill arches are also shown.

The numbers of secondary lamellae/mm for all arches taken togther were 72.06, 38.23 and 43.34 respectively for 1,100 and 55g fish (Table 4.9).

Relationship between Body Weight and Total Number of Secondary Lamellae

As was to be expected from results for the two previous sections, body weight and number of secondary lamellae were highly correlated with one another. The correlation coefficients → were all greater than 0.983 (Tables 4.7 and 4.8). Results of regression analysis for individual arches are given in Table 4.7. The weight specific numbers of secondary lamellae in this species were 40,023.2 for a 1-g fish, 1513.6 for a 100-g fish and 2206.7 for a fish of average weight (about 55 g).

Relationship between Body Weight and the Average Bilateral Surface Area of a Secondary Lamella

Results of the regression analyses are set out in Tables 4.7 and 4.8 and Figure 4.2. Slopes for the 1st, 2nd, 3rd and 4th arches were respectively 0.3156, 0.3028, 0.2973 and

0.2943; for all gill arches combined it was 0.3043. The correlations coefficients all exceeded 0.9465 (Tables 4.7 and 4.8). Weight specific lamellar areas were respectively 0.0117, 0.00048 and 0.00069 mm^2 for fish of 1,100 and 55 g.

Relationship between Body Weight and Gill Surface Areas

The total surface are for each of the gill arches increased with body size. The average value for the total gill area for all arches ranged from 1428.12 to 8134.56 mm^2 for fish of body weights from 5.67 g to 105.67 g. The average value 4826.3 mm^2 for the gill area of a fish of 55g. Results of the regression analysis of bilogarithmic plots are given in Tables 4.7 and 4.8. The slope for all the gill arches was 0.592 and for individual arches, it ranged between 0.572 and 0.614, the area of the 3rd arch increasing most rapidly with body size. Correlations between the figures for gill surface areas and body weight ranged between 0.95 and 0.98 but those for surface area of all gills were more highly correlated (r=0.9837).

Relationship between Weight Specific Gill Area and Body Weight

As shown in Figure 4.4 and the result of the regression analysis in Tables 4.7 and 4.8, the gill area/g body weight decrease with increase in body weight both for individual gill arches and for the total area of the gills. Correlations for these relationships were high, r values ranging from 0.9507 to 0.9763.

Morphometrics of the Air-Breathing Organs

Relationship between Body Weight and Volume of the Suprabranchial Chambers

Average surface area for a fish of about 55g was 24.709 cm^2. Results from regression analysis of log-log plots (Table 4.8) showed that the slope of the straight line was 0.696. The weight specific surface area for fishes of 1,100 and 55 g was respectively 1.533, 0.397 and 0.53 cm^2/g.

Relationships between Body Weight and the Total Respiratory Surface Area

The combined surface area of the gills and accessory organs increases with body weight and analysis of the bilogarithmic plots showed the relationship;

Total respiratory surface area = 6.2976W$^{0.689}$.

The correlation coefficient for these combined data was 0.993. Thus the respiratory surface area/gm decreased from 3.378 cm^2 in a 5.67 g fish to 1.159 cm^2 in a 105.67 g fish average values being 1.626 cm^2/g for a 55 g fish.

It is of interest that for a fish of 5.67 g the gills provide 74.55 per cent of the total respiratory surface area whereas for fish of 105.67 g average weight, the gill contribution was reduced to 66.4 per cent.

Diffusing Capacity of the Respiratory Organs

Calculations of diffusing capacity of the tissue barrier (D$_t$) were calculated for the different organs as shown in Table 4.10. D$_t$ for the lining of the suprabranchial chamber was greater than that of the gills being respectively 0.07533 and 0.05298 ml/min/mmHg/kg. Values for the different gill arches decreased gradually from the 1st-4th gill because of th corresponding change in surface area. No significant differences were found in measurements of the thickness of the water/blood barrier.

Table 4.13: Summary table of slope (b) and intercept values (a) for the relationship between body weight and gill area of different fishes.

Species	A	B	Reference
Small-mouthed bass (*Micropterus dolomieu*)	865.0	0.7800	Price, 1931
Gray's Intermediates	1392.0	0.8200	Ursin, 1967
Yellow fin tuna (*Thunnus albacares*)	3151.0	0.875	Muir and Hughes, 1969
Bluefin tuna (*Thunnus thynnus*)			
Skipjack tuna (*Katsuwonus pelamis*)	5218.0	0.8500	Muir and Hughes, 1969
Toadfish (*Opsanus tau)*	560.7	0.7900	Hughes, 1970a and Hughes and Grey, 1972
Dolphin fish (*Coryphaena hippurus*)	5208.0	0.7100	Hughes, 1970a
Mackerel (*Scomber scombrus*)	424.1	0.9970	Hughes, 1970b
Shanny (*Blennius pholis*)		0.8500	Milton, 1971
Dogfish (*Scyliorhinus canicula*)	262.3	0.9610	
Tench (*Tinca tinca*) (1969)		0.5520	Hughes, 1970b

Drowning Experiments

Fish prevented from coming to the surface to obtain air ultimately died because of inability to satisfy their oxygen requirements. The observations below give some indication of the critical levels of oxygen content for survival by these fish, but no detailed investigations were made.

Chance observations made when the flow of water to the reservoir ceased because of failure of the pumps indicated that when the oxygen content was 4.4 mg/l there was little change in the fish apart from a slight increase in ventilatory frequency.

On 16 August 1974 at about 6.15 a.m., water flow had again stopped and the fishes were very restles. They moved rapidly and were dashing at the lid of the aquarium in attempts to reach the surface. In some fishes the condition appeared very precarious as their eyes were bulging and the ventilatory rhythm was very rapid. Oxygen content of the reservoir was found to be 1.9 mg/l. Following restoration of the circulation to 3.1/min the oxygen content rose to 6.2 mg/l by 7.30 a.m. Four of the fishes died on this day and it would appear that 2.0 mg/l must be close to the critical level of oxygen when surfacing is prevented in this species.

The remaining seven fishes survived the hypoxic stress and were observed until 22 August. The fishes which died were carefully dissected while they were still submerged under water and it was ascertained that air was absent from the suprabranchial chambers.

The increase in body weight of an animal is accompanied by an increase in the living material requiring oxygen for its metabolic processes and is associated with a larger gas exchange surface. The general relationships between metabolism and body weight are expressed by the exponential equation $\dot{V}_{O_2} = aW^b$ where $\dot{V}_{O_2}$ is the rate of oxygen consumption; a is the intercept or value for a 1 g animal of body weight W; b is the slope of the regression line.

This relationship for many fishes was summarized by Winberg (1956) as $\dot{V}_{O_2} = aW^{0.81}$. The b value has been found to vary in different fishes depending on body weight, age, activity, nutritional state, maturity and other factors. Temperature is very important in controlling different reactions of the metabolic chain and consequently the higher the temperature the greater the rate of metabolism; there may also be effects on the slope depending on acclimatization.

The power relating gill surface area to body weight was at first thought to be close to that for the metabolism/weight relationship (Muir and Hughes, 1969). However, more detailed analysis (Hughes, 1970b, 1977) have indicated that these relationships vary according to the species. The data collected in this paper provide information on such relationships for an air-breathing fish; analyses of the relationships for different arches within the gill system have also been made.

Gills

AS has been found the for other species of fish (Muir and Hughes, 1969; Hughes *et al.*, 1974a) the sum of the slope values for total filament length, bilateral surface area of an average secondary lamella, and the number of secondary lamellae/mm gave a value (0.592) which is almost identical to the slope obtained for the regression line of the total gill surface area in relation to body weight (0.5919). These component values are compared with those for other species in Tables 4.12 and 4.13. Similarly the sum of the slope values for total number of secondary lamellae and average lamellar area is equal to thst for gill area (Table 4.11). The slope for the gill area is lower in *Channa punctata* than that reported for *Anabas testudineus* (0.615). Thus, although the gills provide most of the respiratory area they have a comparatively low rate of growth. In younger fish the gills provide about 75 per cent of the total surface for gas exchange but in higher weight groups it is only 66.4 per cent. This low slope is probably associated with the particular air-breathing habits of this species. At all stages of growth the surface area for a fish of the same body size is significantly lower than that for *Anabas, C. punctata* has more sluggish habits and is generally less active than Anabas when observed in an aquarium. *Channa* shows a greater slope (0.4253) for total filament length than does *Anabas* (0.335) and indicates the greater dependence of *Channa* on gill respiration. The frequency of secondary lamellae along the filaments decreases with body weight as in all other fishes investigated. The number estimated from the regression lines for a 1 g fish is 73/mm in Anabas (Hughes, Dube and Munshi, 1973) and 72/mm in *Channa*. Figures for *Saccobranchus* are lower (63/mm) but all these values are much less than that obtained for the fast-swimming tunny where extrapolation of the regression lines gave a value of 134/mm (Muir and Hughes, 1968). All values are for both sides of a filament.

Comparison between different gill arches has shown some changes in surface area distribution during the growth of this species. The area for the 3rd arch increases more rapidly than those of the other arches. The increase in filament number varies very slightly between arches and this was also found to be true for the slope of the regression lines relating numbers of secondary lamellae/mm to body weight.

The slopes of the lines relating total number of secondary lamellae to body weight are greater in *Channa* (0.289) than in *Anabas* (0.177) and again can probably be correlated with the generally greater water-breathing habits of *Channa*, whereas *Anabas* is an obligatory air-breather. However, the b value for bilateral surface area of an average secondary lamella is less for *Channa* than *Anabas* (0.3043, 0.426) indicating that the secondary lamellae are smaller in specimens of *Channa* of the same body weight.

As has been indicated above, the surface area of the air-breathing organs provides 25 per cent of the total respiratory area in small fish (5.67 g) and about 33 per cent in larger specimens (105.67 g). The greater slope of the relationship between body weight and surface area of the suprabranchial chamber (0.696) as against that for the gills (0.592) summarized this relationship and indicates that as the fish grow the need for aerial oxygen becomes more important.

The volume of the suprabranchial chamber increases by a power function of 0.855 which is greater than the rate of increase of surface area, as would be expected. The rate of increase in volume of the suprabranchial chamber in *Anabas* (0.861) is very similar to that found for *Channa*. The average surface area of the suprabranchial chamber for unit volume of air is 3627.73 mm^2 in *Channa*, whereas that reported for *Anabas* (Hughes, Dube and Munshi, 1973) is 2226 mm^2 and for the air-sacs of *Saccobranchus ossilis* its value was even less) 1066 mm^2; Hughes *et al.*, 1974a).

The values for Channa probably exceed those for the frog, *Rana*, but are very much less than corresponding values for the human lung (10-30,000 mm^2). Nevertheless, it is clear that *C. punctata* has developed an efficient mechanism for utilizing the oxygen content of the air that is taken into the air-beathing organ.

The diffusing capacity of the respiratory organs are compared with those of other fish in Table 4.10, D$_t$ for *Channa* gills is much greater than for either *Anabas* or *Saccobranchus*, and the same is also true for the air/blood barriers of the suprabranchial chamber.

Finally, relationships between body weight and total respiratory surface area and those for oxygen uptake by this species are very similar. The power functions are respectively 0.622, 0.625 and 0.629 for total respiratory surface, and oxygen uptake at 20° and 31°C when the fish were prevented from surfacing. These slope values for oxygen uptake are extremely low. However, the relationship for gill surface area (a=0.592) has a lower slope than that obtained for the metabolic rate when the fish is confined to water breathing.

It may be concluded that the respiratory organs of this fish show a fairly homogeneous development within the gill system and that the increase in surface area of the respiratory organs corresponds to that of the metabolic requirements. The increasing development of the air-breathing surfaces relative to those involved in aquatic gas exchange does not seem to reach a stage where the fish becomes dependent upon these organs for such a significant proportion of its oxygen that it becomes an obligate air breather.

Chapter 5

Evolutionary Transformations of the Respiratory Islets of Air-breathing Organs in Teleostean Fishes

In *Monopterus* (=*Amphipnous*) *cuchia* (order Synbranchiformes), *Channa gachua* (order Ophiocephaliformes) and *Anabas testudineus* (order Perciformes) the vascular papillae composing the Respiratory islets (RI) though look alike, differ in their structural details, distribution and microcirculation. Evidences have been obtained that in *Monopterus* they develop from the coiled configurations of the marginal channels of gill lamellae. In this species the rosette of vascular papillae form RI of varied shape and sizes at the base of gill filaments and surrounding the gill clefts. They are also distributed extensively in the buccopharynx, hypo-pharynx near the oesophagus and respiratory airsacs. In *Channa gachua, Channa striata, Channa marulius* they are very well distributed in the buccopharynx, tongue and suprabranchial chambers. In *Anabas* the biserially arranged transverse capillaries with their septate endothelial valves give beaded appearance in corrosion replica of the methyl methacrylate resin preparations under SEM; each beaded structure represents a vascular papilla.

It is postulated that in *Anabas* the vascular papillae in series were laid down over each other and got fused to form the transvers channels of RI, in the same plane which remain in contact with the respiratory surface making the whole structure very much compact and efficient in gas exchange. In *Monopterus* and *Channa* the capillaries forming the domes of vascular papilla are arranged either in spiral or wave-like fashion as a result of which only 10-15 per cent of the total capillary surface come in contact with the respiratory surface making them haemodynamically inefficient.

The respiratory islets act as resistances box stepping down the arterial pressure to venous levels (Abstract from Journal of Morphology June, 1997, Vol.232, No.3, Bristol ICV97, Munshi, J.S.Datta)

The structure and function of the gills and accessory respiratory organs of air-breathing fishes are generating a lot of interest in recent years partly because of an increasing amount of research in fish respiration, and more precisely because of their interesting adaptations and many convergences to the ancestral vertebrates which made the transition from water to land, (Munshi and Hughes 1992; Graham 1997; Mittal, Eddy and Munshi, 1999). In this book I have attempted to survey our present knowledge about the structure and function of gills and accessory respiratory organs of common air-breathing fishes of India, with special reference to the work done in our laboratories.

The transition from water to air-breathing in vertebrates required many structural and functional modifications to their external gas exchange system. In all probability these adaptive changes in the system have evolved by different ways in nature over an extended period of vertebrate evolution. It has been speculated that those animals which succeeded in evolving themselves became the stem forms of our amphibian ancestors and are no longer found today (Hughes, 1996; Rahn and Howell, 1976). Other animals unable to adapt themselves completely to an aerial mode of respiration were caught in the air-water interface where many of their descendants can be found even today. The air-breathing fishes of India represent an agglomeration of such a group of animals. They have developed varied types of bimodal gas exchange mechanisms for using both air and water as respiratory medium. Attempts to obtain oxygen directly from the air seem to be a continuing evolutionary process. There is plenty of evidence to suggest that in these air-breathing forms the gill-skin system in these fishes still remains the main organs for removal of carbon dioxide (Hughes, 1996). It is seen that the handling of the accessory respiratory organs can easily cope with the oxygen demand but is unable to function efficiently in the elimination of carbon dioxide.

In two other genera of siluroid fishes, *Clarias magur* (*batrachus*) and *Heteropneustes fossilis*, the evolutionary transformations of respiratory islets (RI) from gill filaments and lamellae were quite evident (Munshi, 1958, 1961, 1962; Hughes and Munshi, 1978). Respiratory islets have been transformed from the gill lamellae (Munshi, 1958). In the mud eel, *Monopterus* (= *Amphipnous*) *cuchia* and *Channa* (=*Ophiocephalus*) *punctata*, *C. gachua, Channa marulius, C. striata* the respiratory islets are composed of rosettes of papillae. It was postulated earlier that the respiratory islets of *Anabas* have evolved in a different manner – by the transformation of the marginal channels of gill lamellae. The biserial arrangement of transverse capillaries of respiratory islets of *Anabas* were thought to have been developed by the abbreviation of two rows of gill lamellae (Munshi and Hughes, 1991). In recent years there have been extensive studies of the respiratory islets by scanning electron microscopy, which have shown that the transverse capillaries are not simple capillaries but are composed of small units of bead-like structures which resemble the vascular papillae of *Channa*, and *M. cuchia*. The application of methyl methacrylate resin corrosion technique facilitated the understanding of the nature of these transverse capillaries. The transverse capillaries gave the appearance of strings of bead-like structures, each bead representing a vascular papilla. Recently we have made more extensive studies of different regions of the respiratory islets of *Anabas* especially of the marginal areas, and clear evidence was obtained to show that the transverse capillaries of RI of *Anabas* have evolved from the vascular papillae-like structures. For this review we have made a comparative study of the respiratory islets of *Heteropneustes fossilis, Clarias magur (batrachus), Monopterus cuchia, Channa* species and *Anabas testudineus*. We have now established how the so-called transverse channels of RI have evolved with their septate endothelial valves.

Representative Genera of Air-Breathing Teleosts

Amongst the teleostean fishes the air-breathing species are more than 140 in number and are found in sixteen families (Bertin, 1958). They include genera ranging from fish which spend the greater part of their life on land to fish which resort to air-breathing only at times of stress. In India these fishes mainly live in wetlands where water pools and tanks dry up in the summer and become marshy. These marshy waters abound in the eastern zone of India that especially includes the Gorakhpur division of Uttar Pradesh, Sharsa, Purnea, Darbhanga and Bhagalpur in Bihar and most parts of Assam, Bengal and Orissa.

The following species of air-breathing fishes belonging to different families are common in these waters: 1) Fam: Notopteridae (Feather backs) – *Notopterus chitala* (Ham), *Notopterus notopterus* (Pallas); (2) Fam: Cobitidae (Loaches) – *Lepidocephalus* (=Lepidocephalichtys) *guntea* (v); (3) Fam Clariidae (Cat fishes) – *Clarias magur* (=*batrachus*) (Linn.); (4) Fam: *Heteropneustidae* (=Saccobranchidae) (Cat fishes) – *Heteropneustes fossilis* (Bloch); (5) Fam: *Channidae* (=Ophicephalidae) (Snake-heads and Murrels) – *Channa punctata* (Bloch), *Channa striata* (Bloch), *Channa marulius* (Ham), *Channa gachua* (Ham), (6) Fam: Symbranchidae (Finless eels) – *Symbranchus bengalensis* (McClell); (7) Fam: Amphipnoidae (Mud eels) – *Monopterus* (=*Amphipnous*) *cuchia* (Ham); (8) Fam Anabantidae (Climbing perches) – *Anabas testudineus* (Bloch), *Colisa* (=*Trichogaster*) *fasciatus* (Bl. and Schn); (9) Fam: Gobiidae (Gobies) – *Periopthalmus vulgaris* (Eggert), *Boleopthalmus boddaerti* (Pallas), *Pseudopocryptes lanceolatus* (Bl. and Schn.); and (10) Fam: *Mastacembelidae* (Spiny eels) – *Mastacembelus pancalus* (Ham), *M. armatus* (Lac.), *Macrognathus aculeatum* (Bl.). In this paper we will deal with those genera and species which have developed respiratory islets and are involved in oxygen uptake from air.

Habitat Conditions

A. testudineus, C. magur, H. fossilis, M. cuchia and the four species of *Channa* live in different trophic zones of the swamps and derelict ponds. *P. vulgaris* and *B. boddaerti* are true amphibious fishes spending most part of their lives in the soft muds of the estuaries. The shallow derelict and marshy waters of the tropical world may be taken as the normal habitat of these air-breathing fishes. But most of the species are found in all types of waters.

While the two catfishes *Clarias* and *Heteropneustes* inhabit muddy bottoms of weed infested swamps, subsisting on rich detritus of decaying organic matter, *Anabas* and *Channa* spp., are column feeders with high predatory habits. During drought the *Channa* spp., live under mud in a torpid state. *Monopterus, Clarias magur* and *Heteropneustes fossilis*. During the day, it lies in muddy shallows or burrows at the waterside, emerging now and then to the surface to gulp in air. It mainly subsists on worms, bottom biota, insect larvae and shrimps, which they mostly hunt in the night.

Recent studies (Rai and Datta Munshi, 1979; Rai and Munshi, 1979, Munshi, 1990; Munshi *et al.,* 1993) on the environmental features of North Bihar swamps have indicated periodic exposure of water to extreme drought conditions under high atmospheric temperature (30-35°C), which causes rapid evaporation of water leaving a rich humous loamy soil with little or no water. In general, these swamps have very poor O_2 conditions. The lower layers are often deoxygenated. Dissolved free CO_2 up to 45 ppm has been recorded (Rai and Munshi, 1979). The bicarbonate alkalinity varies considerably from 75 to 330 ppm in different swamps, tending to produce variations of pH from 7.5 to 9. In certain weed-infested swamps the chloride content varies from 4.30 to 6.02 ppm. H_2S gas (16.5 to 19.34 mg/L) has often been encountered in these swamps. The diurnal fluctuations in relation to seasonal variations of physico-chemical factors were studied (Dehadrai and Tripathi, 1976; Rai and Munshi, 1979).

Structure of Gills

The gill filaments bear an alternating series of secondary lamellae on both of their surfaces. In some air-breathing fishes like, *Channa striata* there are tertiary lamellae (Munshi, 1960; Hughes and Mittal). In most of the air-breathing teleosts the filaments interdigitate amongst themselves and the interbranchial septa are much reduced (Hughes and Munshi, 1979).

Blood Circulation

In *Anabas*, the fourth branchial arch has no secondary lamella and the efferent and afferent arteries are joined by broad vessels (Munshi, 1968; Munshi *et al.,* 1986; Olson *et al.,* 1986). In *Channa punctata* and *Channa striata* the last two branchial vessels pass directly up to the dorsal aorta (Lele, 1932; Munshi *et al.,* 1994; Olson *et al.,* 1994). But in *Channa argus* the afferent and efferent vessels of this arch are connected by clustered loops of small blood vessels (Wu and Chang 1947). In both the species and *M. cuchia* (Munshi and Singh, 1968a,b; Mishra *et al.,* 1977; Munshi *et al.,* 1990) and *Fluta alba* (Liem 1963) the fourth branchial arch forms a wide diameter shunt vessel connecting the ventral aorta with dorsal aorta. Oxygenated blood returned to the heart from the air-breathing organs may lose its oxygen to the water as it passes through the gill lamellae. Vascular shunts which short circuit the lamellae can prevent this and enable the oxygenated blood to pass to the dorsal aorta for distribution to the tissues (Satchell 1976, Olson *et al.,* 1986).

In *Clarias* and *Heteropneustes* with their 'in series circulation' like other water-breathing fishes the oxygenated blood passes to the dorsal aorta for distribution to the tissue (Olson *et al.,* 1990; Olson *et al.,* 1995), but in *Anabas, Monopterus, Channa* and

Boleopthalmus, the two circulations are wholly or partially in parallel. From the evolutionary point of view this arrangement seems to be one Nature's early experiments towards the establishment of a double circulation, where there is a clear differentiation of pulmonary (suprabranchial) and systemic circulation in fishes. But there is an admixture of oxygenated blood in its passage through the heart (Satchell 1976; Olson 1994). These fishes therefore had to endure a diminution in the oxygenated level of their blood. Perhaps this enforced partial unsaturation of the blood resulted in a high level of haemoglobin and iron in their blood *Anabas* 19.8 Hb g per cent, Pradhan 1961, *M. cuchia* 18-24.0 Hb g per cent in males and 13.8-26.0 Hb g per cent in females, Mishra *et al.*, 1977, *Heteropneustes* 14-19 Hb g per cent, Pandey *et al.* 1976, 29.4-35.4 mg Fe per cent, *Boleopthalmus dussumeiri*, 30.4-36.6 mg Fe per cent *Osphromemus goramy* 32-36.96 mg Fe per cent (*Channa striata* 32.8-40 mg Fe per cent, Dubale 1961, vide Munshi, 1980).

Respiratory Islets

The respiratory islets were discovered in the respiratory membrane of *Clarias batrachus, Heteropneustes fossilis, Anabas testudineus* and *Channa* spp. by Munshi (1958) and reported by Misra and Munshi (1958) in the XVth International Congress of Zoology, London. They reported that in *Anabas scandens* (Dald) the respiratory membrane enclosing the suprabranchial and the branchial cavities developed respiratory islets which may represent modified primary and secondary gill lamellae of a typical fish. The bellows-action of this membranous sac (or enclosure) is assured by the movements of the branchial muscles that perform this function.

The biserial arrangement of lamellae and transverse capillaries (TC) in the respiratory islets (RI) of Accessory Respiratory Organs (ARO) of different genera of fishes are evident under the scanning electron microscope and their homology with the gill filaments and their secondary lamellae has been established. It was postulated that the two sets of transvers capillaries of the respiratory islets of *Anabas* have been derived from embryonic marginal channels of two sets of lamellae of a gill filament. But it was difficult to explain the septate nature of the endothelial valves. The scanning electron microscopy of the respiratory islets of juvenile and adult has shown the segmented nature of the transverse capillaries (Munshi and Hughes, 1986).

Recently extensive studies of the different regions of the respiratory islets show convincing evidence that the transverse capillaries of RI are formed of vascular papillae-like structures. Concurrent extensive studies of the respiratory islets of *Monopterus cuchia* and *Channa* species gave new ideas of evolutionary transformations of the respiratory islets in these groups of fishes.

Methodology

In the last 20 years with the advent of new technology and methods we could probe deeply into the biological materials giving rise to new ideas.

Material and Methods

Twenty specimens of *Heteropneustes fossilis*, body weight 30-40 g were kept in an aquarium at 25°C and fed on a diet of chopped liver. The fishes were anesthetized in

MS 222 (2 g/litre). The pectoral region was opened and a cannula inserted into the bulbus arteriosus and physiological saline (0.9 per cent NaCl) containing 2500 IU heparin/litre was infused at a constant pressure of 1.5 cm H_2O while the respiratory sacs were kept inflated with moist air, 2.5 per cent glutaraldehyde solution in 0.03 molar potassium phosphate buffer (pH 7.4, 50.0 milliosmoles) was then perfused through the vasculature followed by 1 per cent O_sO_4 (pH 7.4, 350 milliosmoles) and 0.05 molar uranyl acetate solution (pH 5, 350 milliosmoles). After perfusion fixation, the respiratory sacs, gills and gill fans were dissected out and sampled for TEM and SEM. The samples from different regions of the respiratory sac and gills were washed and kept in maleate buffer at 4°C. TEM samples were dehydrated in ethanol and embedded in Epon, whereas SEM samples were processed by critical point drying and sputtered with gold.

In a few (three) fishes, 2.5 per cent glutaraldehyde in potassium phosphate buffer at pH 7.4 was injected through the body wall and mouth into the respiratory sacs and these were immersed. After 2 hr of fixation, the whole fish was transferred to potassium phosphate buffer at pH 7.4. A few samples of the fishes were photographed and others were processed for SEM and TEM (Munshi *et al.*, 1986).

Some specimens of *Anabas testudineus* were injected with India ink in order to study the vascularization of the respiratory organs. Pieces of the membranous covering overlying the accessory respiratory organ in the branchial cavity were fixed in appropriate media and then either mounted for macroscopical studies or sectioned for microscopical investigation. The relation obtained between the membranous covering overlying this specialized organ and the gill-arches and the branchial cavity was also studied. Transparent preparations of the parts concerned were also made by the Spalteholtz method. Tissues were fixed in Zenker, Chrome Bouin, Alcoholic Bouin, Gilson's mercuro-nitric and other well known fluids and the sections were stained with Heidenhain's or Delafied's or Mayer's haemalum haematoxylin in conjunction with eosin. Van Gieson Mallory's triple stain and borax carmine with picroindigo carmine were also used (Munshi, 1968).

After stunning, small pieces of gill filament, labyrinthine plate, and the respiratory membrane lining the suprabranchial chamber, were fixed either in (1) 2.5 per cent, osmium tetroxide buffered to pH 7.2 with 0.1 M phosphate buffers (Millonig, 1961) or (2) 2.5 per cent, glutaraldehyde in 0.1 M sodium cacodylate buffer followed by post-fixation for 1 h in 2 per cent osmium tetroxide buffered with 0.1 M sodium cacodylate. Specimens were dehydrated in alcohol and embedded in Araldite. Sections were cut on an LKB ultramicrotome. They were mounted on carbon-coated grids, and stained with uranylacetate followed by lead citrate (Reynolds, 1963) and viewed with an AEI 6B electron microscope (Hughes and Munshi, 1973a, b).

Several specimens of *Monopterus cuchia* (100-300 g) were anaesthetized in MS 222 (1 g/500 ml) and fixation was done by vascular perfusion. Compressed air was fed into the air sacs at a pressure of 20 cm H_2O and a cannula inserted into the bulbus arteriosus and *perfusion* initiated with oxygenated physiological saline (0.9 per cent NaCl) containing 2500 IU heparin/litre. Perfusion was subsequently continued with a 2.5 per cent glutaraldehyde solution in 0.03 molar potassium phosphate buffer (pH

7.4, 500 milliosmoles). Tissue samples were taken from the air sac, gill region, hypopharynx and buccopharynx following the glutaraldehyde perfusion. Tissue blocks were post-fixed in 1 per cent osmium tetroxide solution (pH 7.4, 350 milliosm.), stained in a 0.05 molar uranyl acetate solution (pH 5,350 milliosm.) dehydrated in a graded series of ethanols and finally embedded in epon for transmission electron microscopy (TEM). Post-fixation and staining were performed by perfusion and tissue samples removed from the air sacs and gills, dehydrated and embedded. Some samples were critical point dried with CO_2 after dehydration and sputtered with gold (50-60 nm) for scanning electron microscopy (SEM). Respiratory organs of a few mud eels were fixed, following anaesthesia solution (pH 7.4, 350 milliosm.) through the mouth. Although attempts were made to seal the opercular slits, some fluid was constantly leaking out which resulted in the pressure head varying between 15 and 55 cm H_2O. The fish was then kept submerged in the same glutaraldehyde solution for 3 hours. Afterwards tissue samples were taken from the same positions and processed for TEM and SEM as described above. In order to preserve the mucus lining of the respiratory surface, the anaesthetized fish was decapitated and the cranial portion of the head sectioned in a sagital plane. The two halves were then submerged in glutaraldehyde to which ruthenium red (Ruthenium hexamine trichloride – RHT: Johnson Mathey Chemicals) was added at a concentration of 0.7 per cent (final pH 7.1). Tissue samples were then taken from the same position and post-fixed in O_sO_4 in phosphate buffer at pH 7.4.

The emphasis on gill vascular anatomy and microcirculation has been made only recently (Olson *et al.*, 1986, 1990, 1994; Munshi *et al.*, 1986b, 1990, 1994). The India ink perfusion technique was successfully employed by Munshi (1961, 1962). The application of methyl methacrylate (Mercox) vascular corrosion replication techniques were initially pioneered by Murakami (1971) and subsequently applied to the fish gill by several workers (Gannon *et al.*, 1973; Olson, 1980, 1991; Munshi *et al.*, 1990). Laurent and Dunel (1976) have successfully used Silicone resins as vascular substitution material. Mercox, has low viscosity, low interfacial tension and rigidity. This technique has permitted three dimensional resolution of even the finest capillary networks under SEM. However, success of these techniques depend on the conditions under which the resins are perfused. The gills and ARO (Air-breathing Organ) are to be pre-perfused with appropriate Ringer saline for the removal of blood under appropriate pressure and temperature (Olson, 1991).

Structure of Respiratory Organs

Structure of the Respiratory Air Sac of *Heteropneustes fossilis*

The gills of *Heteropneustes fossilis* (Plates 5.1 and Plate 5.2) are well developed with typical secondary lamellae. The area of an average secondary lamella is, however, relatively small, when compared to other water-breathing fishes (Hughes *et al.*, 1974). The lamellae float freely in a denser medium of water. But if the fishes are taken out of water, the large secondary lamellae cannot remain separate from one another but tend to adhere together, and as a result, the greater part of their surface will become ineffective in respiratory gas exchange. The shorter secondary lamellae of the

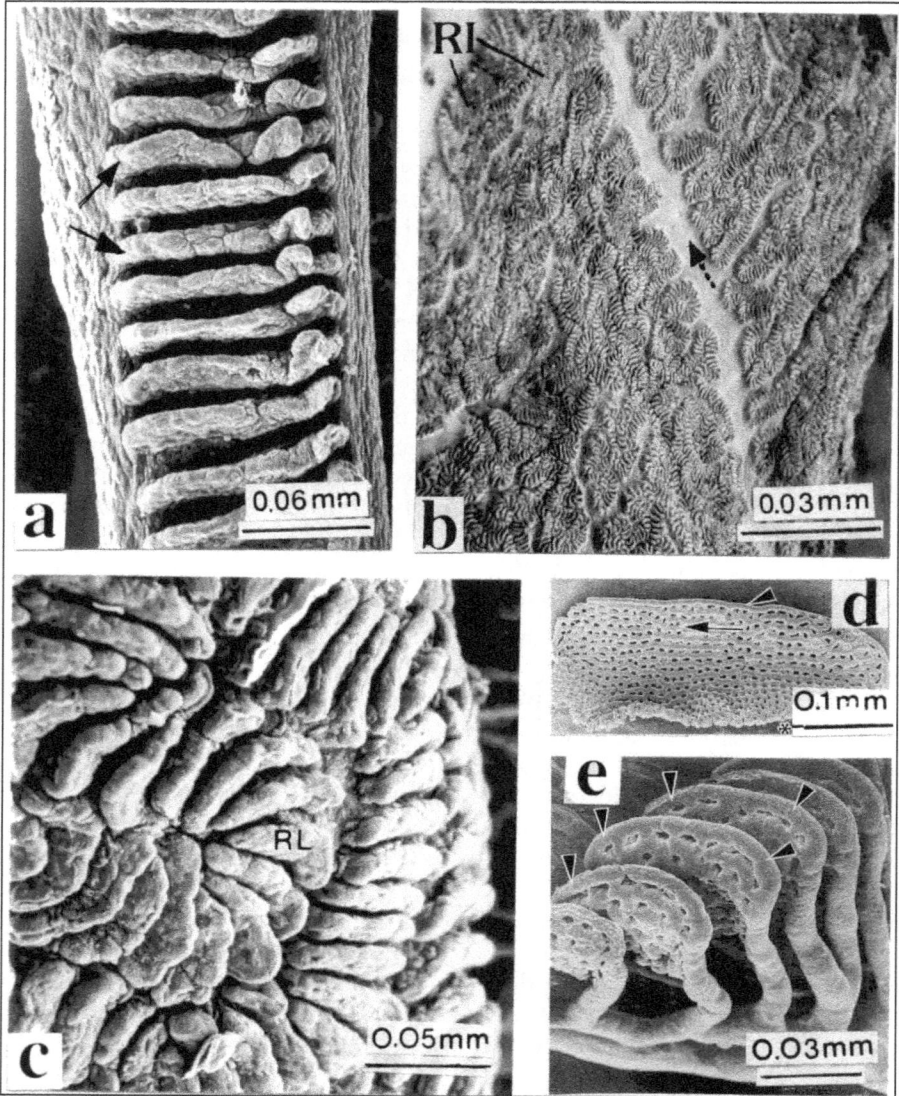

Plate 5.1

(a) SEM photomicrograph showing part of a gill filament along with secondary lamellae of *Heteropneustes fossilis*. Bar = 0.06 mm; (b) Low power SEM photomicrograph showing respiratory islets (RI) and non-respiratory lanes (arrow) of the respiratory air-sac of *H. fossilis*. Bar = 0.03 mm; (c) Higher magnification of part of Figure b showing the biserial arrangement of respiratory lamellae (Rl). Bar = 0.05 mm; (d) SEM of methyl methacrylate corrosion replica (MERCOX) of gill lamella of *H. fossilis*. Note the flow through outer marginal channel (tailed arrow). Inner basal margin (*) is discontinuous. (After Olson *et al.*, 1990) Bar = 0.01 mm; (e) SEM of Mercox preparation of air-sac lamella of respiratory islet outer marginal channel (arrow-heads) and 1-2 adjacent continuous channels are also evident. (After Olson *et al.*, 1990) Bar = 0.03 mm.

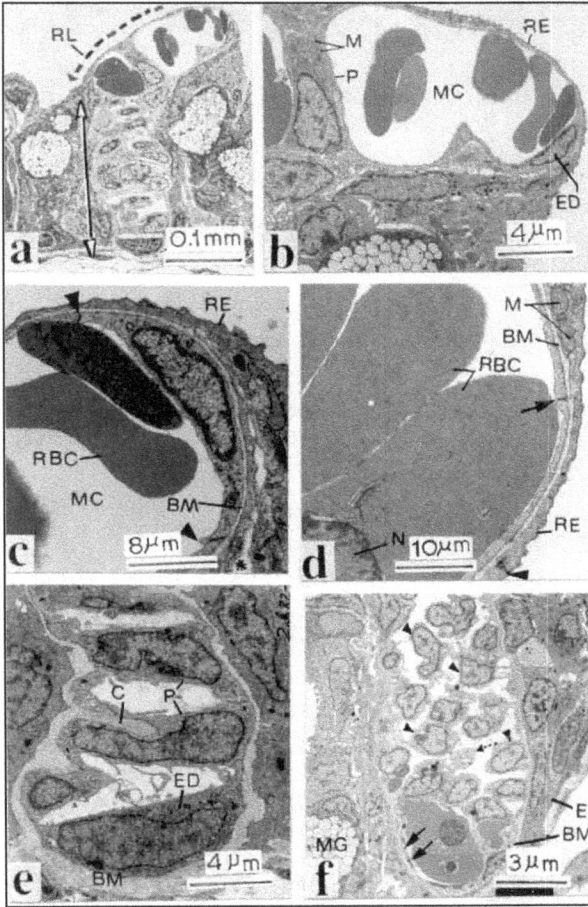

Plate 5.2

(a) TEM of a section of lamella of respiratory islet of *Heteropneustes fossilis* cut at right angle, showing the respiratory (distal broken arrow) and non-respiratory (Proximal) (straight arrow) parts. Bar = 0.1mm; (b) TEM of respiratory distal part of a lamella of respiratory islet as above showing the formation of marginal channel by endothelium (ED) and pillar cell (P) are also seen is basically placed nucleus, many mitochondria (M) and a column. The single layer of epithelium is also clear. Bar = 4 μm; (c) Magnification of distal region of Figure b. Note the junction of endothelial cell with pillar cells flange marked by arrow head and lymphoid spaces (*). Bar = 8 μm; (d) Higher magnification of marginal channels showing the single layer of respiratory epithelium (RE) with mitochondria (M), junction of two epithelial cells (arrow) basement membrane (BM) and arrow-head for junction of the pillar cells flange is also seen. Bar = 10 μm; (e) Proximal (basal) region of respiratory lamella of RI showing the basal endothelial cell with thick basement membrane (BM), Column © and pillar cell (PC). Bar = 4 μm; (f) TEM of an horizontal section of lamella showing the meshwork of pillar cells (PC) separating the respiratory component containing (RBCs) from the basal part containing WBC, broken arrow. Note Weibel Pallade body in the endothelial cell marked by trailed arrow. MG=Mucous Glands. Bar = 3 μm.

respiratory sacs which lie flat on the surface of the membrane are more adapted to air-breathing.

Transformation of Gill Filaments and Lamellae into Respiratory Islets

The fine structure of the respiratory sac of *H. fossilis*, as revealed by SEM and TEM, supports this hypothesis, that fusion of the primary lamellae (gill filaments) and shortening of the secondary lamellae have produced air-breathing devices like gill 'fans' and the 'islets' of the respiratory membrane lining the air-sac (Munshi *et al.*, 1989) (Figure 5.1).

The following steps might be traced in the structural evolution of the air sac:

1. With the advent of aerial breathing, there has been a process of lateral cohesion of successive gill filaments, a process well demonstrated in the case of the structure of the 'fans' where filaments become laterally fused with each other in varying degrees.
2. The secondary lamellae on both sides of the filament shift their positions to come to lie on the same surface.
3. The secondary lamellae diminish in size and lie flat in two rows on the surface of the respiratory membrane of the air-sac giving rise to the typical biserial arrangement of lamellae to form islets.
4. The lamellae retain their position in relation to the afferent and efferent primary filament arteries. The efferent vessel is single and median in position, but the primary afferent vessel now becomes paired instead of remaining single.
5. The main ridge of the respiratory sac containing the IVth afferent and efferent vessels represents the gill arch region of branchial arch.
6. There is a loss of gill rays.
7. The cucullaris muscle innervated by a lateral branch of the occipitospinal nerve originates from the posterior region of the auditory capsule and encircles the entire respiratory sac (Munshi, 1962). This muscle may be involved in the contraction and expansion of the respiratory sac.

The present ultrastructural study confirms the earlier findings of Munshi (1958, 1962) and Hughes and Munshi (1978) that the respiratory islets of the air-sac in *H. fossilis* have been derived by modifications of the lamellae of the gills.

Structural Differentiation of Lamellae

The distal and basal (proximal) blood channels of the lamellae of respiratory islets are lined by endothelial cells. These cells are characterized by the presence of darkly-staining granules resembling the endothelial granules of Weibel and Palade (1964), their functions are not very clear, but in mammals they have been shown to contain clotting factors (Wagner *et al.*, 1982).

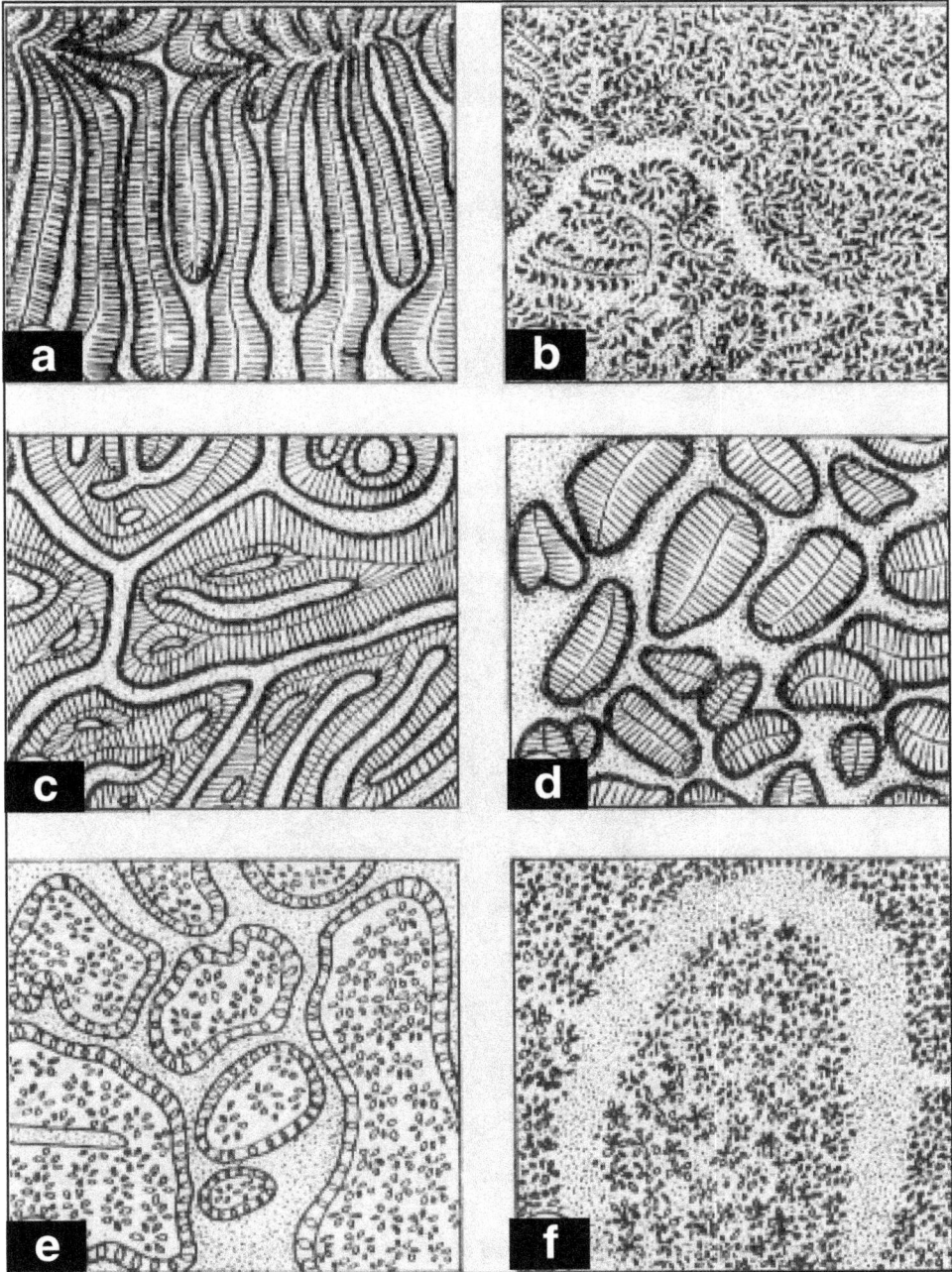

Figure 5.1(a-f): Respiratory islets of (a) *Clarias batrachus*; (b) *Heteropneustes fossilis*; (c) *Anabas testudineus*; (d) *Colisa fasciatus*; (e) *Amphipnous cuchia* and (f) *Channa punctatus.*

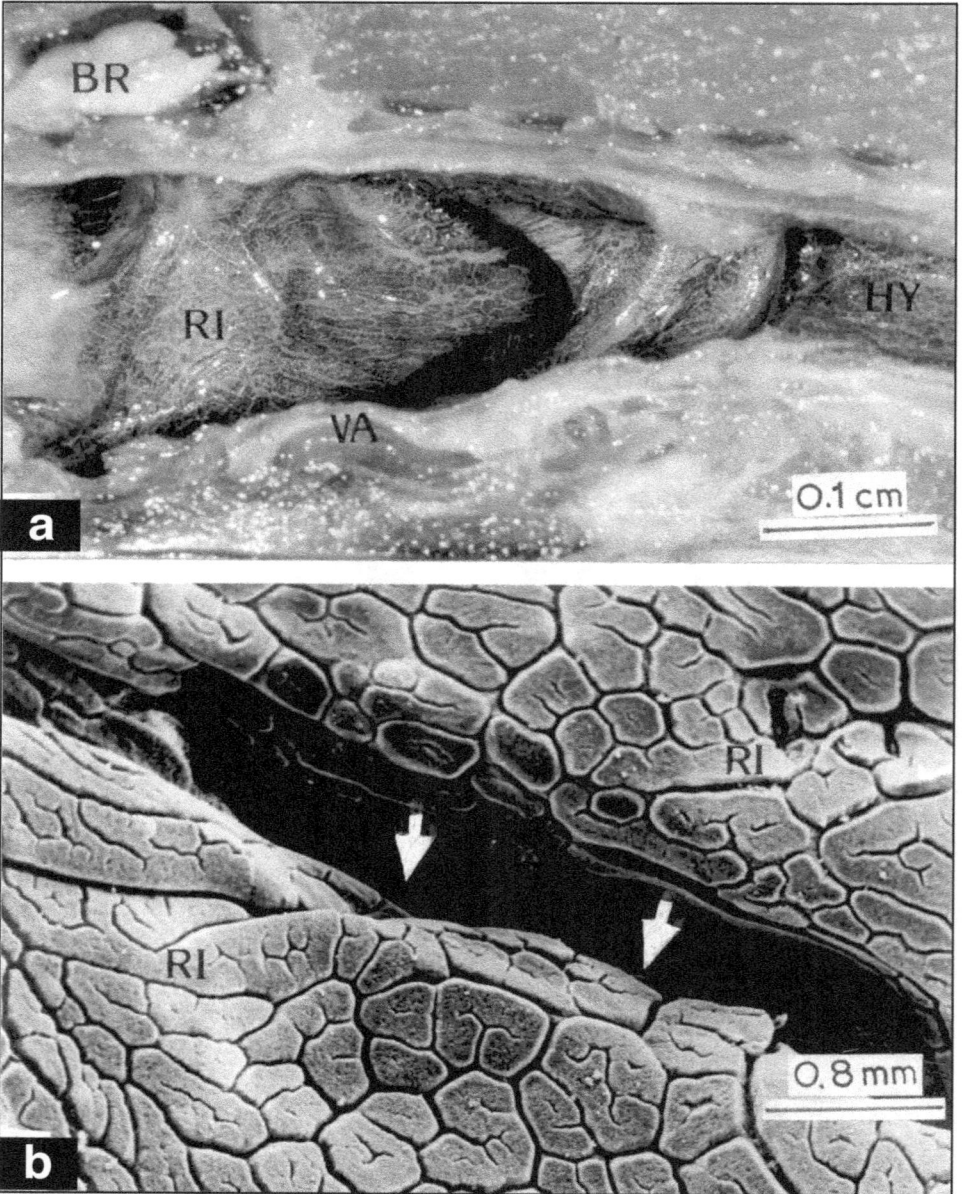

Plate 5.3

(a) Photomicrograph of a dissection of head of *M. cuchia* showing respiratory islets (RI) of buccopharynx, inhalant aperture of respiratory air-sac guarded by a membranous flap like shutter attached with first branchial arch, the position of brain (BR), ventral aorta (VA) and hypopharynx (Hy.) Bar = 0.1 cm; (b) SEM of pharynx showing cleft (arrow) between successive arches through which water moves (Arrow). Note the respiratory islets (RI) on both sides of the cleft. Bar = 0.8 mm.

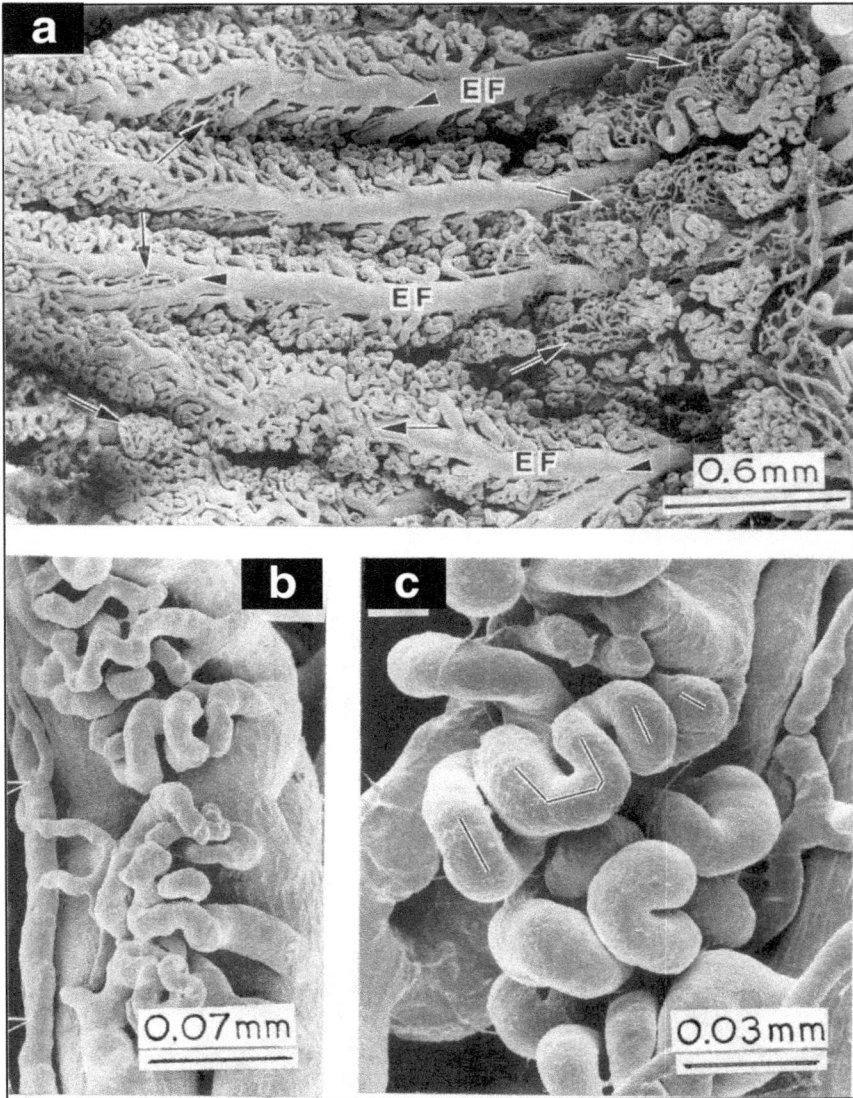

Plate 5.4

(a) SEM photomicrograph of Mercox corrosion replica of part of second gill arch of *M. cuchia*. Efferent artery (EF), coiled arteries of lamellae of gill filaments rosette of vascular papillae formation at bases of filaments and on gill arch, net work of nutrient vessels. Bar = 0.66 mm; (b) SEM photomicrograph of Mercox preparations of gill filament of second arch with coiled marginal capillary of lamellae. Blood flows from afferent filament artery (left) through the tortuous capillaries to efferent filament artery (right), a small vein runs along the afferent filament artery (arrow heads) bar = 0.07mm; (c) Higher magnificence of lamellar capillaries from second arch filament of *Monopterus cuchia*. Solid lines indicate peripheral marginal loops of one capillary which is apposed to epithelium. The bendings of the capillaries give rise to vascular papillae. X640.

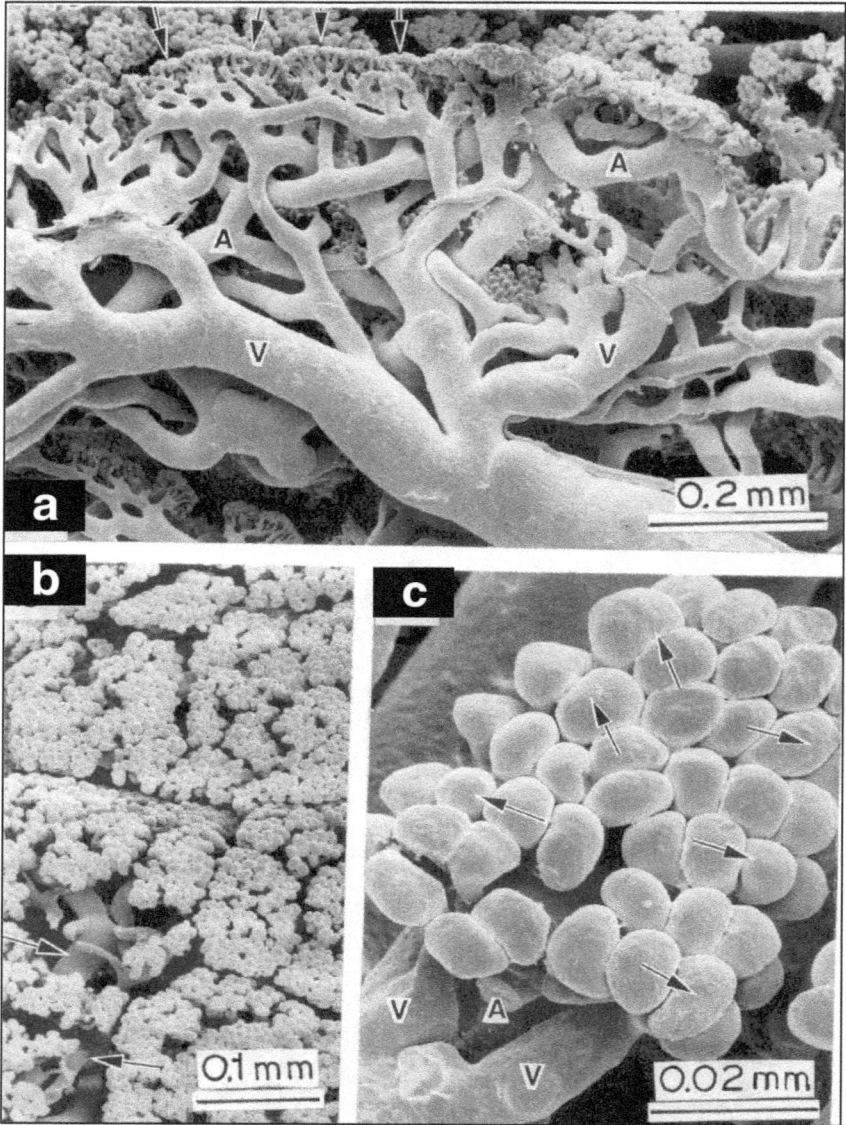

Plate 5.5

(a) SEM photomicrograph of vascular replica of respiratory epithelium seen in a transverse view. Large arteries (A) and vein (V) interwine to form a vascular lattice supplying the respiratory capillaries (arrow) forming rosette of vascular papillae. (After Munshi *et al.*, 1990) Bar = 0.2 mm; (b) SEM photomicrograph of corrosion replica of respiratory islet showing many rosettes of vascular papillae. (Note several distributing vessels are also visible beneath capillaries) (arrow) (Munshi *et al.*, 1990) Bar = 0.1 mm; (c) SEM photomicrograph of surface appearance of respiratory rosette vasculature from pharyngeal epithelium. The central artery (A) supplies respiratory rosette the vascular papillae toward peripheral vein (V). (Munshi *et al.*, 1990) Bar = 12 μm.

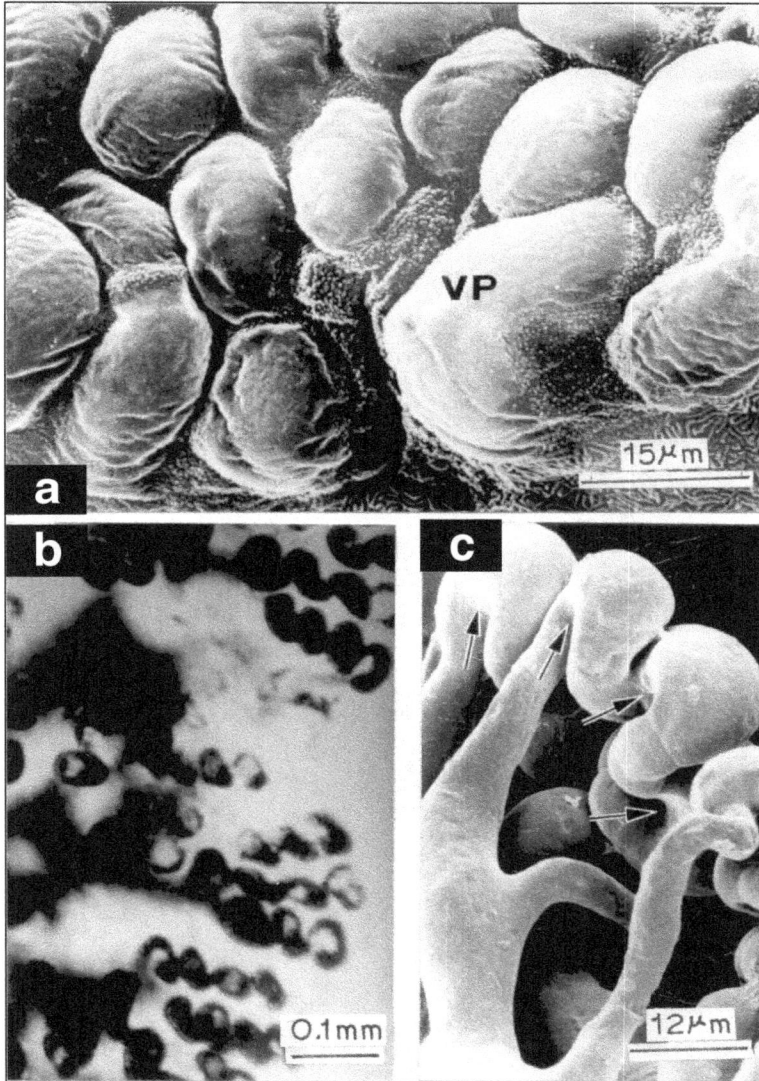

Plate 5.6

(a) SEM high power view of the respiratory islet of respiratory air-sac showing the formation of vascular papillae (VP). Note the tortuous course of one blood capillary, to form the vascular papillae (VP) and microvilli (MV) of the epithelial cell junction. (After Munshi *et al.*, 1989). Bar = 15 µm; (b) Photomicrograph of a part of whole mount of an India ink injected preparation of the respiratory membrane lining the air-sac of *M. cuchia.* Note the spiral nature of vascular capillary forming a series of vascular papillae. The white spots show the position of endothelial valves (marked by arrow). (Munshi and Singh, 1968). Bar = 0.1 mm; (c) SEM photomicrograph of Mercox preparation of respiratory capillary showing spiral nature of capillary forming vascular papillae in *M. cuchia.* Sphincter at the neck of one of the vascular papillae is seen (arrow head) (Munshi *et al.*, 1990). Bar = 12 µm.

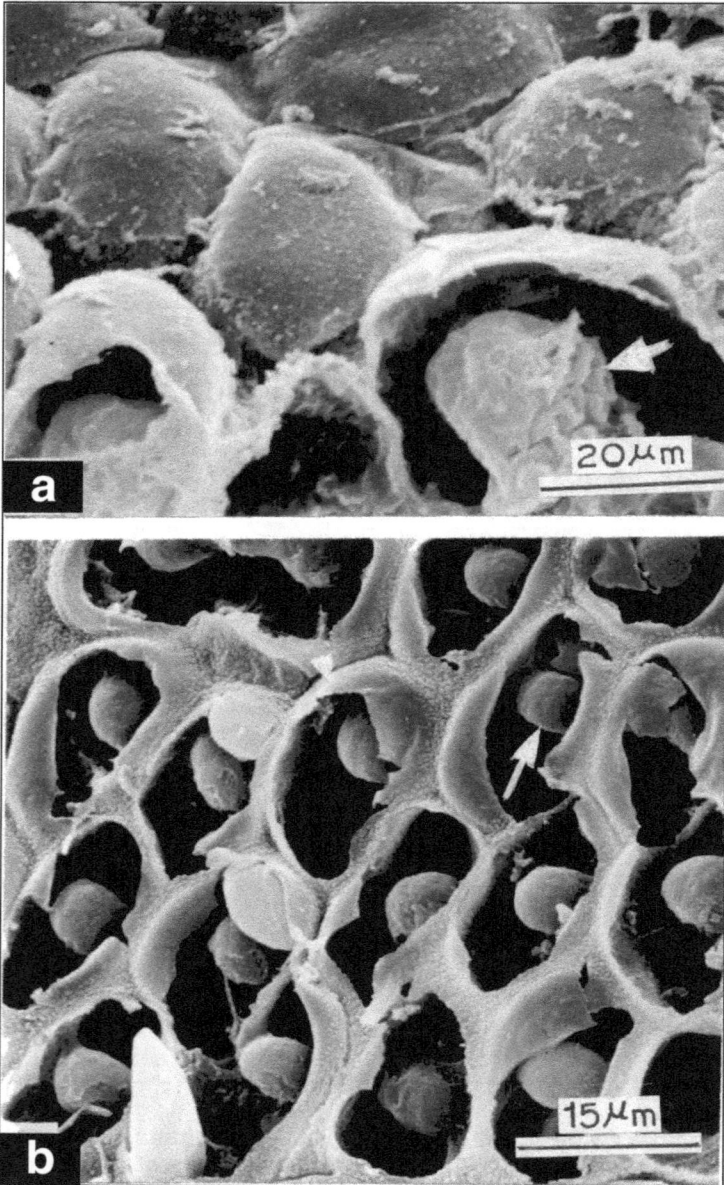

Plate 5.7

(a) SEM photomicrograph of vascular papillae showing scattered mucus on the roof of the papilla. A vertical section through a vascular papilla has exposed the surface of an endothelial valve in tissue material marked by white arrow. Bar = 20 µm; (b) SEM of respiratory islets of hypopharynx showing the ruptured papillae (VP) white arrow head with their endothelial valves (EDV) (tailed arrow), giving a honey comb like pattern. Each valve forms the Fulcum base of the vascular papilla leading the afferent pathway to the efferent side. Bar = 15 µm.

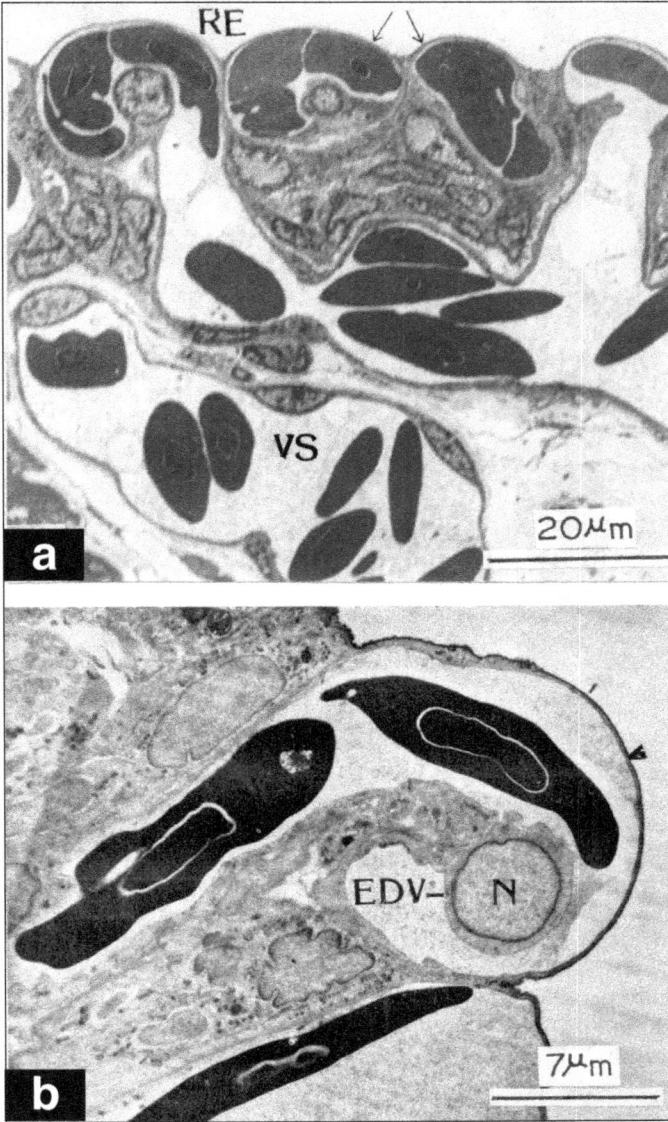

Plate 5.8

(a) TEM photomicrograph of vertical section passing through the respiratory islet of *M. cuchia* connective tissue region with the arterial and venous supply to the vascular papillae (thin arrow). Note endothelial valves (EDV). Respiratory epithelium RE, Red blood cells (RBC), and vascular papillae. A venous (VS) sinus like structure is seen below the rosette of vascular papillae. Bar = 20 µm; (b) TEM photomicrograph of vertical section through the vascular papilla showing respiratory membrane of the air-sac, see the position of endothelial valves (EDV) in the vascular papilla (arrow head). Note the deflected position of the valves with its prominent nucleus (N). (Hughes and Munshi, 1973). Bar = 7 µm.

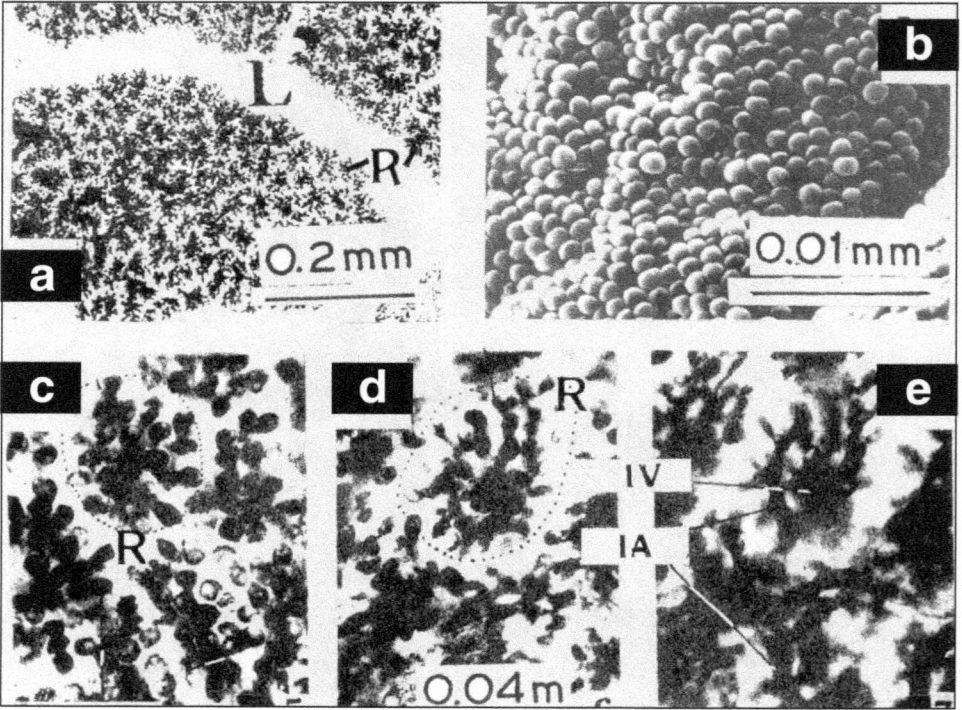

Plate 5.9

(a) Photomicrograph of a whole mount of the respiratory membrane lining the suprabranchial chamber of *Channa punctata*, injected with Indian ink, showing the vascular patches representing respiratory islet and the white non-respiratory 'lanes' (L). The vascular papilla are in the form of numerous 'rosettes' (R) (After Munshi, 1962); (b) SEM micrograph of a respiratory islet of suprabranchial chamber of *Channa marulius* showing numerous vascular papillae. Bar = 0.01 mm; (c), (d) and (e) Photomicrograph of a few rosettes at a different level of focus showing their vascularisation with arterial supply in the form of a Ý and central venous drainage of group of vascular papillae. (Munshi, 1962). Bar = 0.04 mm.

Typical pillar cells with 3-4 columns have been found in the lamellae of respiratory islets. These columns are not intracellular but remain out of contact with the blood in the folding of the pillar cell membrane (Hughes and Grimstone, 1965).

There appears to be some structural differentiation of the lamellae. The exposed part that lies over the surface of the respiratory membrane has regularly arranged pillar cells, whereas, the embedded region shows an irregular meshwork of pillar cells. The marginal channels are broad and appear to show a lower resistance to blood flow. The distal exposed part contains whole blood, while the proximal basal parts contain plasma and some white blood cells only; hence there appears to be some sort of filtration mechanism which separates the flow of RBC's and WBC of the blood in these lamellae. Differences in haematocrit of blood are well known in different parts of the lamellar circulation of the gills of fishes especially in relation of blood flow to the marginal channels (Hughes, 1984). The outer exposed parts of the lamellae have not more than 2-3 distinct channels and may be adapted for aerial respiratory exchange by allowing greater blood flow.

In *Clarias batrachus* the same sort of structural transformations of gill filaments and secondary lamellae have been observed (Munshi, 1961).

Structure of Respiratory Islets of *Monopterus cuchia*

The air-sacs in *Monopterus cuchia* (Plates 5.3–5.8) are in the form of a pair of lung-like structures situated along the lateral sides of the head, partly masked by the operculum and connected with the pharynx through an opening that serves as an inhalant and exhalant apertures (Munshi and Singh, 1968b). Morphologically, they are dorso lateral extensions of the pharyngeal cavity. The mucosa lining of the respiratory pharynx, hypopharynx and airsacs consists of vascular and non vascular areas. The vascular areas are small and large respiratory islets formed of intra-epithelial blood capillaries which are extended spirally. The blood capillaries penetrate the epithelium at many points and emerge on the surface to form vascular papillae. The rosettes are separated from each other by micro air-pockets bearing microvilli. The papilla wall is formed by the extended flanges of the endothelial cells with their characteristic round nucleus. The axis of the vascular papilla is formed by the folding of the basement membrane in which some of the supporting epithelial cells are tucked in. They work as small valves. Each valve projects freely into the papilla lumen. The valve seems to be associated with regulating the flow of blood through individual vascular papillae. The air-blood pathway is extremely thin, the total thickness being 0.435 μm in *M. cuchia*. It is formed of a thin layer of respiratory epithelial cells, basal lamina and thin endothelium.

Respiratory Islets of *Ophiocephalus* (=*Channa)*

In *Ophiocephalus punctata*, *Ophiocephalus striata* and *Ophiocephalus marulius* (Plates 5.9–5.12) the main air-breathing organs are in the form of a pair of suprapharyngeal chambers, which develops dorsal to the gill arches above the pharynx and lateral to the auditory capsule (Munshi, 1962). The respiratory chambers almost freely communicate with the pharynx through inhalant apertures. A shutter of dendritic plates, located on the first gill-arch, guards the posterior – most inhalant aperture. The morphological basis of the breathing mechanism has been worked out fairly well. Alternate replacement of air and water during air ventilation have been observed in *Channa argues* and *Channa punctata* (Ishimatsu and Itazawa, 1981; Liem, 1980). These air-breathing chambers also take part in aquatic respiration, when the fish is submerged in the water for a long period of time. The respiratory mucosa has many respiratory islets studded with vascular papillae. The valve that projects into the lumen of a papilla is formed of connective tissue bounded by a strong basement membrane. The surface of the valve, however is covered by the endothelial cells.

In *Channa* (=*Ophiocephalus*) spp. The respiratory islets are composed of vascular papillae, which are specialized parts of the intra-epithelial arterioles. The respiratory islets are supplied with mixed blood by the labyrinthine arteries that originate from the first and second pairs of afferent branchial arteries. The oxygenated blood is

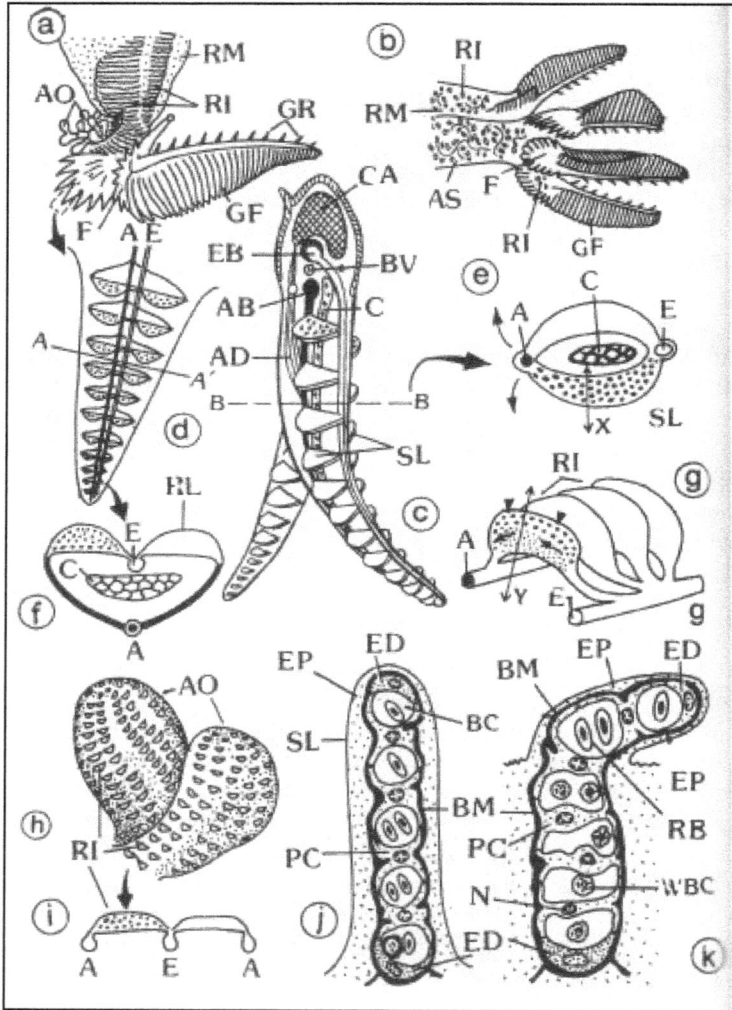

Figure 5.2: Schematic diagram of morphological transformations of gills into accessory respiratory organs in *Heteropneustes fossilis* and *Clarias batrachus*.

(a) Lateral view of the second gill arch of *Clarias batrachus* showing the formation of the fan, Arborescent organ (AO) and parts of the respiratory membrane (RM) lining the suprabranchial chambers, the latter have been derived from the gill lamellae of the gill arch. Note gill raker (GR) and gill filament (GF); (b) The respiratory sac of *Heteropneustes fossilis* opened out by means of a mid-ventral incision to show the relationship between the 'fans' (F), the gill arches (GA) (I-IV) suprabranchial chamber and respiratory sac. Note the arrangement of respiratory islets on fans, respiratory membrane of air-sacs; (c) Three dimensional figure of typical gill filament showing the biserial arrangement of secondary lamellae, afferent (AB) and efferent branchial artery (EB), cartilage, gill arch (CA), adductor muscle (AD) and branchial vein (BV). Note pillar cells in secondary gill lamellae shown by black dots; (d) The diagrammatic view of a part of the fan of *Clarias batrachus* showing the position of secondary gill lamellae arranged in two rows on the surface of the fan formed by the fusion of the primary gill lamellae and their vascular supply. The primary afferent and efferent vessels lie at different levels. (e) A diagrammatic representation of a transverse section of the gill filament at the level marked B-B in figure "C" showing the position of the secondary gill lamella and their vascular supply. (f) A diagrammatic transverse section of a part of the Figure d passing through A-A showing the position of the secondary gill lamellae, on both sides of the primary efferent artery, the gill ray (C) and the primary and secondary afferent arteries; (g) A diagrammatic view of SEM of vascular corrosion replica of air-sac lamellae of *Heteropneustes fossilis*. Outer marginal channel (arrow heads) and 1-2 adjacent channels and afferent and efferent lamellar arteries are seen (After Olson *et al.*, 1990); (h) Diagramatic view of dendritic bulb of arborescent organ (AO) showing the nature of respiratory islet and lane; (i) A diagrammatic transverse section of the Figure h passing through two lamellae of one islet showing their vascular supply; (j) Diagramatic representation of a TEM section of secondary lamellae of *H. fossilis* showing blood channel (BC), Epithelium (E), pillar cell (PC), basement membrane (BM), endothelial cell (ED); (k) Diagramatic representation of TEM section of respiratory lamellae of respiratory islet air-sac of *H. fossilis* at right angle, showing the respiratory (distal) and non-respiratory (basal) along with epithelium (E), Basement membrane (BM), Endothelial cell (ED), RBC, WBC and pillar cell (PC).

drained from the suprabranchial chamber by suprabranchial veins into the jugular veins.

The vascular papillae of the buccopharynx are of specialized nature, as they are retractile. There is some mechanism to shunt blood from the respiratory islets of this region when the fish manipulates its prey into the buccal cavity. As a result, the vascular papillae are retracted into cup-like depressions of the buccopharyngeal epithelium. This mechanism protects the vascular papillae from being bruised when the fish preys on other aquatic animals. There are prominent epithelial band like structures connecting one papilla with another (Hughes and Munshi, 1986).

Accessory Respiratory Organs of *Anabas testudineous*

The air-breathing organs of *Anabas testudineous* (Plate 5.13–5.18) are labyrinthine plates borne on the epibranchials of the first pair of). The Labyrinthine organs and the inner surface of these chambers bear hundreds of respiratory islets (RI), which are the functional units of gas exchange. They are surrounded by non-vascular areas called 'gill arches lodged in suprabranchial chambers (SBC) on either side of the head (Misra and Munshi, 1958; Munshi, 1958, 1968; Hughes and Munshi, 1968 (lanes)'. The transformation of gill filament lamellae to respiratory islets is illustrated in Plate 5.13, Figure 5.3 and Plate 5.15a.

Blood Circulation

The labyrinthine artery (LA) originates as a branch from the 1st efferent branchial artery and supplies blood to the RI of the labyrinthine plates. Likewise the respiratory membrane of the SBC gets its blood supply from the 2nd efferent branchial artery. The blood is drained via a series of veins (LV) leading to the jugular vein (JV).

Respiratory Islets

Blood Circulation

There is a typical tripartite arrangement of the major vessels comprising a central artery and paired lateral veins for each unit of respiratory islets (Plate 5.16a) (Munshi *et al.,* 1986). The artery supplying an islet, the median islet artery, will branch 2-4 times, before forming many transverse capillaries biserially arranged over the islet. Blood drains into venules through short arteriovenous anastomoses (Plate 5.16b). The venules travel away from the respiratory surface, anastomose, and open into the lateral islet veins which dilate to form the central venous sinus (Plate 5.16 and Figure 5.3J). The relationship between the transverse capillaries and CVS is depicted. The CVS gives shape to the respiratory islets and are the storehouse of oxygenated blood.

Microscopic and Submicroscopic Structure

As viewed by light microscopy and SEM, the labyrinthine organ is seen to be a complex structure having many crenulated plates. Each plate is supported by a thin lamina of bone lying within the connective tissue (Plate 5.18a). The first epibranchial bears labyrinthine plates instead of gill filaments. Respiratory islets of varied shapes and dimensions are found distributed on both surfaces of the labyrinthine plates Plate 5.13a. At higher magnifications the biserial arrangement of capillaries that run

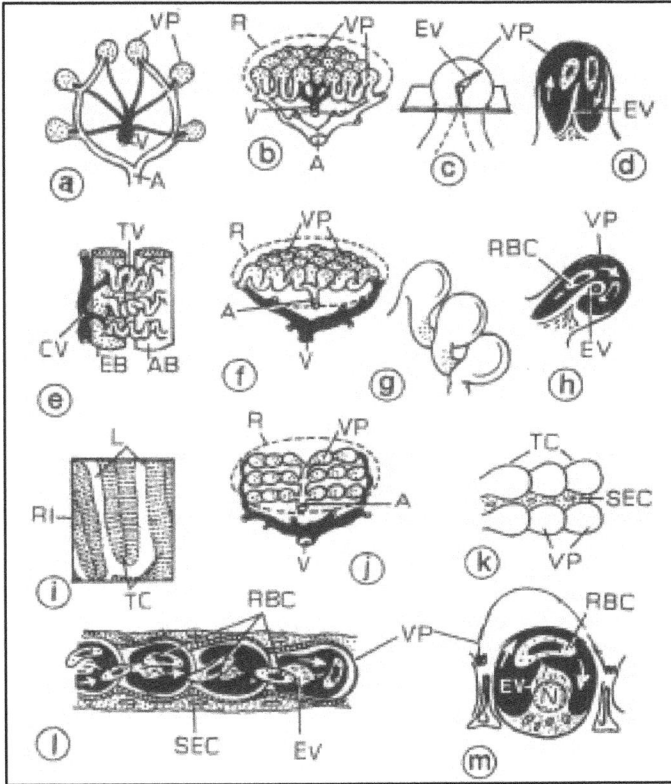

Figure 5.3: The vascular papillae of *M. cuchia, Channa* sp. and *Anabas* are arranged in biserial fashion on both sides of a vascular rosette which form the respiratory islet in all these advance percomorph group of teleostean fishes.

(a) Circulatory design of vascular papillae of respiratory rosette in *Channa* showing arterial (A) and venous (V) drainage; (b) One unit of respiratory rosette (R) of *Channa* with their arterial (A) and venous drainage (V); (c) Formation of vascular papillae in *Channa* from the top bended part of the capillary; (d) TEM section of vascular papilla of *Channa* showing the endothelial part of the capillary; (e) Formation of vascular papillae in *Monopterus cuchia*. Collateral vein (CV) which runs parallel to efferent filament artery which will take up the work of venous drainage of the rosette of respiratory islet; (f) One Unit of respiratory rosette of *M. cuchia* and their arterial (A) and venous (V) drainage; (g) Spiral nature of vascular papillae in *M. cuchia*; (h) TEM section of vascular papilla showing endothelial valve (EV), arrow indicates the pathway of RBC; (i) Part of a respiratory islet of *Anabas* showing vascular and non-vascular areas (Lane); (j) Higher magnification of a part of respiratory islet of *Anabas* showing the arrangement of vascular papillae (VP) forming transverse channels (TC) along with arterial and venous supply; (k) Showing the supporting epithelial cells (SEC) between two transverse channels (TC); (l) Horizontal section of transverse channel (TC) in *Anabas* to show the fusion of a series of vascular papillae (VP) along with their endothelial valves (EV), arrow indicates the movement of RBC; (m) Vertical section of a vascular papilla in *Anabas* to show the endothelial valve (EV), RBC and supporting epithelial cells (SEP) on both sides.

transversely across the surface of the respiratory islets becomes apparent. The capillaries are arranged in parallel lines in two rows on the surface of the respiratory islet. These transverse capillaries have a segmented appearance with a series of bulges when they are full of red blood cells (RBCs). The epithelial cells covering the respiratory surface of the capillaries at these bulging regions remain smooth, microvilli being present only at the marginal non-vascular areas. It is in this part of the respiratory surface of the islets, a large number of rod-shaped bacteria are found (Plate 5.17a).

Relationships between the epithelium, the capillary system, and the underlying subepithelial organization of the respiratory islets become clear in a low power TEM of a vertical section of respiratory islets of the labyrinthine plate (Plate 5.18a,b). The blood capillaries are arranged in a series in the same plane, separated from each other by thin sheets of supporting epithelial cells. These cells in vertical section have broad triangular bases with long, forked apical ends. The nuclei are basal in position. Vesicles and mitochondria are discernible in the cytoplasm. The thin extensions of the cells give pillar-like support to the walls of the contiguous capillaries. The apical parts at times forked into two terminal processes to accommodate the extended processes of respiratory epithelial cells (Munshi and Hughes, 1991).

The respiratory epithelial cells are also triangular in vertical section, with long, thin extensions spread over the capillary surface. Sometimes extensions of these cells enclose narrow spaces over the blood capillaries, which occasionally contain lymphocytes. The supporting epithelial cells are seen in the spaces between the capillaries. The cytoplasm of these cells are electron dense and finely granular in nature. The cells bear short microvilli on their outer surface in the inter-capillary region, but the extensions which cover the capillaries remain chiefly smooth.

The transverse capillaries of the respiratory islets are formed by specialized cells surrounded externally by the basement membrane (lamina) of the epithelium. The main bodies of the endothelial cells lie on a regular sequence (Plate 5.13b). They are characterized by (1) large, round nuclei and (2) thin perinuclear cytoplasm forming septa-like structures extending halfway across the lumen of the channel and functioning as valves between the series of pocket like widening of the capillary lumen (Plate 5.17b), and by (3) thick, tongue-like cytoplasmic extensions, protruding freely in the direction of blood flow in the channel. The tongue like processes have a cytoskeleton of keratin-like fibrillae. Moderate numbers of mitochondria are discernible in the cytoplasm of the main body of the endothelial cells (Plate 5.18b).

The Respiratory Membrane

The respiratory membrane covering the labyrinthine organ and bounding the suprabranchial chamber consists of (a) vascular and (b) non-vascular areas.

(a) Vascular Area

The vascular part of the membrane comprises small and large islets distributed over its entire surface. The elevation of the islets above the general surface varies according to their situation. In the posterior contractile and saccular section of the membrane, the islets are comparatively smaller in size, but elevated so as to impart a folded appearance to the membrane are irregular and are roughly polygonal in outline.

In the remaining part of the membrane, the islets are less pronounced in elevation and are differently patterned. The same kind of structural pattern is also obtained on the surface of the labyrinthine organ (Plate 5.13).

Underlying the epithelium there is a thick layer of connective tissue between the bony labyrinthine plate and the vascular layer of the respiratory islets. Large paired sinusoids are found in the connective tissue layer just below the vascular system of the respiratory islets. The venous sinusoids are somewhat flat, bag-like structures having very thin walls formed by endothelial cells. They contain blood corpuscles, but sometimes they are seen to be empty (Plate 5.18a).

Structural Evolution of Vascular Papillae (Figures 5.3a,b,c,d,e,f)

In Monopterus (=Amphipnous) cuchia, Channa (=Ophicephalus) punctata, Channa striata, Channa marulius, Channa gachua the respiratory islets are distributed extensively on the floor and roof of buccopharynx, tongue, suprabranchial chambers, and dendritic nodules of labyrinthine organs. In *Monopterus cuchia* the respiratory islets are found almost everywhere in the buccopharynx, hypopharynx in front of oesophagus and in the respiratory airsacs (Munshi and Singh, 1968; Munshi, 1976, 1985; Hughes and Munshi, 1986; Munshi *et al.*, 1989). Both in *Monopterus cuchia* and in all the *Channa* spp. The respiratory islets are formed of hundreds of rosettes of vascular papillae. In contrast in *Anabas testudineus* belonging to the Anabantidae group of fishes, the respiratory islets are restricted to the suprabranchial chambers which extend into the roof of buccopharynx. The RI are formed by biserially arranged transverse channels (Munshi, 1958, 1968, 1980, 1985; Hughes and Munshi, 1973a,b; Hughes, Munshi and Ojha, 1986, Munshi and Hughes, 1991). The vascular corrosion replica technique using Mercox was used extensively by Olson *et al.* (1986); Munshi *et al.* (1986b); Olson *et al.* (1990, 1994, 1995) to study the vascular organization of the RI of *Anabas testudineus, Monopterus cuchia, Channa punctata, Channa striata, Channa marulius, Channa gachua, Heteropneustes fossilis* and *Clarias batrachus*.

These studies have revealed the possible steps involved in the structural evolution of the respiratory islets of the accessory respiratory organs of fishes. In this chapter we have made a comparative study of the respiratory islets of *Monopterus cuchia, Channa* spp. and *Anabas testudineus*. All the specialized features of the transverse channels of respiratory islets of *Anabas* is now explicable. Evidence is given to show that the transverse channels of RI of labyrinthine organs have been derived due to the coalescence of a series of vascular papillae-like structures (Figure 5.3j-m).

The vascular papillae are sections of capillaries which connect the arterial system of the branchial region with those of the veins of the jugular system and which are specialized for gas exchange (Figure 5.4). Similar types of vascular papillae are also found in the respiratory islets of *Monopterus cuchia*, an air-breathing fish belonging to the order Symbranchiformes (Munshi and Singh, 1968). In both *Channa* and *Monopterus*, the respiratory islets are distributed not only in the respiratory sacs but also in the buccopharynx. In *M. cuchia* the first pair of gills have been transformed into the respiratory air sacs. These sacs represent the embryonic branchial pouches during development of gills (See Goodrich, 1930, Chapter-IX). It seems that the gill slits did not form and the branchial pouches enlarge into the air-sac in the adult stage

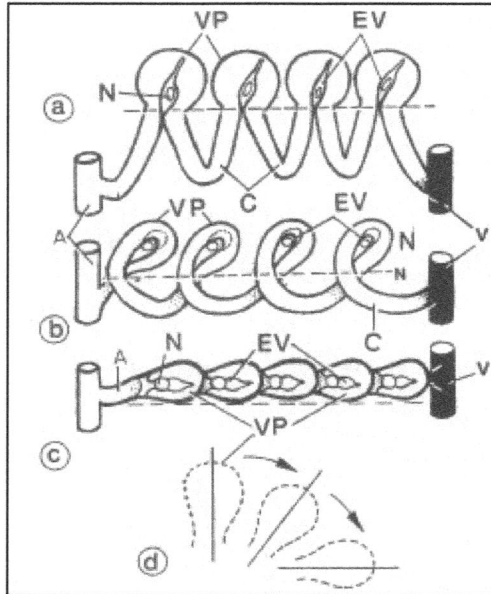

Figure 5.4: Three conceptual models of respiratory islets of *Channa, Monopterus* **and** *Anabas* **showing the arrangement of vascular papillae in biserial fashion on both sides of a vascular rosette which form the respiratory islet in all these advance group of Percomorph teleostean fishes.**

(a) *Channa model* – Upper part of the capillary forming the vascular papillae (VP) covered over by thin respiratory epithelium for gaseous exchange. The nature of their capillary is wave like; (b) *Monopterus model* – Most part of the capillary remains embedded in the epidermal tissue of the buccopharynx; (c) *Anabas model* – Vascular papillae fused with each other to form a transverse channel of respiratory islet to make whole structure efficient for gaseous exchange from air; (d) Orientation of vascular papilla in relation to the horizontal plane of the respiratory membrane in *Channa, Monopterus* and *Anabas.*

of *M. cuchia*. The 1st branchial arch supply blood to the respiratory air-sac. The second branchial arch only bears a few gill filaments; the respiratory islets have developed extensively on the floor and roof of the pharynx as well as in the hypopharynx. In the climbing perch, *Anabas testudineus* (Perciformes), the respiratory islets are restricted to the labyrinthine organs and epithelial lining of the suprabranchial chambers (Hughes and Munshi, 1973a). In this, the respiratory islets are composed of biserially arranged transversely oriented blood capillaries. These transverse capillaries are formed of series of vascular papillae which coalesced over one another. Each unit of the capillary is provided with valves with broad collar-like extensions and supported by large prominent nucleated endothelial cells. These endothelial cells have tongue-like processes that have a cytoskeleton with coarse keratin fibrillae. The tongue works like a 'spring board' of a swimming pool. It seems that they are movable due to the cardiac pressure of blood flow. As RBCs fall over them they are pushed up towards

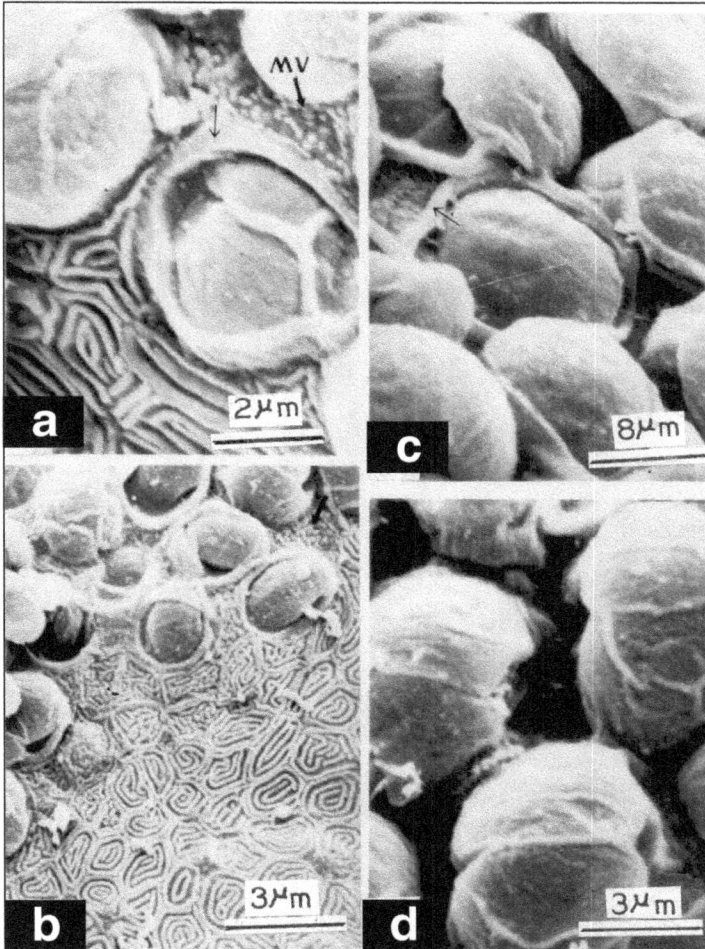

Plate 5.10

(a) SEM of the buccopharynx of *Channa striata*. Enlarged view of vascular papillae of buccopharynx, one of which is retracted within the receptacle (thin arrow). Prominent, band-like epithelial cell boundaries are seen on the surface of the vascular papillae. Prominent microridged epithelial cells are found in the non-vascular area. Microvilli (MV) are visible at the base between the vascular papillae. (After Hughes and Munshi, 1986). Bar = 2 µm; (b) View of the vascular and non-vascular area of buccopharynx of *Channa* showing whorla microridges on the epithelial cell surface. In the transitional area, the microridges break-up into microvilli (Hughes and Munshi). Bar = 3 µm; (c) Higher magnification showing the alterate arrangements of vascular papillae in different stages of retraction. The prominent rims of the receptacles are seen as band-like structures connected strongly with each other (thin arrow). (Hughes and Munshi, 1986). Bar = 8 µm; (d) Fully protracted vascular papillae showing the loop-like formations of blood vessels. Prominent band-like cell boundaries of the epithelial cells and the alternate arrangement of vascular papillae are clearly discernible. (Hughes and Munshi, 1986). Bar = 3 µm.

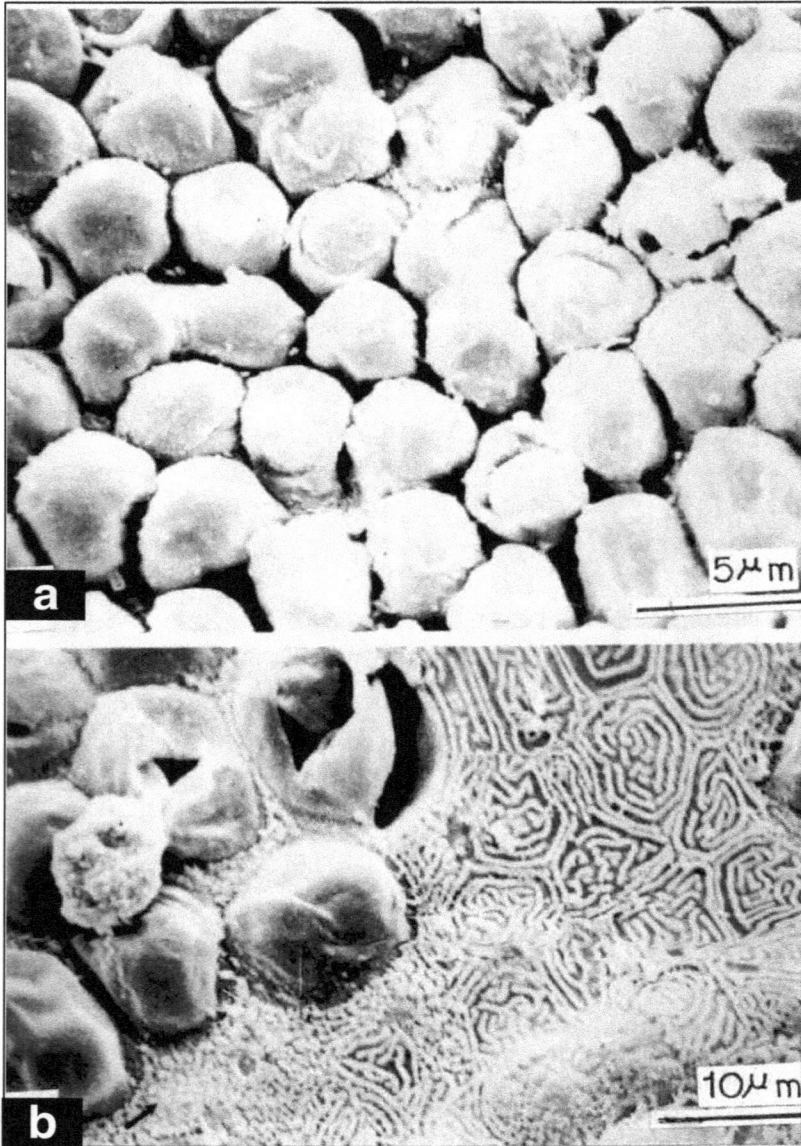

Plate 5.11

(a) SEM of surface of respiratory air-sac (=suprabranchial chamber) of *C. striata*. A bunch of fully protracted vascular papillae of air-sac without receptacles and bands. See the fusion of vascular papillae at several places. (After Hughes and Munshi, 1986). Bar = 5 µm; (b) Higher magnification of vascular and non-vascular areas of respiratory air-sac showing vascular papillae with microvilli at their bases and microridges bearing epithelial cells in the non-respiratory area. The microridges become irregular and then break-up into microvilli in the transitional area. (Hughes and Munshi, 1986). Bar = 10 µm.

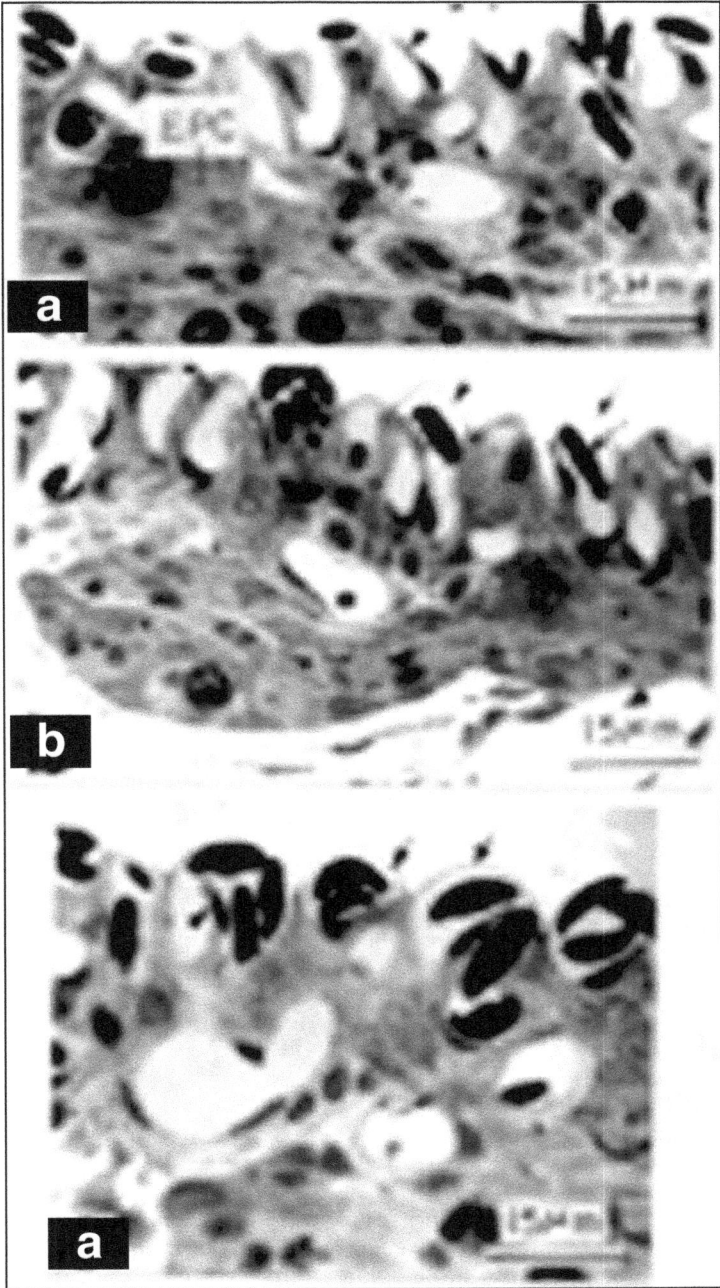

Plate 5.12

(a-c) Photomicrograph of vertical sections of air-sac of *C. striata.* Vertical section of wall of air-sac showing vascular respiratory (EPC), vascular papillae (tailed arrow) and dermis. Note endothelial valve are also shown by arrow head. Bar - 15μm.

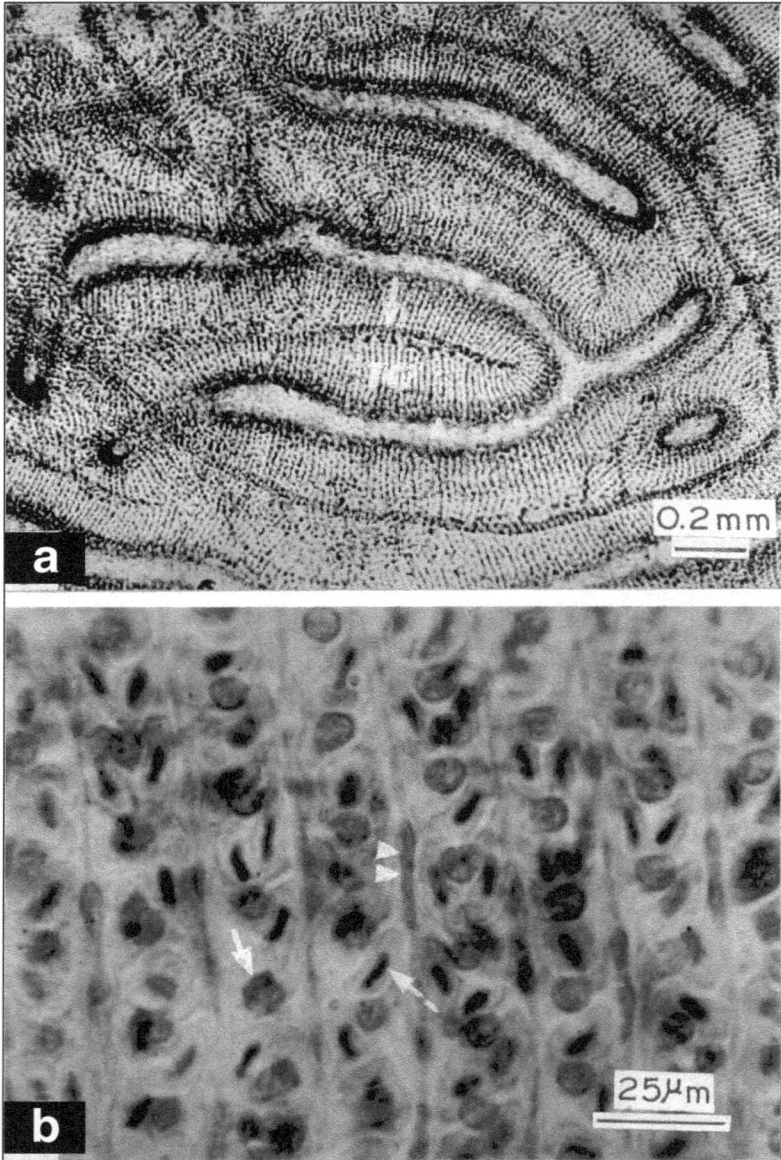

Plate 5.13

(a) Photomicrograph of a toto preparation of the respiratory membrane of *Anabas* showing the structure of a respiratory islet. The biserial arrangement of two rows of transverse blood channels with the median afferent arteries marked by arrow and arrow heads are seen. Bar = 0.2 mm; (b) Higher magnification of a part of above respiratory islet of *Anabas* showing transverse blood channels alongwith supporting epithelial cell (two arrow heads) in between the transverse blood channels, round nuclei of the endothelial valves (tailed arrow) and red blood corpuscle by broken tailed arrow. Bar = 25 μm.

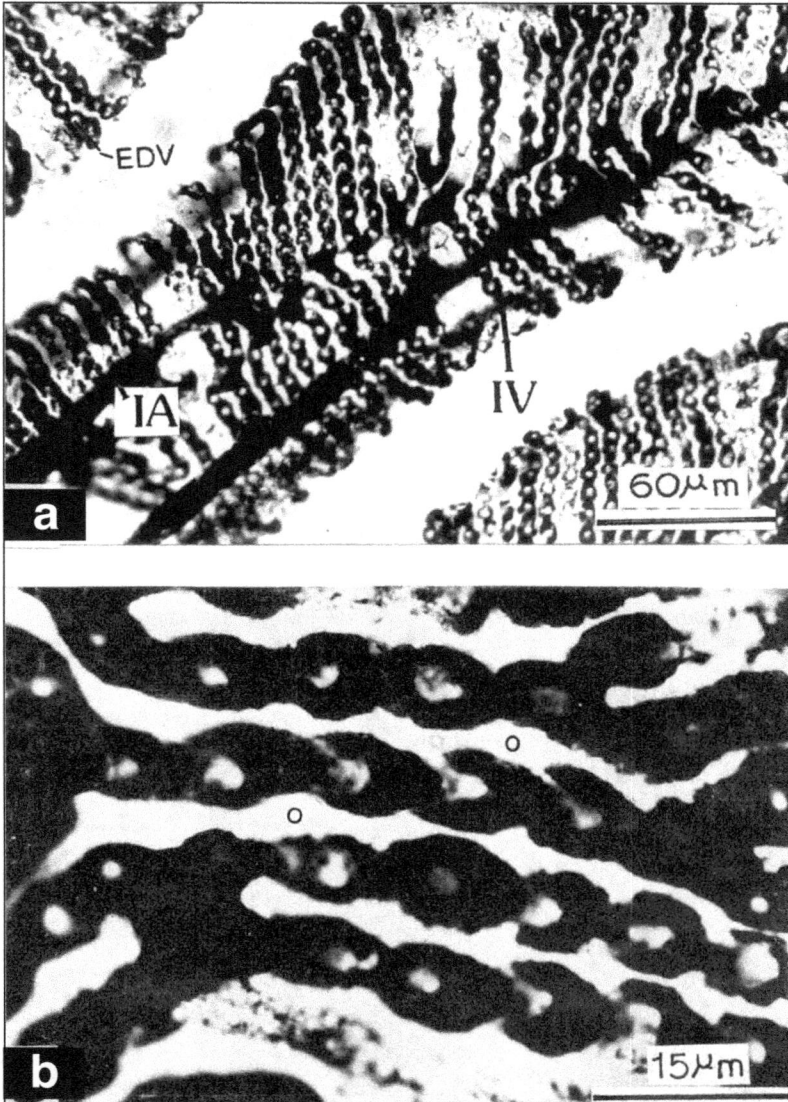

Plate 5.14
(a) Photomicrograph of India ink injected preparations of respiratory membrane of *Anabas* showing the pattern of the blood capillaries. The blood channels look like a string of rings with white spots at regular intervals. The white spots represents the positions of endothelial valves. Note (1A) islet artery and islet vein (IV) are also shown. Bar = 60μm. (After Hughes and Munshi, 1973); (b) Higher magnification of the above preparation to show the nature of the internal space of vascular papillae like structure. The position of endothelial valve (arrow) is also seen. The formation of the blood channels due to the coalescence of vascular papillae is apparent. The white space between the transverse capillaries represent the supporting epithelial cells between the transverse channels marked by (o). bar = 15 μm (After Hughes and Munshi, 1973).

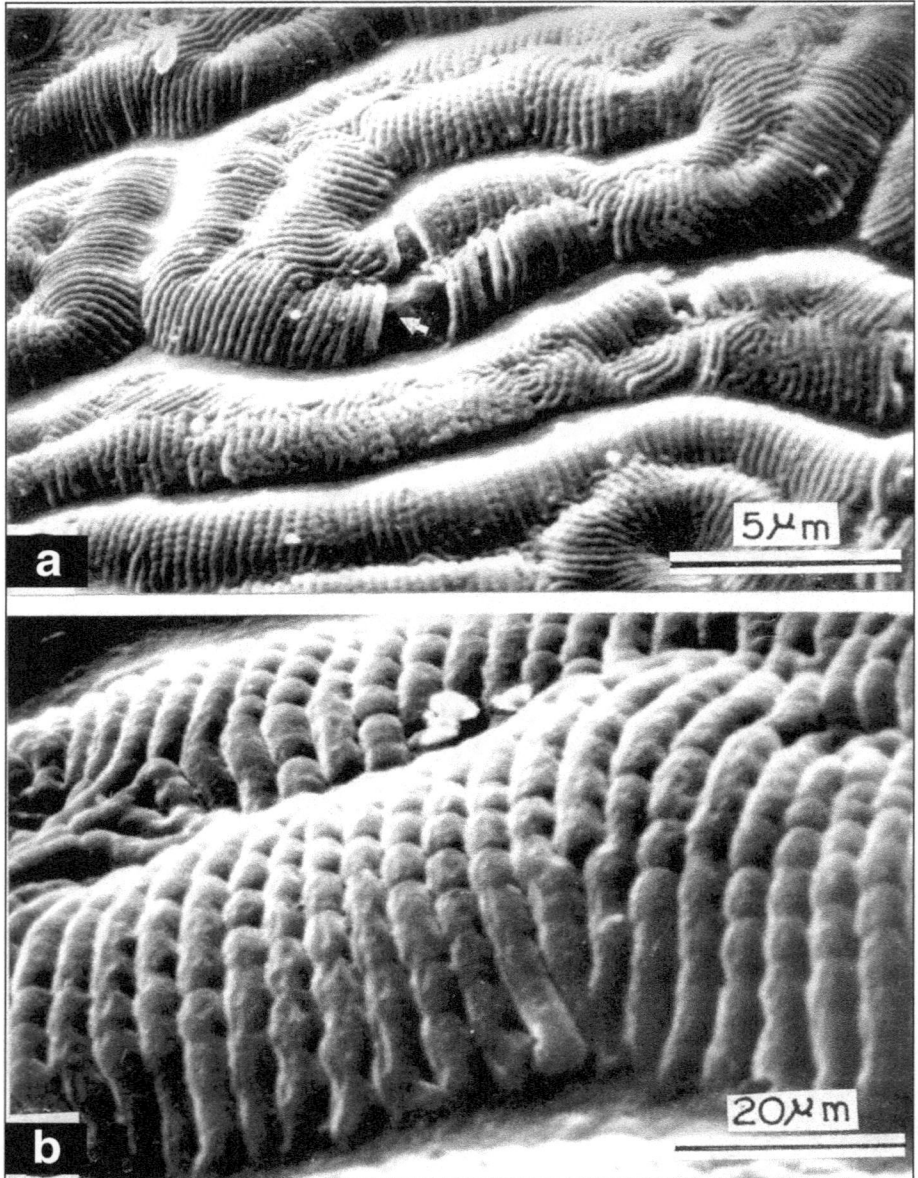

Plate 5.15

(a) SEM photomicrograph of methyl methacrylate (mercox) corrosion replica of
respiratory islets of *Anabas* showing the arrangement of transverse blood channels.
The blood channels curve over the venous sinus (arrows) of the respiratory islets bar
= 5 μm; (b) SEM of resion cast corrosion vascular replica of a respiratory islet showing
biserial arrangement of transverse capillaries. They give the appearance of strings of
beaded necklace. Each bead represents replica of a vascular papilla. Bar = 20μm.
(After Munshi, 1996).

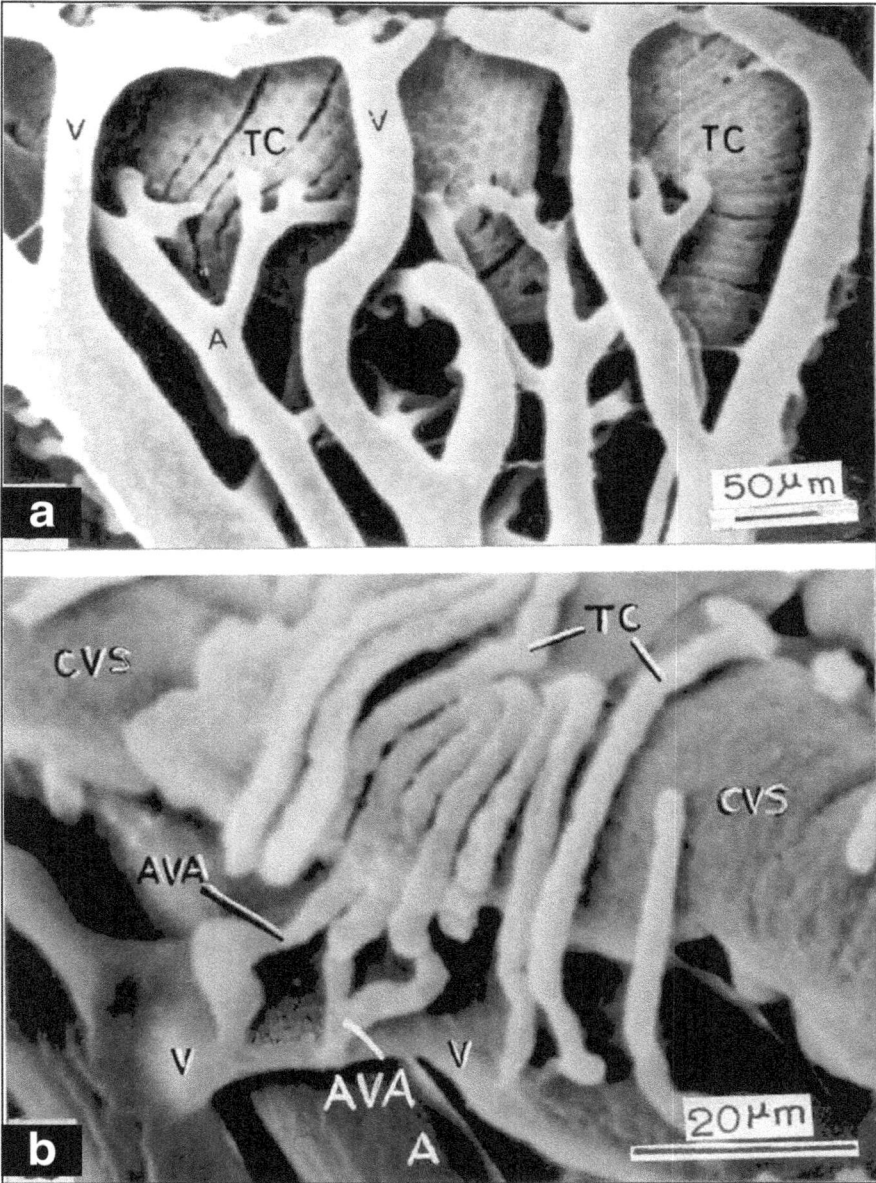

Plate 5.16

(a) SEM of photomicrograph of mercox preparation of respiratory islet of *Anabas* from the tissue side showing the arterial supply (A) of the transverse channels and venous drainage (V) of the system. The beaded appearance of the transverse channels is very much apparent. Bar = 50 μm. (After Munshi *et al.*, 1986); (b) Dorsolateral view of portion of corrosion replica respiratory islet of *Anabas*. Transverse capillaries to show the structure of Arterio-venous Anastomosis (AVA), see the position of Sphincters at the junctions of AVA to control the flow of blood. Position of CVS is also seen. Bar = 20μm.

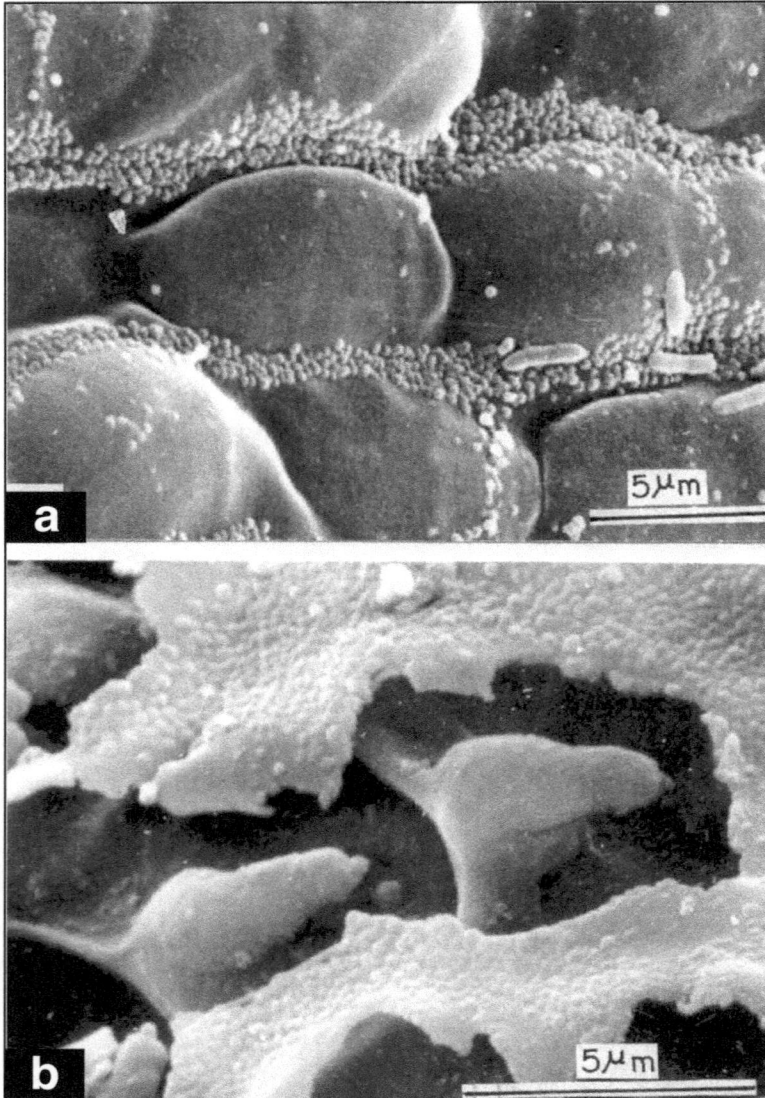

Plate 5.17

(a) SEM photomicrograph of highly magnified view of a series of vascular papillae lying one above the other forming blood channels. The peduncle (arrow heads) of one the vascular papilla is clearly visible with its curved margin showing how the vascular papillae fall over one another to form transverse channels. The position of microvilli of the supporting epithelial cells are seen with entangled rod shaped bacteria lying over the surface. Bar = 5 μm; (b) SEM photomicrograph of respiratory membrane partly removed to show the position and structure of Endothelial valves. The septate nature of the valve is very much apparent with their extended tongue like structure freely positioned in the blood space. Note microvilli on the surface of the respiratory epithelium (After Munshi *et. al*, 1986) Bar = 5 μm.

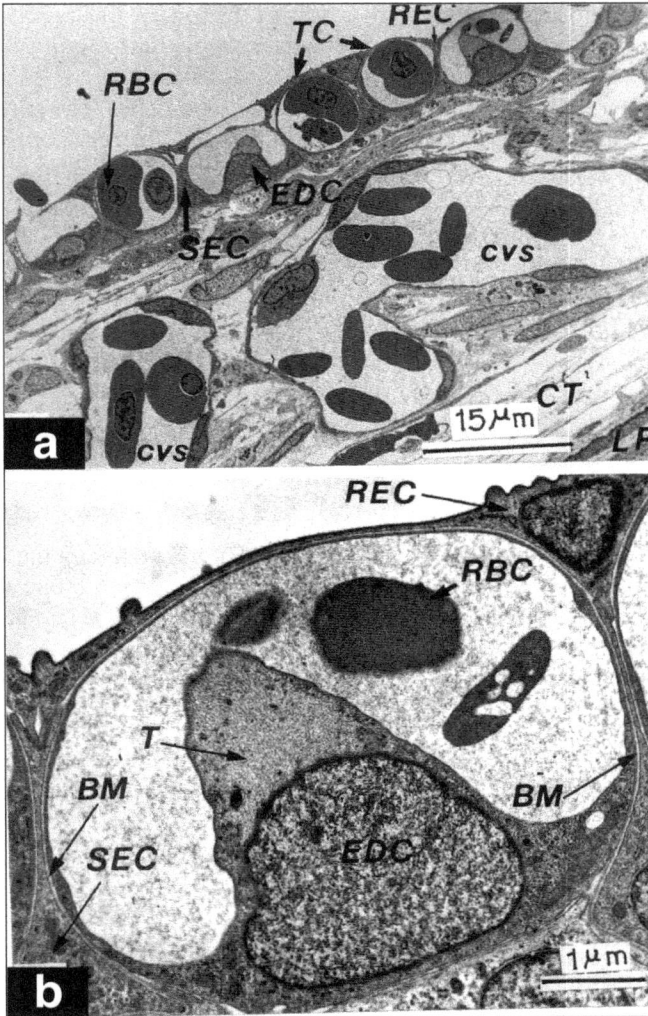

Plate 5.18

(a) TEM of a section passing through the respiratory islet of a labyrinthine plate, showing cross section of a group of transverse capillaries (TC) separated from each other by a single layer of supporting epithelial cells (SEC), Respiratory epithelial cells (REC) with microvilli on their surface. One lymphoid space is seen over the vascular papilla (arrow). Two venous sinuses (CVS) are seen underneath the respiratory islet. Connective tissue (CT) and part of the bony labyrinthine plate (LP) are also seen. RBC red blood cell. Bar = 15µm. (After Munshi and Hughes, 1991); (b) TEM of one vascular papilla of an islet, showing structure of its endothelial valve consisting of an enlarged endothelial cell (EDC) with tongue like process (T), with coarse keratin like fibrillae, the supporting epithelial cells with apical forked ends (SEC), their endothelium and basement membrane (BM), and transverse capillary extended edge of a respiratory epithelial cell (REC). Respiratory epithelium has low, microvillous protuberances at capillary margins. RBC red blood cell. Bar = 1 µm. (After Munshi and Hughes, 1991).

the respiratory membrane by the spring action of the tongue like processes. The SEM photomicrographs do not show any deformation of the tongue like endothelial processes (Plate 5.17a,b). This *Anabas* model seems to represent the last stage in the evolution of air-breathing organs among these groups of fishes. The parallel biserially arranged capillaries are separated from each other, not by pillar cells but by supporting epithelial cells. This *Anabas* model shows how thousands of vascular papillae have been packed in a small area thus affecting an economy of space in the small head of the fish. The aerial respiratory process has been completely separated from the feeding process. In *Channa* species and *Monopterus cuchia* the same pharyngeal chambers are used for feeding, aquatic and aerial respiration.

In *Channa striata*, the vascular papillae of pharynx are retractile but they are not retractile in *Monopterus cuchia*. In *M. cuchia* we have seen how the afferent-efferent connecting loop vessels become convoluted to form a rosette of vascular papillae. On the gill head regions and on both sides of the gill clefts the vascular papillae form RI where the arterio-venous anastomoses are involved.

Secondary Vascular System and Accessory Respiratory Organ

The secondary vascular system originates from the post lamellar circulation in the form of hundreds of tortuous vessels in air-breathing teleosts, commonly from the efferent branchial arteries (Munshi, 1996). They form the nutrient capillary web along the outer border of the efferent filament artery to nourish the filament abductor and adductor muscles. Some of the nutrient vessels originate from the medial wall of the efferent filament artery in the body of the filament.

The glomeruli-like tortuous vessels of the gills unite to form nutrient vessels. Two physiological processes seem to be involved *viz.* (i) plasma skimming of the circulatory blood of efferent branchial vessels before they supply blood to the accessory respiratory organs with high haematocrit value (Olson, 1984; Hughes, Roy and Munshi, 1992); and some sort of ultrafiltration of vascular fluid at high cardiac pressure take place.

Arterio-Venous Anatomoses and Phylogeny of Air-Breathing Fishes

Vogel *et al.* (1998) and Hughes (1998) have drawn attention to the special significance of the arterio-venous anastomoses in establishing possible phylogenetic relationships between tetrapods and the main groups of living bony fishes. In particular it has been emphasized (Hughes, 1999) that the special situation in Dipnoi cannot be associated with their air-breathing habit as the Indian air-breathing teleosts have a secondary gill circulation which is essentially similar to that of the other major groups of living fishes including *Latimeria*.

Conclusions

The examples discussed in this review provide further examples of the way in which different parts of the alimentary canal of air-breathing fishes are adapted to their combined functions in nutrition and respiration. The three genera (*Anabas, Channa, Monopterus*) mainly studied here also confirm another general finding with air-breathing fishes, namely that the precise ways in which parts of the alimentary canal have become modified to show a remarkable diversity. But in this case as our studies have become more detailed it has been possible to discern similarities which further illuminate our understanding of all three genera. Clearly we are dealing with end products of different evolutionary lines and the comparative approach has been invaluable but cannot be used too rigidly in defining phylogenetic relationships. Neverthless all studies of air-breathing fishes continue to help our comprehension of the many integrated changes involved in all water/air transitions.

Chapter 6

Chloride Cells in the Gills of Freshwater Teleosts

Specialised cells of several kinds, namely (a) mucous cells (b) large bi- or trinucleate glandular cells, and (c) mast cells occur in the gills of some species of freshwater teleosts.

The typical 'goblet' type of mucous glands are present in large number in freshwater species, such as *Catla catla, Labeo rohita, Ophiocephalus punctatus* and *Mastacembalus armatus*. In *Catla catla* these glands respond to the chloride test. This indicates that besides discharging mucous, they also play some part in the elimination of chloride. In *Ophiocephalus punctatus, Clarias batrachus* and *Heteropneustes fossilis* only some of the mucous cells give a positive reaction with the $AgNO_3/HNO_3$ test for chloride. This may mean that a few of them are in a state of active secretion of chloride, while others are in a non-secretory phase.

Large eosinophil glands with 2 or 3 nuclei also occur, chiefly in the siluroids. The function of these hypertrophied multicellular glands is not clear.

Judged by the $AgNO_3/HNO_3$ test, *Catla catla* possesses the highest number of chloride cells. In *Hilsa ilisha, Rita rita, Ophiocephalus striatus* and *Mastacembalus armatus*, only a network of silver exists on the surface of the primary and secondary lamellae.

The 'chloride cells' are said to be a characteristic of marine fishes in which an extrarenal mechanism for the elimination of excess of salts from the body fluid is necessary for correct osmo-regulation. These cells are also said to make their appearance in freshwater fishes that have been experimentally subjected to a saline medium. The presence and occurrence of 'chloride cells' in freshwater fishes living in their natural habitat is, therefore, interesting. The discovery of the presence of actively secreting chloride cells in certain species of freshwater teleosts and their mucoid nature are new results reported here (Munshi, J.S.D., 1964)

It has been claimed that the bulk of the sodium, potassium and chloride absorbed in the gastro-intestinal tract of marine teleosts is excreted by the 'chloride secreting' cells in the gills (Smith, 1930; Keys, 1931, 1933; Keys and Willmer, 1933); but Bevelander (1935, 1946) belived that the supposed excretory cells are nothing but intra-epithelial mucous glands and that the general respiratory epithelium might be a site of chloride excretion.

Liu (1942) attempted to acclimatize the freshwater air-breathing fish, *Macropodus opercularis*, to different concentrations of salt solutions. He stated that even an exclusively freshwater fish can tolerate a salt solution nearly as saline as sea-water by virtue of the enormous development of latent 'chloride secreting cells'. From this experiment he concluded that freshwater teleosts posses 'chloride-secreting cells' in the gills in a dormant condition.

Copeland (1948a,b) noted cytological changes in the Chloride cells of *Fundulus heteroclitus* during adaptation to varying degrees of salinity and by using the Leschke test demonstrated the presence of a copious amount of chloride in the secretory cells of fishes adapted to the salt-water condition, but only a limited amount of it at the free ends of the cells in fishes adapted to freshwater life. Getman (1950) working on *Anguilla rostrata* came to a similar conclusion.

More recently Vickers (1961) has said, with reference to the gills of *Lebistes reticulates* (a freshwater teleost), that after subjection to hypertonic-salt solutions of varying concentrations, its mucous cells become functionally transformed into chloride cells.

The present studies on gill epithelia of freshwater teleosts were undertaken in order to remove some of the ambiguities and deficiencies in our knowledge.

Materials and Methods

Fishes were collected from the river Ganges and the local ponds. Pieces of their gills were immediately fixed in Zenker, Helly, chrome-Bouin, alcoholic Bouin, Gilson's mercuro-nitric, or Carnoy's fluid after gently removing the adhering mucus with a cotton swab and rinsing them in river water. Sections of the gills were stained with Heidenhain's or Delafield's haematoxylin or Mayer's haemalum, and counterstained with eosin. Mallory's triple stain, borax-carmine followed by picro-indigo-carmine, and other stains were also employed.

Parts of the gill lamellae were dissociated in several different fluids such as sodium-chloride solution, Ranvier's alcohol, and borax solution. The dissociated cells were stained with methylene-blue and examined with the microscope.

Thionin was used to demonstrate the presence of mucous and mucous producing cells and mast cells (the latter by their chromotropic reaction). Sections were stained in a 0.2 per cent aqueous solution of thionin, quickly passed through acetone to xylene, and mounted in balsam. The PAS technique was used to demonstrate the mucous-producing cells. Fresh tissues of the gill lamellae were also studied supravitally with neutral red.

The Leschke silver technique, as modified by Copeland (1948a,b) was also used to detect the presence of the chloride in the so called branchial glands. Histological studies were made on the following species of fishes.

Sl.No.	Species	Family	Result of Chloride Test
Group – A			
1.	*Hilsa ilisha* (Ham.)	Clupeidae	Done
2.	*Gadusia chapra* (Ham.)	Clupedae	Not done
Group – B			
3.	*Catla catla* (Ham.)	Cyprinidae	Done
4.	*Labeo rohita* (Ham.)	Cyprinidae	Not done
5.	*Rohtee cotio* (Ham.)	Cyprinidae	Not done
Group – C			
6	*Clarias batrachus* (Linn.)	Clariidae	Done
7.	*Heteropneustes fossilis* (Bloch)	*Heteropneustes*	Done
8.	*Wallago attu* (Bloch and Schn.)	Siluridae	Not done
9.	Rita rita (Ham)	Bagridae	Done
10.	*Mystus aor* (Gunther)	Bagridae	Not done
Group – D			
11.	*Ophiocephalus punctatus* (Bloch)	Ophiocephalidae	Done
12.	*Ophiocephalus striatus*	Ophiocephalidae	Done
Group – E			
13.	*Anabas testudineus* (Bloch)	Anabantidae	Not done
14.	*Trichogaster fasciatus* (Bl. & Schn.)	Osphronemidae	Not done
Group – F			
16.	*Mastacembalus armatus* (Lacep.)	Mastacembelidae	Done

The chloride test was only applied to nos. 1,3,6,7,9,11,12 and 15 in the above list.

Results

The principal findings are summarized in Tables 6.1 and 6.2 and illustrated in Figures 6.1 and 6.2.

The structure of the gills of *Labeo rohita, Hilsa ilisha, Rita rita* and *Ophiocephalus striatus* has been recently described in detail by myself (1960). In the following pages the branchial glands of *Catla catla, Heteropneustes fossilis, Ophiocephalus punctatus* and *Mastacembalus armatus* are described in detail.

Table 6.1: Summary of the tests

Sl.No.	Species	Specialisation		Response to Chloride Test
		Mucous glands	Multimucleated Eosinophil and Glandular Cells	
1.	Hilsa ilisha	+	−	Only silver network present
2.	Gadusia chapra	+	−	Test not carried out
3.	Catla catla	++	−	Mucous glands responsive
4.	Labeo rohita	++	−	Test not carried out
5.	Rohtee cotio	+	−	Test not carried out
6.	Clarias batrachus	+	+	Some of the mucous glands are responsive
7.	Heteropneustes fossilis	+	+	Some of the mucous glands are responsive
8.	Rita rita	+	++	Only silver network present
9.	Walago attu	+	+	Test not carried out
10.	Mystus aor	+	+	Test not carried out
11.	Ophiocephalus striatus	++	−	Some of the mucous glands are responsive; silver network present
12.	Ophiocephalus punctatus	++	−	Some of the mucous glands are responsive
13.	Trichogaster fasciatus	−	−	Test not carried out
14.	Anabas testudineus	+	−	Test not carried out
15.	Mastacembalus armatus	++	−	Only silver network present
	General conclusions	Generally present in most of the species	Present only in the siluroids	

Catla catla

In this species only two kinds of specialized cells are present, the mucous glands and the acidophil mast cells.

Mucous glands are present in large numbers in the epithelium covering the gill arches and the primary and the secondary gill lamellae. In a horizontal section of the epithelium of the head of the gill arch, the mucous glands are found to be unicellular and rounded, and to possess flattened nuclei owing to the presence of the secretory substance. The glands are scattered on the general surface of the primary gill lamellae and in the interlamellar space close to the base of the secondary lamellae (Figure 6.1E). A few of them may occur on the secondary lamellae as well.

A paper on the mast cells of this and other species was communicated by myself to the Second All-India Congress of Zoology held in Varanasi (Munshi, 1962).

Table 6.2: Result of the chloride test

Sl.No.	Species	Mucous cells	Result of the Chloride Test Applied to the Mucous Glands	Silver Network Reaction	Remarks
1.	Hilsa ilisha	Present	No response	Present	Mucous glands do not respond to the chloride test
2.	Catla catla	Abundantly present	Most of the mucous glands repond	Present	Mucous glands respond to chloride test
3.	Clarias batrachus	Present	Some of the mucous cells respond	Present	Some of the mucous glands respond to the chloride test
4.	Heteropneustes fossilis	Present	Some of the mucous cells respond	Present	Some of the mucous glands respond to the chloride test
5.	Rita rita	Present	No response	Present	The mucous glands do not respond to the chloride test
6.	Ophiocephalus striatus	Abundantly present	A few of them respond	Present	A few of the mucous glands respond to the chloride test
7.	Ophiocephalus punctatus	Abundantly present	A few of them respond	Present	A few of the mucous glands respond to the chloride test
8.	Mastacembalus armatus	Abundantly present	No response	Present	Mucous glands do not respond to the chloride test

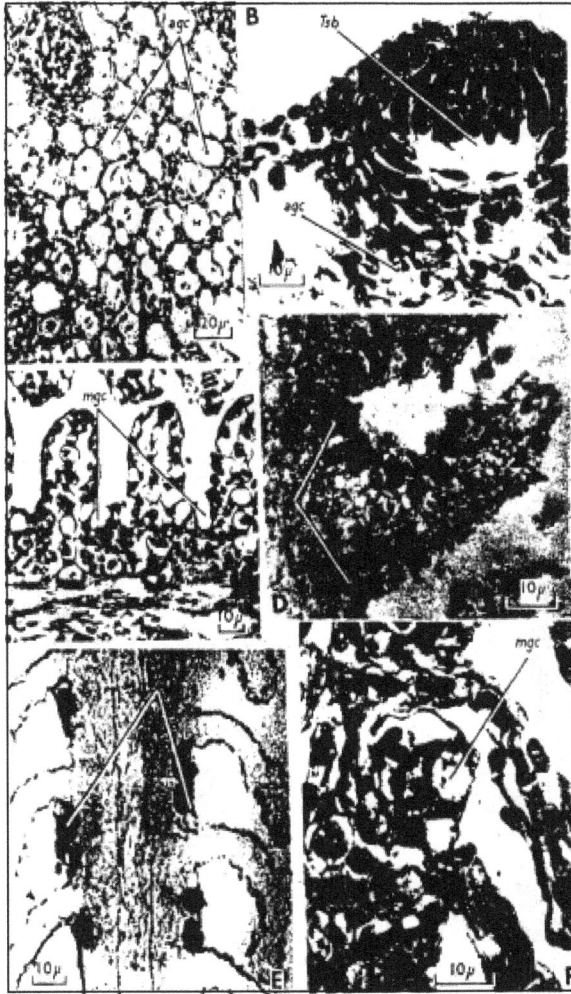

Figure 6.1

A: a horizontal section passing through the epithelium covering the head of the gill of *Heteropneustes fossilis* showing the detailed structure of the acidophil glandular cells (age) and a taste-bud cut across; B: A transverse section of the epithelium covering the head of the gill of *Wallago attu*, showing the large acidophil-glandular cells (age), epithelial cells, and a taste bud (tsb); C: A horizontal longitudinal section of a primary gill lamella of *Ophicephalus punctata* showing the presence of mucous glands (age) in the inter-lamellar areas as well as in the epithelium covering the secondary lamellae; D: A horizontal longitudinal section of a primary gill lamella of *O. punctata*, treated with the $AgNO_3/HNO_3$ test for chlorides, showing the chloride secreting cells (clc) at the base and on the secondary gill lamellae; E: A horizontal longitudinal section of a primary gill lamella of *Catla catla*, showing the structure and position of the 'chloride secreting cells (clc) in the inter-lamellar areas, after treatment with $AgNO_3HNO_3$; F: A horizontal longitudinal section of a primary gill lamella of *Catla catla*, showing the mucous glands (age) in the inter-lamellar area between and secondary gill lamellae.

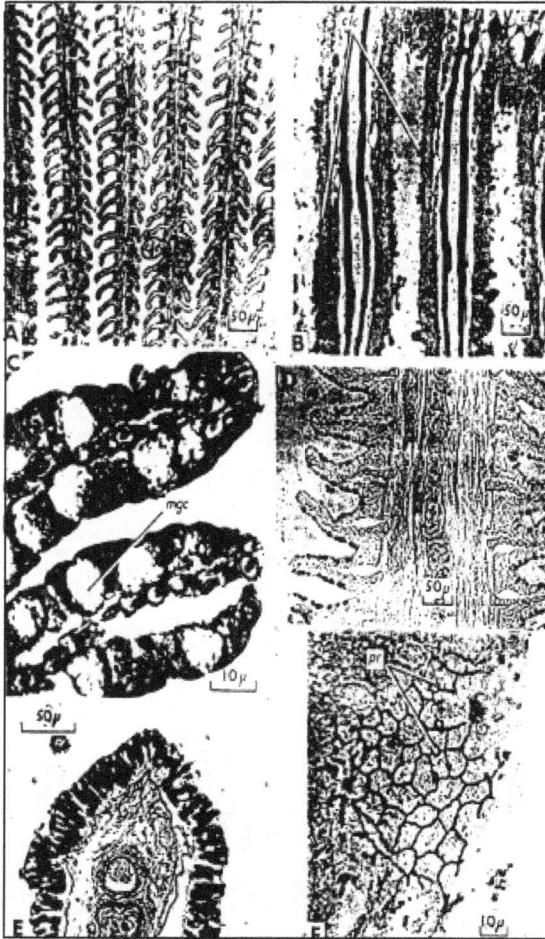

Figure 6.2

A: A horizontal longitudinal section of the primary gill lamellae of *Catla catla* passing through the secondary lamellae of both sides, showing the presence of the chloride cells after the application of the AgNO$_2$/HNO$_3$ test for chlorides; B: A longitudinal section passing through several primary gill lamella of *Heteropneustes fossilis*, showing the distribution of the 'chloride secreting cells.' The chlorides extruded by the secretory cells have reacted to the AgNO$_2$/HNO$_3$ test and are seen as black dots; C: A horizontal longitudinal section of a primary gill lamella of *Mastacembelus armatus*, showing the presence of hypertrophied mucous glands (mgc) on the secondary lamellae; D: A longitudinal section of a primary gill lamellae of *Ophiocephalus punctatus*, showing the distribution of PAS – positive cells on the secondary gill lamellae; E: A transverse section of a primary gill lamella of *Ophiocephalus punctatus*, showing the distribution of PAS positive cells on the secondary gill lamellae; F: A section passing superficially across the surface of a primary gill lamella of *Hilsha ilisha* showing the silver network formed after treatment gill lamellae of *Hilsha ilisha* showing the silver network formed after treatment with AgNO$_2$/HNO$_3$ for testing chlorides. The pores (pr) of the mucous glands are also clearly seen.

In *Catla catla*, a number of epithelial cells of the primary lamellae react on the $AgNO_3/HNO_3$ test for chlorides. In Figure 6.2A the black marks present in the inter-lamellar area, at the bases of the secondary lamellae, represent epithelial cells that have responded to this test. Under higher magnification the outline and shape of these cells can be made out (Figure 6.1E). Some of them react strongly to this test, proving thereby the presence of chlorides in them. These darkened cells occupy the same position as that occupied by the mucous glands.

It seems that the mucous glands present on the primary gill lamellae and those lying between the bases of the secondary lamellae are concerned with the excretion of chlorides also. The exact parallelism between the position of the mucous glands in Figure 6.1F and the incidence of the chloride-secreting cells in Figure 6.1 E is a pointer in this direction. These cells react metachromatically with thionin and give a positive reaction with PAS.

Heteropneustes fossilis

In this species large acidophil-gland cells and mucous glands are present. The large acidophil-gland cells are present in the epithelium covering the surface of the gill arches. They are larger than the other epithelial cells of the gill (Figure 6.1A). Most of the cells show 2 zones, the central one clear and containing the nuclei and the outer one staining more darkly. Similar acidophil glands cells also abound in the epidermis of the skin of *Heteropneustes fossilis*.

The mucous glands are abundant in the epithelium covering the head of the gill arches and occur in smaller numbers on the primary gill lamellae. However, these glands are absent from the secondary lamellae.

Result of Chloride Test

A number of epithelial cells of the primary-gill lamellae respond to the chloride test. Figure 6.2B shows the general distribution of the responsive cells borne by the gill lamellae. But all the epithelial cells do not react to the chloride test, and this indicates that certain of the epithelial cells are specially concerned in eliminating chlorides.

Ophiocephalus punctatus

In this species only two kinds of specialized cells, namely mucous glands and acidophil mast cells are present in the gills.

Mucous glands are present in the epithelium covering the heads of the gill arches and the primary and the secondary gill lamellae (Figure 6.1C). Each mucous gland is a unicellular, flask-shaped structure, opening to the exterior through a small pore. These glands are also found on the primary lamellae at the bases of the secondary lamellae and may occur even in the secondary lamellae (Figure 6.1C). They give a strong reaction with PAS, showing thereby that the secretory substance is a mucopolysaccharide (Figure 6.2 D, E.)

Some of the epithelial cells of the primary and the secondary lamellae respond to the chloride test (Figure 6.1D). Those occurring at or near the bases of the secondary lamellae are prominent in form and react sharply to the $AgNO_3/HNO_3$ test. Sometimes they appear to be vesicular, with a spout-like opening. It seems probable that some, if not all, of the mucous glands are capable of acting as chloride excreting cells (compare Figure 6.1C with Figure 6.1D)

Mastacembalus armatus

In this species only the mucous type of cells are found. Mucous glands are present in the epithelium covering the heads of the gill-arches and the primary and the secondary gill lamellae (Figure 6.2C). They are of the usual goblet type. They are abundant and lie close to one another, forming almost a continuous layer over the head of the gill-arch. In the primary, and more particularly in the secondary lamellae, they exist in a hypertrophied condition. Each is a unicellular gland containing finely granular cytoplasm and a nucleus displaced to one side. The cytoplasm of these glandular cells is eosinophobe.

Neither the epithelial cells nor the mucous glands respond to the $AgNO_3/HNO_3$ test for chlorides. This shows that the specialized mucous glands present in the gill region are not concerned in the work of chloride excretion. Occassionally, a silver network is formed on the surface of the primary and the secondary gill lamellae. Obviously this is due to the presence of chloride in the inter-cellular matrix (Figure 6.2F).

Discussion

It may be useful to list here the characters of the chloride cells according to the description of the authors already mentioned.

1. The cells are ovoid, or sometimes columnar (Copeland, 1948 a,b).
2. They are large.
3. Their cytoplasm is finely granular.
4. There is a marked affinity for eosin.
5. The nucleus is nearly spherical, and frequently eccentric (Liu, 1942, 1944).
6. The cells are present in the epithelium of the secondary and the primary lamellae. They are closely packed together in the epithelium of the filament, between the bases of the respiratory platelets (Getman, 1950). They may reach the sub-epithelial connective tissue layer.
7. They are close to the vascular layer. The correlation between the number of chloride cells and the degree of vascularity is very well marked (Burns and Copeland, 1950)
8. They respond to the chloride test.
9. They are secretory (Keys and Willmer, 1932; Liu 1942; 1944).
10. They give a strong reaction to the PAS test (as is shown in the present paper).
11. They give a chromotropic reaction with metachromatic dyes.

In a comparative study of the branchial epithelium of fishes, Bevelander (1936) found several types of intra-epithelial gland cells' which according to him belong to unicellular, multicellular and transitional types. In the present study of the branchial glands of the freshwater teleosts only 3 types of specialized cells have been found in the gills: the mucous glands of goblet type, the mast cells, and the large eosinophil, multinucleate gland cells. The multicellular type of branchial glands were described by Bevelander either as crescentic patches of cells, definitely oriented, with well differentiated tall columnar cells, or as typically pear-shaped or flask-shaped glands composed of closely packed, tall columnar cells. These (multicellular) branchial glands of Bevelander closely resemble the taste buds found invariably on the surface of the heads of the gill-arches (Figure 6.1B) and in the branchial regions of the pharynx in fishes. As these organs are definitely sense organs, they have been omitted from consideration here. Branchial glands, possibly multicellular in composition, are represented by the large eosinophil glands that are bi- or tri-nucleate and are found chiefly in the siluroids (Figure 6.1A). Such glands also abound in the integument of the siluroid fishes. The function of these hypertrophied multicellular glands is not clear.

Mucous glands of the ordinary goblet type are present in large numbers in the gill epithelium of the freshwater fishes examined, except in *Trichogaster fasciatus*. They generally occur on the head of the gill arches and may even extend on to the surface of the primary gill lamellae. In certain cases, such as *Catla catla*, *Labeo rohita*, *Ophiocephalus punctatus*, *O. striatus* and *Mastacembalus armatus*, they extend into the epithelial covering of the secondary gill lamellae. In the gill rakers of *Labeo rohita* a modified type of these glands occurs, which may be looked upon as having been derived from the usual kind of mucous glands. The mucous glands of the primary lamellae of *Catla catla* respond to the 'chloride test' and this indicates that these glands besides discharging mucus, also play a part in the process of elimination of the chlorides. In *Ophiocephalus punctatus*, *Clarias batrachus* and *Heteropneustes fossilis* only a few of the mucous glnads react positively to the $AgNO_3/HNO_3$ test.

The $AgNO_3/HNO_3$ test for chlorides was applied to several species of fishes, *Catla catla* showed the greatest number of chloride cells; the other species examined, in which the mucous gland cells reacted positively to a limited extent, were *Ophiocephalus punctatus*, *Clarias batrachus* and *Heteropneustes fossilis*. In *Hilsa ilisha*, *Rita rita*, *Ophiocephalus striatus* and *Mastacembalus armatus* only a network of silver is formed on the surface of the primary and the secondary lamellae. Shelbourne (1957a,b) also obtained a similar kind of silver network in the integument of the marine plaice larva, which he took to mean that the excretion of chlorides also takes place through the general integumental surface. In the fishes reported here, the silver network is evidently caused by the presence of chlorides in the intercellular cementing substance. The 'openings' of the mucous glands can be seen distinctly in the interistices of the network. Occassionally a little of the secretory matter containing chlorides is detectable in the small pores in the silver network. In *Catla catla* it is seen beyond doubt that ordinary mucous glands are capable of copiously excreting the chlorides in addition to mucus, *Ophiocephalus punctatus*, *Heteropneustes fossilis* and *Clarias batrachus* also react like *Catla catla* to the $AgNO_3/HNO_3$ test, but to a lesser degree.

The specificity of the $AgNO_3/HNO_3$ test as a histochemical index of the presence of chlorides has been doubted by Lison (1936) and others. Copeland (1948a,b), however, believe the $AgNO_3/HNO_3$ test for chlorides to be dependable. Shelbourne (1957) has attempted to confirm the presence of silver chloride by using solvents of silver chloride.

According to the theory of osmo-regulation, extra-renal excretion of chloride is not to be expected in purely freshwater teleosts like *Catla catla, Ophiocephalus punctatus, O. striatus, Clarias batrachus* and *Heteropneustes fossilis*. The occurrence of chloride cells in the gills of the freshwater fishes is therefore paradoxical. Copeland (1948a,b) explains this away by saying that the cell reverses its polarity and serves as a physiological mechanism to absorb chloride ions from freshwater and Krogh (1937) thinks that a mechanism for the absorption of chlorides from the surrounding medium exists in freshwater fishes; but these authors have not provided satisfactory evidence for this view. From the fact that the chloride cells are abundantly present in some of the freshwater teleosts, it is difficult to escape the conclusion that extrarenal excretion (*i.e.* excretion of chlorides by the gills) occurs in freshwater teleosts.

Liu (1942) has suggested that the occurrence of the supposedly dormant chloride-secreting cells in the gills of freshwater teleosts probably indicates that the progenitors of the freshwater fishes once inhabited the seas. I have found chloride cells to be present in certain Indian species of freshwater fishes in an active state and not in a dormant condition.

Smith (1951) and Bevelander (1935) have questioned the nature of the cells that perform this 'electrolyte excretion' in the gills of fishes and think that the whole of the respiratory epithelium of the gill lamellae might be involved in this work (of excretion). But histochemical tests have disclosed the presence of only a limited number of epithelial cells concerned in this 'electrolyte excretion'. Histological studies of freshwater fishes by the present author have shown that mucous-gland cells excreting chlorides occur in the gill epithelium of some species of freshwater fishes. The chloride tests, when applied to 8 Indian species of freshwater fishes, belonging to different genera and families, succeeded in the case of 5 speciess only (see Table 6.2), in which the mucous glands responded positively to the test. This result is contrary to that obtained by Keys and Willmer (1932), Liu (1942) and Copeland (1948a,b), who found the chloride-secreting cells to be essentially non-mucoid in character. The discovery of the presence of actively secreting chloride cells in certain species of freshwater teleosts and their mucous nature are, therefore, new results.

According to the theory of osmo-regulation, extra-renal excretion of chlorides is to be expected only in marine fishes and an interesting situation is created by the discovery of 'chloride cells' in certain freshwater teleosts. The test for chloride is very positive in some freshwater species of teleosts; in others it is less marked, but a silver netwok was found to exist in all the species examined. These obvious differences can be reconciled by assuming that the hypertonicity of the blood and of other body fluids of the different species of freshwater fishes varies within a certain range, so that chloride cells are called into play according to the needs of the fish. It seems probable that the silver network is due to the presence of chlorides in the intercellular cementing substance.

H.W. Smith (1931) was aware of chloride excretion in freshwater fishes and remarked that it is a very primitive process which is being common to fresh and salt water teleosts'.

Chapter 7

Respiratory Surface Area Metabolism Relationship in Air-Breathing Fishes of India

With increases in body size, changes in structure and functional ability of the animal occur and this instigates biologists to make a comparative study on functional anatomy and physiology of particular organs. The best studied problem is the process of aerobic metabolism, *i.e.*, oxygen uptake and extent of respiratory surface. Aerobic oxygen uptake of animals ($\dot{V}_{O_2}$) increases less rapidly than increase in body mass and has a long history of study by physiologists, and it has been recently reviewed by Heusner, 1982; Peters, 1983; Hughes, 1984; Schmidt Nielsen, 1984. On the basis of these studies the best known generalized exponent value for the relationship between resting metabolism and body mass is given as 0.75; however, it varies from less than 0.67 to more than 1.0. In fishes, the dimension of the respiratory surface of gills increases by a power function of 0.8 of body mass. The studies on individual fish species have indicated the variation of 'b' value from 0.5-1.0 (Hughes, 1977; Roy and Munshi, 1995).

As $\dot{V}_{O_2}$ is the amount of oxygen uptake per unit time through a respiratory surface of 'A' cm^2 and barrier of 't' μm thickness under the influence of pressure gradient of ΔP_{O_2} mmHg, it can be estimated by modified Fick's equation (Hughes, 1972).

$$\dot{V}_{O_2} = \frac{K.A.\Delta P_{O_2}}{t}$$

where, K is Krogh's permeation coefficient for oxygen through medium. For a fish, if K. ΔP_{O_2} and 't' are considered as constant, then $\dot{V}_{O_2}$ will be directly proportional to the area. It is, therefore, of considerable interest to establish the relationship between area and oxygen uptake in water-breathing fishes, the regression coefficient of the relationship of $\dot{V}_{O_2}$ and Respiratory Area (RA) is almost similar, so the slope value for RA and $\dot{V}_{O_2}$ relationship can be considered to be unity or nearer to one (Roy and Munshi, 1984). But it is more interesting and essential to investigate the relationship between oxygen uptake and respiratory area among air-breathing fishes, which have developed air-breathing organs and have both options, to extract oxygen from water and air.

Air-Breathing Fishes

Air-breathing fishes constitute a group of fishes from different taxonomic positions, which have the ability to obtain oxygen from different media Graham (1970) and Bridges (1988) differentiated the freshwater and marine air-breathing fishes on the basis, that the former has developed accessory respiratory organs which take care of oxygen extraction, but not CO_2 excretion, while in the latter, both purposes are served by the gills and skin. Indian air-breathing fishes live in freshwater chaurs* swamps, wetlands usually uninhabitable for purely gill breathers. These are growing carnivorous and hardy fishes sustaining their lives in extremely adverse ecological conditions such as very low dissolved oxygen (DO), high free carbon dioxide (Free CO_2), and temperature fluctuations. Obligatory air-breathing species are represented by *Anabas testudineus*, *Channa striatus*, *Channa marulius* and *Monopterus cuchia* which rely mainly on air-breathing and die when prevented from access to air (Figure 7.1). Facultative air breathers are capable of sustaining life without air-breathing and are represented by *Clarias batrachus*, *Heteropneustes fossilis*, *Channa punctatus* and *Channa gachua*.

Body Weight, Respiratory Area and Oxygen Uptake Relationship

Bimodal oxygen uptake of fishes has been measured by respirometry at both air and water interfaces. To avoid the effects of photoperiod, time and temperature, the experiments were carried out at about the same time period (08-10h) in a light and temperature regulated room. The dimension of respiratory surface of different respiratory organs was taken from previous studies of Hughes *et al.* (1973, 1974a,b). Hakim *et al.* (1978). Munshi *et al.* (1980), Choudhary (1992), and Dandotia (1978), and relationship to the form of allometric equations were established after logarithmic transformation of the data. Aquatic $\dot{V}_{O_2}$ values were analysed from gill and skin area measurements, while aerial $\dot{V}_{O_2}$ values were analysed with respect to accessory organs.

Aquatic Oxygen Uptake and Gill area

Regression coefficients of aquatic $\dot{V}_{O_2}$ and body weight at both winter and summer temperatures is higher than that of gill and body weight relationships in *Anabas tstudineus* and *Channa punctatus* (Figure 7.2). In *Clarias batrachus* and *Heteropneustes fossilis*, the regression line of $\dot{V}_{O_2}$ at $30 \pm 1°C$ with body weight has lower slope value than that of gill area. But at $20.0 \pm 1°C$, the slope value of $\dot{V}_{O_2}$ (0.736) is lower than the slope value of gill and skin in *Clarias batrachus*, but it is higher in *H. fossilis*.

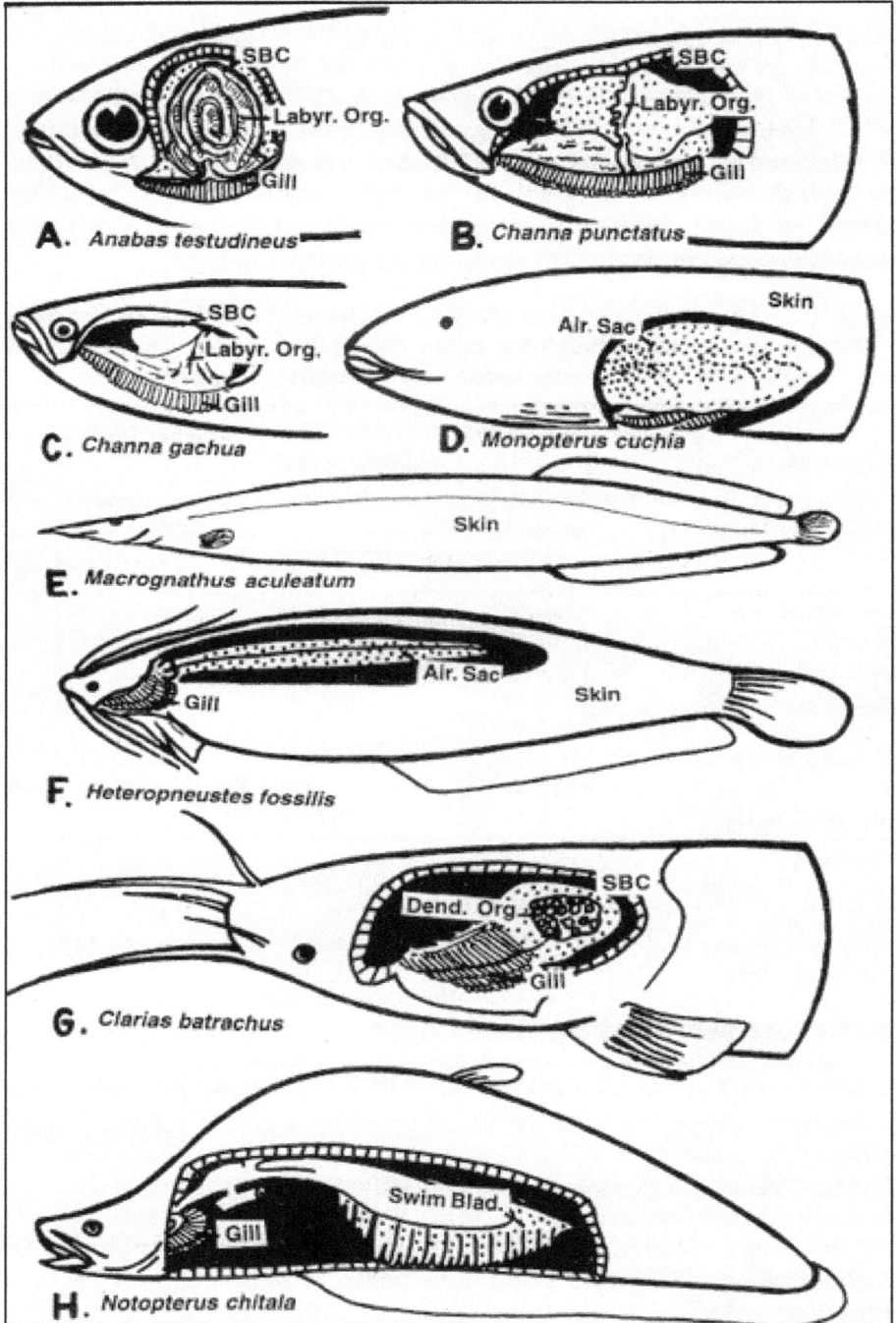

Figure 7.1: Air-breathing fishes of India showing their accessory respiratory organs.

Table 7.1: Allometric equations showing relationship between Body weight (g), area of water-breathing organ (mm²) and aquatic oxygen uptake (mIO₂/h) of air-breathing fishes at two seasonal temperatures.

Fish Species	Allometric Equations for Relationship between Body Weight Vs.		
	Respiratory Area (mm^2)	Aquatic Oxygen Uptake (mIO_2/h)	
		$30 \pm 1^\circ C$	$20 \pm 1^\circ C$
Anabas testudineus	$278.0\ W^{0.615}$	$0.1117\ W^{0.809}$	$0.0774\ W^{0.851}$
Channa punctatus	$470.4\ W^{0.592}$	$0.13001\ W^{0.822}$	$0.0852\ W^{0.863}$
Channa gachua	$148.8\ W^{0.757}$	$0.2075\ W^{0.454}$	
Clarias batrachus	$227.5\ W^{0.781}$	$0.6137\ W^{0.316}$	$0.132\ W^{0.736}$
	$564.9\ W^{0.743}$		
Heteropneustes fossilis	$186.1\ W^{0.746}$	$0.2882\ W^{0.520}$	$0.0755\ W^{0.840}$
	$851.1\ W^{0.684}$		
Monopterus cuchia	$877.4\ W^{0.706}$		

When aquatic $\dot{V}_{O_2}$ values were analysed with water-breathing area, two slope values were obtained at each seasonal temperature (Figure 7.3). The regression coefficient values were more than one for *Anabas* and *Channa punctatus* at both the temperatures. The values were less than one for *Channa gachua, Clarias batrachus, Heteropneustes fossilis* and *Monopterus cuchia* at $30 \pm 1^\circ$C but at $20 \pm 1^\circ$C the value approximates one in *C. batrachus* (Table 7.2).

Table 7.2: Allometric equations showing relationship between water-breathing area (A: mm²) and aquatic oxygen uptake (mIO₂/h) in different air-breathing fishes at two seasonal temperatures.

Fish Species		$\dot{V}_{O_2}$ at $30 \pm 1^\circ C$	$\dot{V}_{O_2}$ at $30 \pm 1^\circ C$
Anabas testudineus	Gill	$0.000068\ A^{1.315}$	$0.000032\ A^{1.363}$
Channa punctatus	Gill	$0.000026\ A^{1.387}$	$0.000011\ A^{1.456}$
Channa gachua	Gill	$0.0087\ A^{0.632}$	
Clarias batrachus	Gill	$0.0680\ A^{0.405}$	$0.000091\ A^{1.215}$
	G+S	$0.0376\ A^{0.419}$	$0.000197\ A^{0.975}$
Heteropneustes fossilis	Gill	$0.00749\ A^{0.698}$	$0.00021\ A^{1.126}$
	G+S	$0.0164\ A^{0.745}$	$0.000018\ A^{1202}$
Monopterus cuchia	Skin	$0.0014\ A^{0.684}$	

Aerial Oxygen Uptake and Accessory Respiratory Surface

Regression analyses of aerial $\dot{V}_{O_2}$ and body weight, and air-breathing organ area and body weight, indicate that resting $\dot{V}_{O_2}$ values at both the temperatures increase more rapidly than the area in all the fishes except *Channa gachua* (Table 7.3, Figure

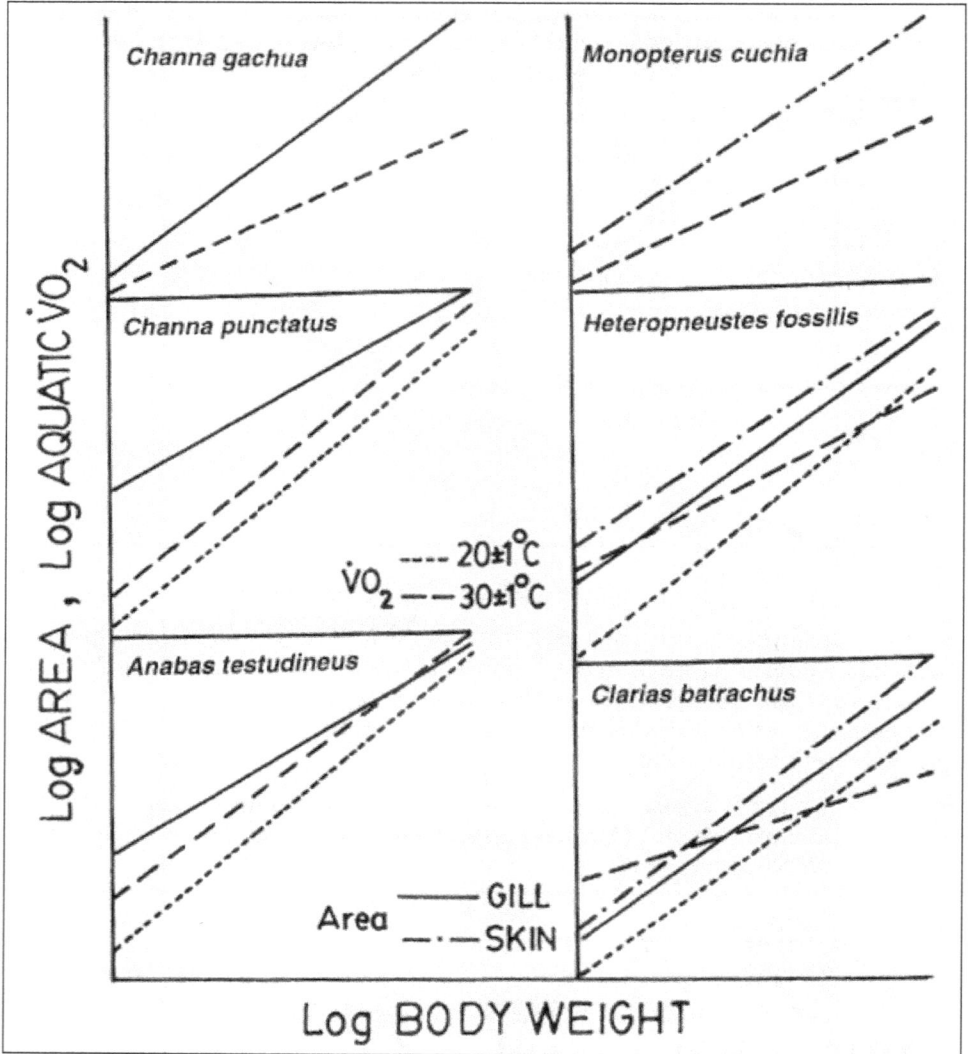

Figure 7.2: Bilogarithmic plots of body weight and gill area to compare with plots of body weight and oxygen uptake at two seasonal temperature.

7.4). The slope values are higher at winter temperatures in all the fishes except *Clarias batrachus*. Analysis of aerial $\dot{V}_{O_2}$ with area of air-breathing organs gave two slope values at the two seasonal temperatures in all the species (Table 7.4; Figure 7.5). The values were more than one in all the species except *Channa gaschua* (0.889). In Monopterus cuchia, the slope value for $\dot{V}_{O_2}$ and air-breathing organ area was 1.897. The values were higher at winter temperatures in all species except *Clarias batrachus* (Table 7.4)

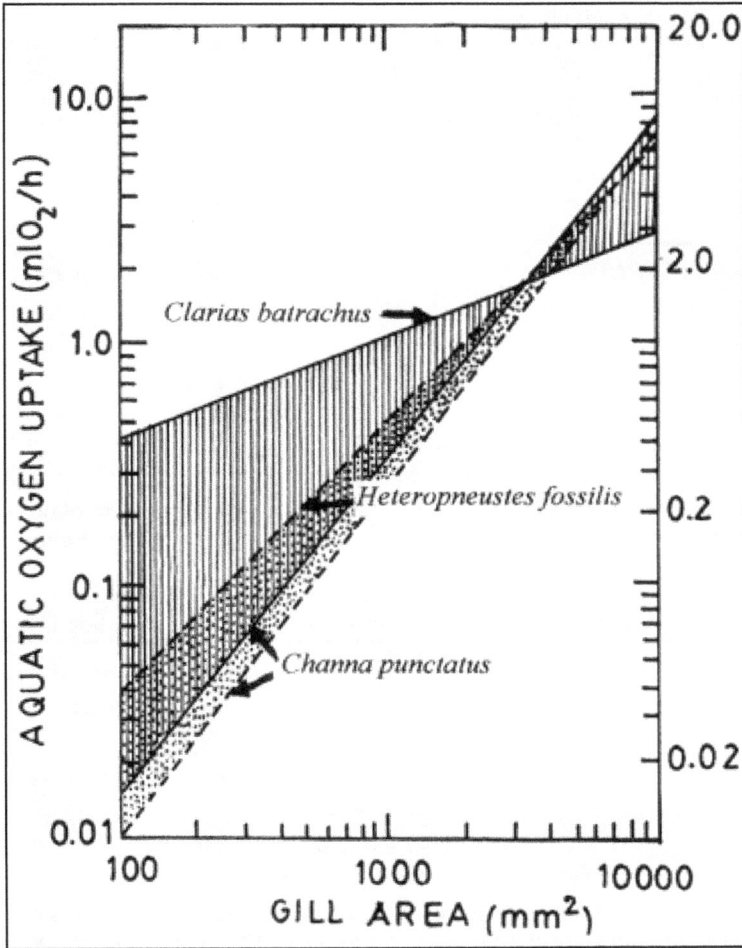

Figure 7.3: Bilogarithmic plots of aquatic oxygen uptake against gill area in different air-breathing fishes. Cross hatched area denotes $\dot{V}_{O_2}$ at 30±1°C, while dotted area represents $\dot{V}_{O_2}$ at 20±1°C. Values for all other fishes lie within the limits.

Bimodal Oxygen Uptake and Total Respiratory Area

Comparison of regression coefficient of bimodal oxygen uptake and body weight and total respiratory area and body weight indicates that with unit increase in body weight $\dot{V}_{O_2}$ increases at a higher rate than the respiratory area (Table 7.5; Figure 7.6). During the winter seasons slope values of oxygen uptake at 20 ± 1°C in both *Channa gachua* and *Heteropneustes fossilis* were found to be lower than those of respiratory areas.

Table 7.3: Allometric equations showing relationship between Body weight (g), area of air-breathing organ (mm²) and aerial oxygen uptake (mlO$_2$/h) of air-breathing fishes at two seasonal temperatures.

Fish Species	Allometric Equations for Relationship between Body Weight Vs.		
	Respiratory Area (mm²)	Aquatic Oxygen Uptake (mlO$_2$/h)	
		30±1°C	20 ± 1°C
Anabas testudineus 1	147.2 W$^{0.713}$	0.09319 W$^{0.929}$	0.03848 W$^{0.851}$
Channa punctatus 2	159.1 W$^{0.606}$	0.12035 W$^{0.888}$	0.04798 W^{1067}
Channa gachua 3	86.0 W$^{0.678}$	0.1556 W$^{0.603}$	
Clarias batrachus 4	124.7 W$^{0.790}$	0.0256 W$^{1.069}$	0.0364 W$^{0.864}$
Heteropneustes fossilis 5	145.9 W$^{0.662}$	0.0644 W$^{0.785}$	0.03105 W$^{0.818}$
Monopterus cuchia 6	12.5 W$^{0.797}$	0.0017 W$^{1.497}$	

Table 7.4: Allometric equations showing relationship between area of air-breathing organ (mm²) and aerial oxygen uptake (mlO$_2$/h) of air-breathing fishes at two seasonal temperatures.

Fish Species	$\dot{V}o_2$ at 30 ± 1°C	$\dot{V}o_2$ at 30 ± 1°C
Anabas testudineus	0.000139 A$^{1.303}$	0.0000317 A$^{1.542}$
Channa punctatus	0.00019 A$^{1.275}$	0.00002 A$^{1.533}$
Channa gachua	0.00297 A$^{0.889}$	
Clarias batrachus	0.000038 A$^{1.351}$	0.000185 A$^{1.094}$
Heteropneustes fossilis	0.00017 A$^{1.186}$	0.0001 A$^{1.238}$
Monopterus cuchia	0.000015 A$^{1.879}$	

Table 7.5: Allometric equations showing relationship between Body weight (g), total respiratory area (mm²) and total oxygen uptake (mlO$_2$/h) in different air-breathing fishes at two seasonal temperatures.

Fish Species		Allometric Equations for Relationship between Body Weight Vs.		
		Respiratory Area (mm²)	Aquatic Oxygen Uptake (mlO$_2$/h)	
			30±1°C	20 ± 1°C
Anabas testudineus 1	G+A	419.2W$^{0.658}$	0.1847 W$^{0.899}$	0.1286W$^{0.933}$
Channa punctatus 2	G+A	429.8W$^{0.625}$	0.2475 W$^{0.859}$	0.1327W$^{0.971}$
Channa gachua 3	G+A	213.5W$^{0.749}$	0.3564 W$^{0.536}$	
Clarias batrachus 4	G+A	351.2W$^{0.787}$		
	G+S+A	916.2W$^{0.760}$	0.1894 W$^{0.763}$	0.1616W$^{0.781}$
Heteropneustes fossilis 5	G+A	328.2W$^{0.715}$		
Monopterus cuchia 6	G+S+A	1181.0W$^{0.694}$	0.3145 W$^{0.624}$	0.3069W$^{0.833}$
	S+A	889.8W$^{0.708}$	0.0082 W$^{1.264}$	

G: Gill; S: Skin; A: Air-breathing organ.

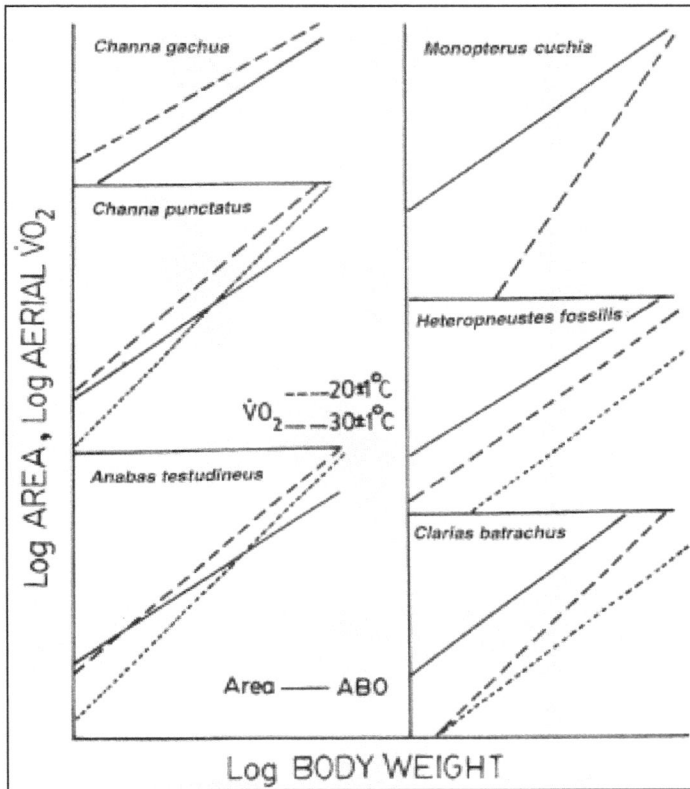

Figure 7.4: Bilogarithmic plot of accessory respiratory organs area and aerial oxygen uptake against body weight to compare the regression lines.

Table 7.6: Allometric equations showing relationship between total respiratory area (mm²) and total oxygen uptake (mlO$_2$/h) in different air-breathing fishes at two seasonal temperatures.

Fish Species		$\dot{V}o_2$ at 30 ± 1°C	$\dot{V}o_2$ at 30 ± 1°C
Anabas testudineus	G+A	0.000048 $A^{1.367}$	0.000025 $A^{1.417}$
Channa punctatus	G+A	0.000034 $A^{1.378}$	0.000006 $A^{1.559}$
Channa gachua	G+A	0.00768 $A^{0.716}$	
Clarias batrachus	G+A+S	0.000181$A^{0.016}$	0.000067 $A^{1.111}$
	G+A	0.000643$A^{0.970}$	0.000263 $A^{1.063}$
Heteropneustes fossilis	G+S+A	0.00054$A^{0.809}$	0.000022 $A^{1.200}$
	G+A	0.0020 $A^{0.8735}$	0.000126 $A^{1.164}$
Monopterus cuchia	S+A	0.00000045$A^{1.784}$	

G: Gill; S: Skin; A: Air-breathing organ.

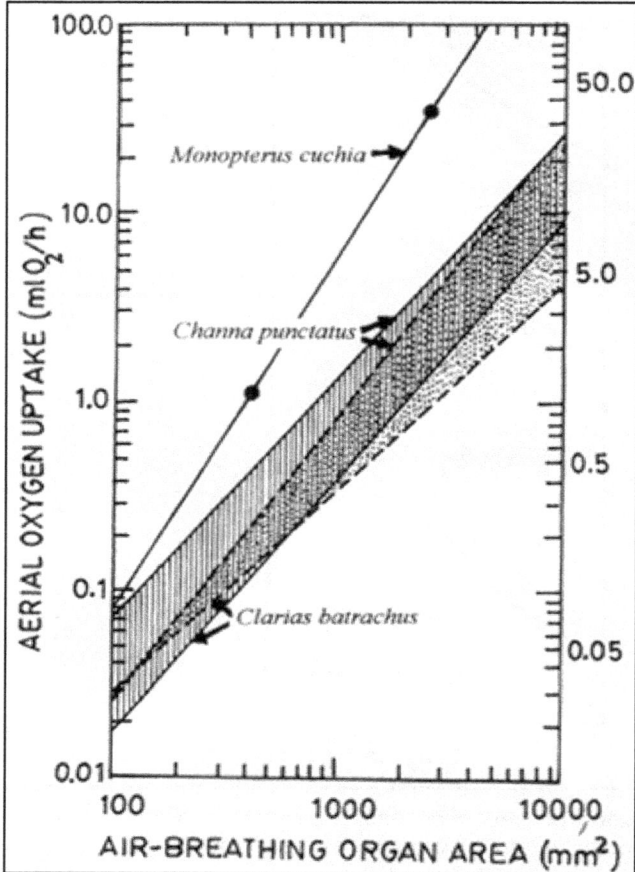

Figure 7.5: Bilogarithmic plots of aerial oxygen uptake against air-breathing organ area in different air-breathing fishes. Cross hatched area represents $\dot{V}o_2$ at 30±1°C. A separate line has been drawn to show aerial $\dot{V}o_2$ at 30±1°C in *Monopterus cuchia* having developed pharyngeal air-sac. Dotted area represents $\dot{V}o_2$ at 20±1°C. Value for all other fishes lie within the limits.

When total $\dot{V}o_2$ was analysed in respect to total respiratory surface area, two exponents were obtained in both seasons. The values were more than one in *Anabas testudineus, Channa punctatus, Clarias batrachus, Monopterus cuchia* at both summer and winter temperatures (Table 7.6; Figure 7.7). It was, however, lower than one in *C. gachua* and *H. fossilis* at 30.0 ± 1°C.

Functional Analysis

Respiratory Area as Determination of Metabolism

The relationship between gill and/or skin area and aquatic $\dot{V}o_2$ suggests that with unit increase in gill area, the oxygen uptake increases by an exponent value of

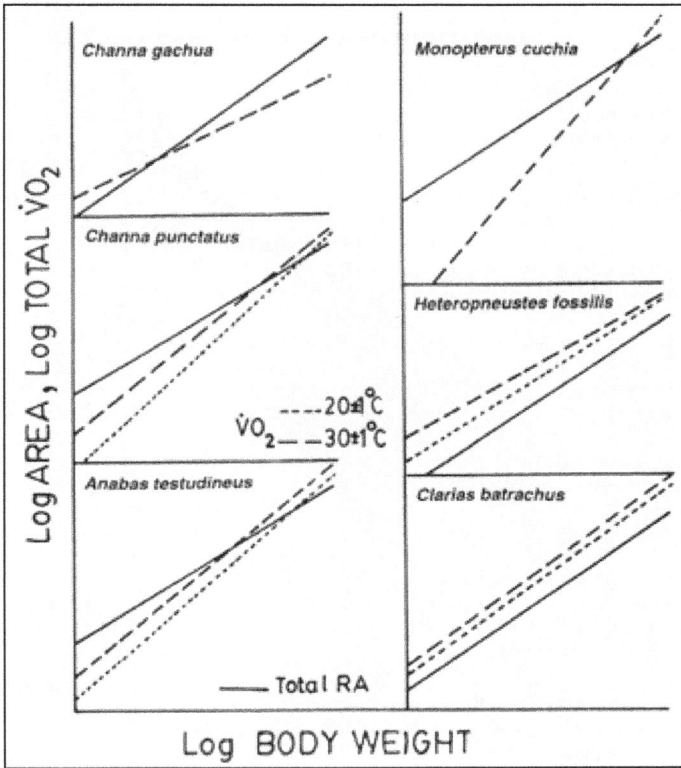

Figure 7.6: Bilogarithmic plots of total respiratory area and against body weight in different air-breathing fishes. Regression line of total oxygen uptake against body weight is also drawn for comparison.

more than one especially at winter temperatures in all the species and at summer temperatures in *Anabas* and *Channa* punctatus. At summer temperatures the aquatic $\dot{V}_{O_2}$ value does not increase with growth rate of gills in *C. gachua, Clarias batrachus* and *H. fossilis*. It means that these fishes do not utilize gills and skin areas to the maximum during the summer season. Interestingly, it has been found that higher weight groups of *Clarias, Heteropneustes* and *Channa gachua* have higher exponential growth rates of gill than the juveniles (Roy and Munshi, 1995).

Regression analysis of aerial $\dot{V}_{O_2}$ and aerial respiratory surface area indicate that aerial O_2 uptake increases more rapidly than the air-breathing surface, in almost all the fishes except *Channa gachua*. It means that with increase in air-breathing surface, the fishes want to rely more on aerial breathing. At winter temperatures the aerial oxygen uptake increases at a faster rate than that at summer temperatures in species like *Anabas testudineus, Channa punctatus* and *Heteropneustes fossilis*. However, for *Clarias batrachus* the 'b' value at winter temperatures (1.094) is lower than that obtained at summer temperatures (1.351). Scaling of air-breathing organ areas at different stages of life suggested higher scaling coefficient for adult groups than that in juveniles of

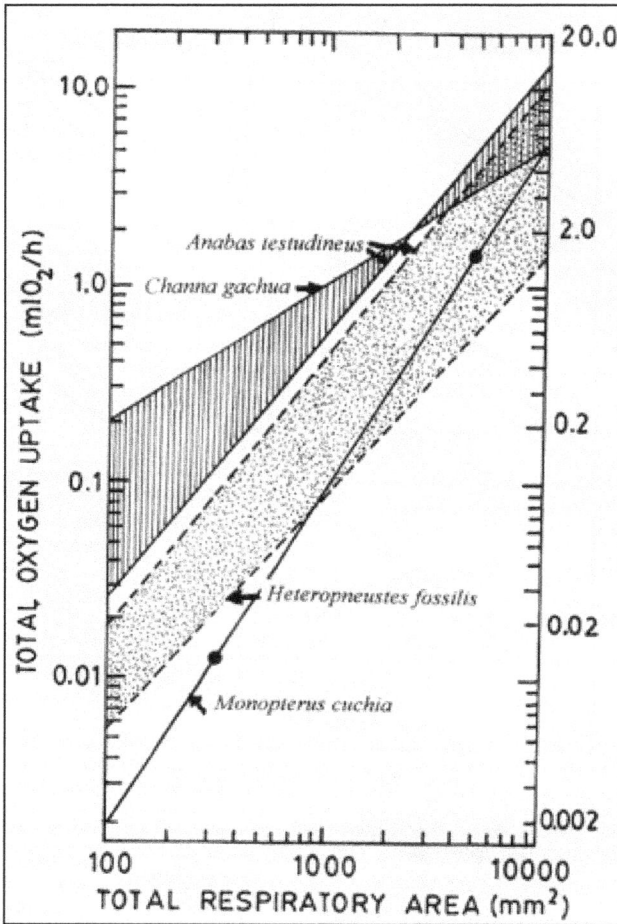

Figure 7.7: Bilogarithmic plot of total oxygen uptake against total respiratory area in different air-breathing fishes. Cross hatched area represents $\dot{V}o_2$ at 30±1°C. Separate line for *Monopterus cuchia* without functional gill has been drawn. Dotted area represents $\dot{V}o_2$ at 20±1°C. Lines for all other fishes fall within the limits.

Anabas, Heteropneustes, Clarias and *Monopterus cuchia* (Roy and Munshi, 1995). In *Channa punctatus* and *Channa gachua*, the scaling coefficients are not very different. Total oxygen uptake also increases with exponent values of more than one with unit increase in total respiratory area in all the species except *Channa gachua* and *Heteropneustes* at summer temperatures.

Comparison of efficiency of aquatic and aerial oxygen uptake in different air-breathing fishes can be made at two seasonal temperatures by calculating the $\dot{V}o_2$ value through an area of 10 cm^2 by using the relationship equations derived for $\dot{V}o_2$ and respiratory area (Table 7.7). It has been found that the fishes absorb more O_2 at summer temperatures that at winter temperatures. In *Anabas, Channa* species and

Table 7.7: Oxygen uptake by air-breathing fishes through 10 cm² area of different respiratory organs at two seasonal temperatures.

Fish Species	Temp. ± 1°C	Oxygen Uptake (ml O₂/hour) through				
		Gill	Skin	Gill+skin	ABO	Total TA
Anabas testudineus	30	0.5999	–	–	1.1257	0.6057
	20	0.4546	–	–	0.7354	0.4423
Channa punctatus	30	0.3764	–	–	1.2520	0.4624
	20	0.2572	–	–	0.7917	0.2754
Channa gachua	30	0.6920	–	–	1.3767	1.0769
Clarias batrachus	30	1.1164	–	0.6765	0.4299	0.5223
	20	0.4013	–	0.1655	0.3541	0.4064
Heteropneustes fossilis	30	0.9302	–	0.2810	0.6144	0.8298
	20	0.5004	–	0.0728	0.5176	0.3918
Monopterus cuchia	30	–	0.1578	–	6.5027	0.1012

Table 7.8: Area of water-breathing organ, estimated oxygen uptake (E $\dot{V}_{O_2}$) and actual oxygen uptake alongwith per cent value of (E $\dot{V}_{O_2}$) of air-breathing fishes.

Fish Species with Body wt.	Resp. Organ	Resp. Area (mm²) [τ h (μ)]	Diffusing Capacity (mlO₂/min/ mmHg)	E $\dot{V}_{O_2}$ (mlO₂/h)	Actual $\dot{V}_{O_2}$ (mlO₂/h)	Tem.± 1°C	Per cent Value
Anabas testudineus[1] 40g	Gill	2687.3 (10.0)	0.000403	2.419	2.212	30	91
					1.784	20	74
Channa punctatus[2] 40g	Gill	4172.2 (2.033)	0.003082	18.492	2.692	30	15
					2.052	20	11
Channa gachua[3] 40g	Gil	2103.2 (2.4)	0.001315	7.887	1.108	30	14
Clarias batrachus[4] 40g	Gill	4056.9 (7.67)	0.000793		1.970	30	35.6
	Skin	8755.9 (101.6)	0.000129	5.534	1.992	20	36.0
	Total	12812.8	0.000922				
Heteropneustes fossilis[5] 40g	Gill	2916.6 (3.58)	0.00122		1.962	30	24
	Skin	10611.8 (98.0)	0.00016		1.674	20	20
	Total	13528.4	0.00138	8.292			
Monopterus cuchia[6] 200g	Skin	36959.4 (119.0)	0.000466	2.796	1.821	30	65

Monopterus the $\dot{V}_{O_2}$ value is greater through air-breathing organs, while in *Clarias* and *Heteropneustes* $\dot{V}_{O_2}$ is greater through gills. With free access to air *Heteropneustes* obtains 60 per cent of total oxygen demand from water (Hughes and Singh, 1971) while under similar conditions Anabas obtains only 40 per cent of O_2 demand from water (Hughes and Singh, 1970). However, Singh *et al.* (1997) found 60 per cent of oxygen demand is fulfilled form water in both the fishes up to 100 mg body weight. Larvae of Monopterus cuchia meets 50 per cent of its oxygen demand from air and 50 per cent from water (Singh *et al.*, 1997), while adult Monopterus obtains 75 per cent of its oxygen requirements from air and 25 per cent from water (Singh and Thakur, 1979).

The ratio of oxygen uptake and respiratory area (cm^3 O_2 h^{-1} m^{-2} is shown in Table 7.11 for a number of species. This also shows the effect of seasonal temperatures as well as the effect of body weight on the ratio of $\dot{V}_{O_2}$ /A. Hughes (1976) concluded that majority of fishes the ratio is usually about 200-300 ml O_2/h/m^2 and this value may increase or decrease with body mass (Hughes, 1977). The ratio of aquatic $\dot{V}_{O_2}$ / gill area of air-breathing fishes is higher than the reported values. This ratio is higher in adult *Anabas* and *Channa punctatus* than in juveniles, but the ratio is lower for adult *H. fossilis* and *Clarias batrachus*. These indicate that in adult *Anabas* and *Channa*

Table 7.9: Area of air-breathing organ, estimated oxygen uptake (E $\dot{V}_{O_2}$) and actual oxygen uptake alongwith per cent value of (E $\dot{V}_{O_2}$) of air-breathing fishes.

Fish Species with Body wt.	Resp. Organ	Resp. Area (mm^2) [τ h (μ)]	Diffusing Capacity (mlO$_2$/min/ mmHg)	E$\dot{V}_{O_2}$ (mlO$_2$/h)	Actual $\dot{V}_{O_2}$ (mlO$_2$/h)	Tem.± 1°C	Per cent Value
Anabas testudineus[1] 40g	ABO	2042.6 (0.21)	0.01459	87.54	2.866 1.220	30 20	3.3 2.5
Channa punctatus[2] 40g	ABO	2073.5 (0.78)	0.00399	23.94	3.178 2.453	30 20	13.3 10.2
Channa gachua[3] 40g	ABO	1048.8 (0.80)	0.00197	11.82	1.439	30	12.2
Clarias batrachus[4] 40g	Den.Org.	2120.0 (0.45)	0.00707		1.319	30	3.1
	Gill Fan	178.8 (7.9)	0.000034				
	Total	2298.8	0.007104	42.624	0.882	20	2.1
Heteropneustes fossilis[5] 40g	Air Sac	1677.3 (1.6)	0.001573	9.435	1.166	30	12.4 1.9
		2391.5 (0.34)	0.01049	62.94	0.635	20	1.7 1.0
Monopterus cuchia[6] 200g	Air-Sac	850.08 (0.44)	0.0029	17.4	4.733	30	27.2 18.9
		2000.00 (0.72)	0.00417	25.0			

punctatus the gill function requires a positive oxygen partial pressure to fulfil requirements for oxygen and thus they need to evolve more efficient air-breathing organs. The ratio of aerial $\dot{V}_{O_2}$/air-breathing organ area is always higher for adult fishes than the juveniles. In adult *Monopterus cuchia*, the value is five times more than the juveniles. But the ratio of total $\dot{V}_{O_2}$/total respiratory area is approximately same for both *Heteropneustes* and *Clarias*, whereas it is higher for adult Anabas and Channa. These ratios are also helpful in calculation of oxygen uptake, if the area is known, or vice versa in different air-breathing fishes.

The morphometrically estimated oxygen uptake of 40 g weight specimens of *Anabas, Channa, Clarias* and *Heteropneustes* and 200 g specimen of *Monopterus cuchia* were calculated by using modified Fick's equation ($\dot{V}_{O_2} = D_t \times \Delta P_{O_2}$). Here the ΔP_{O_2} was considered as 100 mmHg and comparison was made with actual oxygen uptake values by the fishes under laboratory conditions (Tables 7.8–7.10). It was found that in obligate air-breathers the actual aquatic $\dot{V}_{O_2}$ was more than 60 per cent of the estimated aquatic $\dot{V}_{O_2}$ (Table 7.8), which reflects that gill area (and/or skin area) does not provide enough scope of increasing in $\dot{V}_{O_2}$, especially when ΔP_{O_2} is less during the summer season. But in facultative species, the scope for increase is wider. The morphometric estimate of aerial oxygen uptake is much higher than the experimentally obtained aerial $\dot{V}_{O_2}$ value. The air-breathing organs provide greater scope for oxygen uptake, than the gills in obligate fishes. The ratio of total actual $\dot{V}_{O_2}$/total estimated $\dot{V}_{O_2}$ is less than 20 per cent in all the species except *Monopterus* which seems to have developed more efficient mechanisms of perfusion, and circulation.

Table 7.10: Total area of respiratory organ (mm²), estimated total oxygen uptake (E $\dot{V}_{O_2}$)

and actual oxygen uptake and per cent value of (E $\dot{V}_{O_2}$) in air-breathing fishes.

Fish Species with Body wt.	Resp. Organ	Resp. Area (mm²)	Diffusing Capacity (mlO₂/min/ mmHg)	E$\dot{V}_{O_2}$ (mlO₂/h)	Actual $\dot{V}_{O_2}$ (mlO₂/h)	Tem.± 1°C	Per cent Value
Anabas testudineus[1] 40g	Gills+ ABO	4748.7	0.01499	89.952	5.094 4.019	30 20	5.7 4.5
Channa punctatus[2] 40g	Gills+ ABO	6270.3	0.00707	42.432	5.880 4.778	30 20	13.9 11.3
Channa gachua[3] 40g	Gills+ ABO	3383.3	0.00328	19.707	2.574	30	13.1
Clarias batrachus[4] 40g ABO	Gills+ Skin+	15158.8	0.00803	48.156	3.161 2.882	30 20	6.6 6.0
Heteropneustes fossilis[5] 40g	Gill+Skin+ ABO	13528.4	0.00295	17.715	2.309	20	13.0
Monopterus cuchia[6] 200g	Skin+ ABO	37881.0	0.003366	20.196	6.642	30	32.9

However, the oxygen uptake measurement of fishes were made in laboratory conditions, under temperature and photoperiod controlled conditions. In natural conditions these fishes show circadian fluctuations in oxygen uptake (Patra *et al.*, 1978; Munshi *et al.*, 1979; Ghosh and Biswas, 1980; Ghosh *et al.*, 1990). The aquatic $\dot{V}_{O_2}$ of *Anabas* and *Channa punctatus* was maximum during 08-10 hours, and minimum at dusk (16-18 hours), but in *Anabas* the aerial $\dot{V}_{O_2}$ was maximum at dawn (04-06 hours). In *C. punctatus* the maximum aerial $\dot{V}_{O_2}$ was at 08-10 hours. Almost all the fishes showed variation in their peak period of activities, but it has been found that total $\dot{V}_{O_2}$ was minimum during midday (12-14 hours) and maximum at dawn and dusk. Exceptionally, *Channa gachua* showed maximum oxygen uptake around midnight. Dissolved O_2 and free CO_2 of its natural habitat also show diurnal fluctuation (Munshi and Ghosh, 1993). It is interesting to note that during the midday period the amount of dissolved oxygen (DO) in water is maximum, but all the air-breathing fishes consume minimum oxygen during the time when they rest or sleep taking shelter under macrovegetation (Figure 7.8). During this period the fishes mainly use their gills, for oxygen uptake. Falling DO and rising free CO_2 at night act as signals, originating from the autonomic respiratory centre, modified by reflex processes, directing the fish to gulp air for aerial breathing. During the night the fishes become active, the metabolic rate increases and the accessory respiratory organs

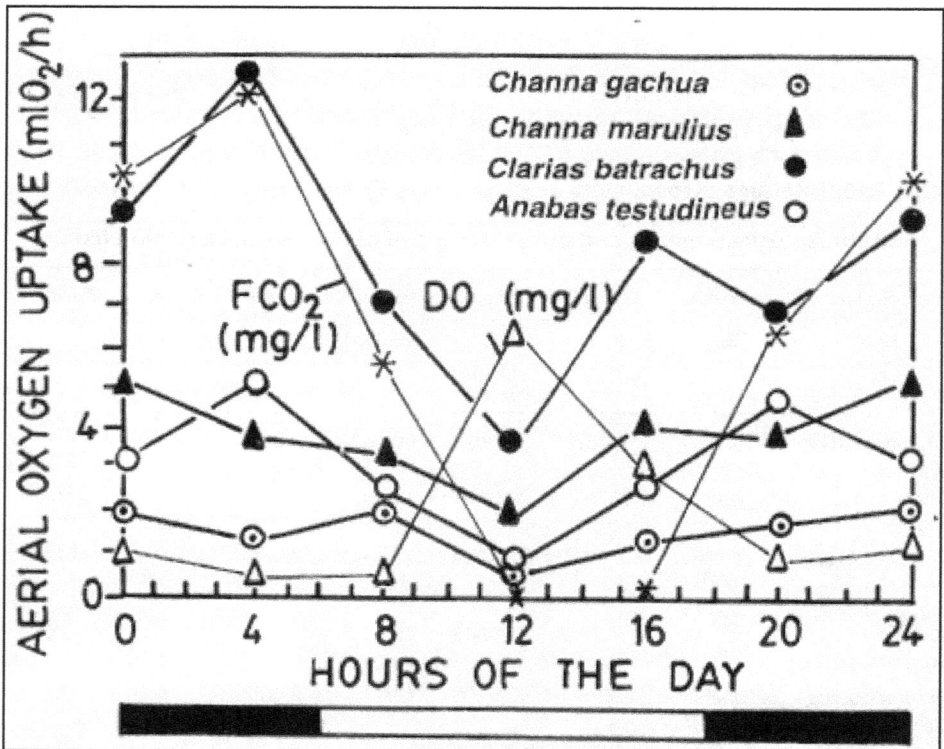

Figure 7.8: Diurnal variation in aerial oxygen uptake of different air-breathing fishes.

play a major part in oxygen uptake (Munshi and Ghosh, 1993). The combined respiratory surface area (gill + accessory respiratory organs) give the maximum limit of bimodal oxygen uptake.

At summer temperatures ($30 \pm 1°C$ for one ml or oxygen uptake (ml O_2/h), the functional water-breathing areas of gills of *Anabas, Channa punctatus, C. gachua, Heteropneustes fossilis* and *Clarias batrachus* were 14.75 cm^2, 20.23 cm^2, 17.9 cm^2, 55.0 cm^2, and 25.2 cm^2, respectively. While at winter temperatures ($20 \pm 1°C$) the functional areas were 17.81 cm^2, 25.41 cm^2, 88.45 cm^2 and 63.3 cm^2 for *Anabas, Channa punctatus, Heteropneustes fossilis* and *Clarias batrachus*, respectively. When air-breathing was taken into account the utilized air-breathing surface area for 1 ml of aerial $\dot{V}_{O_2}$ at summer temperature of $30 \pm 1°C$ was 9.1 cm^2, 8.28 cm^2, 6.98 cm^2, 18.68 cm^2, 15.08 cm^2 and 3.7 cm^2 for *Anabas, C. punctata, C. gachua, Clarias batrachus, H. fossilis* and *M. cuchia*, respectively (Figure 7.9). At winter temperatures the functional area was greater *i.e.*, 12.21 cm^2 for *Clarias batrachus* and 17.02 cm^2 for *H. fossils*. The lung area expressed per unit of oxygen uptake/minute have been calculated to be 800-900 cm^2 for Lepisosteus, 900-1000 cm^2 for Rana and 2000-3000 cm^2 for mammals (Rahn *et al.*, 1971; Hughes *et al.*, 1974b).

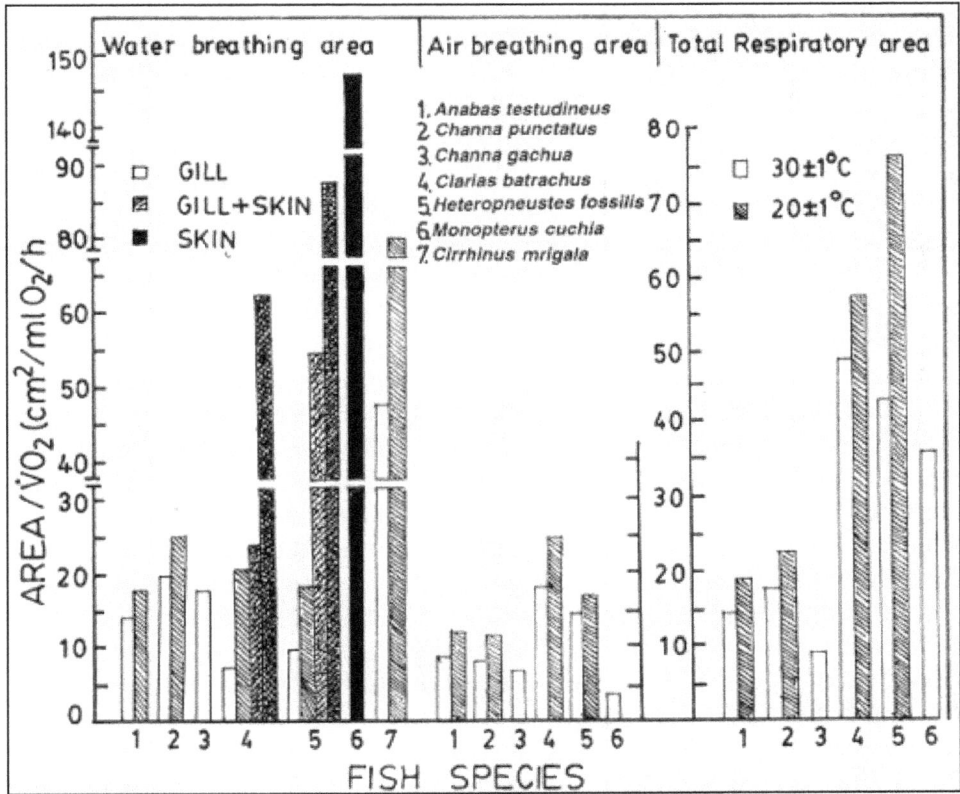

Figure 7.9: Bar diagram showing area of different respiratory organs available per unit mIO$_2$ in different air-breathing fishes of same body weight (40 g).

Table 7.11: Numerical values for the ratio between oxygen consumption and respiratory surface area (cm³O_2/h/m²) for fish and other animals. Values in parantheses indicate the addition of skin area to respective areas.

Species	Temp.°C	Aq. $\dot{V}O_2$ / Gill Area		Aer. $\dot{V}O_2$ /ABO Area		Total $\dot{V}O_2$ /Total Area		References
		10g	*100g*	*10g*	*100g*	*10g*	*100g*	
Ray	—	130	198					After
Torpedo	—	198	210					
Skipjack	—	131	185					Hughes,
Trout	—	481	329					1977
Tench	—	375	172					
Anabus testudineus	30	628	982	1041	1712	767	1337	
	20	479	925	636	1546	578	1088	
Channa punctata	30	469	797	1177	1831	677	1165	
	20	338	631	709	1665	470	1046	
Channa gachua	30	694	346	1522	1281	1022	626	Present work
Heteropneustes fossilis	30	920 (185)	547 (123)	586	778	717 (207)	630 (193)	
	20	504 (101)	625 (141)	305	437	427 (125)	561 (172)	
C. batrachus	30	925 (282)	317 (103)	390	742	510 (208)	483 (210)	
	20	523 (160)	472 (153)	346	410	454 (185)	448 (194)	
M. cuchia	30	(96)	(58)	682	3417	(33)	(119)	
Salamander Lungless	—	159	203	155	178			After
Lunged								Hughes,
Lacerta				197	312			1977
Mammal				534	255			
Bird								

The air-breathing area per unit of oxygen uptake per minute at summer temperature was high for adult silurids (*Clarias*, 809 cm^2; *Heteropneustes*, 771 cm^2), low for adult *Anabas* (351 cm^2) and *C. punctata* (328 cm^2) and lowest for adult *Monopterus* (176 cm^2). But the values are 1537,1024, 576,510 and 880 cm^2, respectively, for juvenile fishes.

This analysis poses a question, whether more respiratory area is functional for $\dot{V}_{O_2}$ during the winter season, or whether a part of the area remains unused during oxygen uptake. By Fick's equation $\dot{V}_{O_2}$ is directly proportional to K.A. and ΔP_{O_2} and inversely proportional to 't' (barrier thickness). So the change in $\dot{V}_{O_2}$ may be due to a change in one or all of these factors. K is called Krogh's permeation coefficient or Diffusion constant. It varies for different media, but does it vary with temperature? K is the product of two material properties: diffusion coefficient and solubility coefficient. Krogh (1941) estimated the value at 20°C, but this value may vary at summer and winter temperatures. The area of the respiratory organ (A) may be modified by change in blood shunting and ventilation patten Singh (1976) has shown that after air gulping, the fish stops gill ventilation for some time, and water is used only for CO_2 elimination. At that time gill area for O_2 exchange becomes minimal. ΔP_{O_2} may change due to variation in concentration of O_2 in water medium, as it depends upon temperature and activities of the biota. Pick's equation may be rearranged to $\dot{V}_{O_2}/A = K. \Delta P_{O_2}/t$).

Change in the ratio of $\dot{V}_{O_2}/A$ with body weight and temperature may be considered due to changes in ΔP_{O_2} or Table 7.11 again induces us to consider "Whether larger *Anabas* and *Channa* and smaller *Heteropneustes* and *Clarias* maintain higher ΔP_{O_2} through gills. Singh *et al.* (1997) observed changes in the water-blood diffusion barrier during development of some air-breathing fishes. Earlier stages of Anabas and Channa species have a much thinner water-blood diffusion barrier than the later stages. But such changes will cause lesser values of oxygen uptake/area relationship in these fishes at adult stages. However, no such data is available on air-breathing organs. Variation in tissue barrier thickness has been proposed due to hypoxic or hyperoxic condition, but does it change with temperature is again a problem for investigation? Further, the effective water/blood barrier increases, as the water film around lamella becomes stagnant for a brief period. For air-breathing organs, the ventilation is tidal and there may be some inherent dead space. Functional area here too may be changed. After gulping air the P_{O_2} in the organ remains high but slowly starts decreasing. This change in P_{O_2} results in change of ΔP_{O_2} causing change in oxygen diffusion rate. The air-breathing organs of different fishes possess varied degrees of adaptation, particularly in blood perfusion pattern and capillary structure – spiral, wavy, lamellar. Blood of these fishes have higher haematocrit values and lower oxygen affinity.

It may be concluded that respiratory area metabolism relationship in air-breathing fishes is not rigid but flexible in different seasons.

Conclusion

The metabolic rate (or oxygen uptake) shows an allometric relationship with body mass and it is principally determined by a few factors like temperature, body size and phylogeny. Similarly the respiratory area is also exponentially related to the body mass and depends mainly on phylogeny and habitat condition. It is suggested that metabolic rate is closely associated with, and perhaps dependent upon, the ability of an organism to obtain oxygen from the external environment, especially on respiratory and circulatory systems to provide O_2 and remove CO_2. This oxygen-exchange capacity of the gas exchanger is determined by : (i) surface area of the gas exchanger, (ii) the effective diffusion barrier, and (iii) the O_2 pressure gradient between the medium and the exchanger.

In Indian air-breathing fishes, the metabolic rate is exponentially related with the respiratory surface area, and these exponent values vary at different seasonal temperatures and with the state of development and with the nature of respiratory organs. In the summer season ($30 \pm 1°C$), both the aquatic and aerial oxygen uptake is higher than that in the winter season ($20 \pm 1°C$), but the exponent values, for the relationship with respiratory area, show opposite trends. At summer temperatures ($30 \pm 1°C$), with unit increase in water-breathing respiratory area the aquatic oxygen uptake increases by powers of 1.315, 1.387, 0.419, 0.745 in *Anabas testudineus, Channa punctatus, Clarias batrachus* and *Heteropneustes fossilis* respectively, while the respective slope values at winter temperatures are higher (1.383, 1.456, 0.975 and 1.238). But the intercept values (O_2 absorbed through 1 mm² area) are always lower at winter temperatures than that at summer temperatures.

At summer temperatures, the aerial oxygen uptake values increases with slope values of 1.303, 1.275, 1.351 and 1.186 respectively, with unit increase of accessory respiratory area in *A. testudineustes, C. punctatus, Clarias batrachus* and *H. fossilis*; while at winter temperature, the respective slope values are 1.542, 1.533, 1.094 and 1.238. The slope value for aerial oxygen uptake and air-breathing organs of *Monopterus cuchia* was calculated to be 1.879 at summer temperatures. It can be summarized that in spite of the same respiratory surface area for a particular species, the intercept value of $\dot{V}_{O_2}$ (mlO_2/mm² area) is higher at summer temperatures, but the $\dot{V}_{O_2}$ increases with higher slope at winter temperatures.

In average sized fish of 40 g body weight, the aquatic oxygen uptake is more than 60 per cent of morphometric oxygen uptake in obligate air-breathers (*M. cuchia* 65 per cent at 30°C and 74 per cent at 20°C), but it is below 40 per cent in facultative air-breathers. But the morphometrically calculated aerial oxygen uptake is much higher (4-15 times) than the actual aerial oxygen uptake. Through a fixed area (10 cm²) of aerial respiratory organ, the $\dot{V}_{O_2}$ is highest for *Monopterus* (6.503 ml O_2/h) at summer temperatures. It may be concluded that the structure of respiratory organs is designed to meet the functional demand of oxygen and flexibility in the exponent values is either due to changes in functional respiratory surface area or permeability factor guided by barrier thickness and pressure gradient. It is clearly indicative that in course of evolution nature has provided a more efficient and flexible mechanism for oxygen diffusion in obligate air-breathers, like *M. cuchia* and *A. testudineus* and the accessory respiratory organs provide a greater scope for $\dot{V}_{O_2}$ during active and stressful conditions. It is essential therefore to study the diffusion barrier thickness of the respiratory organs, pressure gradient at different seasonal temperatures, for better understanding of the mechanism of oxygen uptake in fishes under different physico-chemical conditions of the ambient swampy environment and the physiological status of the fishes.

Chapter 8

Water/Air Transition in Biology

The hydrosphere is an important part of the biosphere. On the basis of its salt concentrations, it is differentiated into marine, estuarine and freshwater ecosystems. The marine ecosystem is comparatively more stable than estuaries and freshwaters. These water bodies have natural inbuilt purification mechanism to maintain adequate ecological standards needed to preserve biodiversity. Biotic communities in turn participate in driving the giant wheel of aquatic ecosystems. The obligate aquatic animals extract oxygen from water which is about 1000 times more dense than its counterpart, the air, and contains about 30 times less oxygen than in the same volume of the latter at normal temperature and pressure (Dejours, 1976). Even though such physical stresses are imposed by the respiratory medium, the gills are meticulously designed to perform their functions effectively in water (Hughes, 1984), while air as a respiratory medium is better suited to terrestrial animals for gaseous exchange. High atmospheric temperature in the summer months may result in the hypoxic conditions of water bodies. High quantities of free carbon dioxide, low precipitation, high temperature and low dissolved oxygen were the impelling forces for the transition of aquatic to terrestrial forms during Devonian, some 350 million years ago. At the end of Devonian, stegocephalian amphibians with true lungs appeared. Once again fluctuations in the atmospheric temperature and oxygen levels during the tertiary period resulted in the evolution of modern air-breathing teleosts (Munshi, 1990). The emergence of land vertebrates is one of the most spectacular events of animal evolution marked by adjustment in respiratory, cardiovascular, acid-base balance and osmo-regulatory processes in the new arrivals of the land.

Respiratory Responses

Obligate water-breathers and air-breathers respectively, use their gills and lungs for gaseous exchange. However, there are about 140 species of teleostean fishes which show different degrees of bimodal gas exchange mechanisms (Rahn and Howell, 1976). These vertebrates use their bimodal machinery effectively to extract oxygen from water and air for their total metabolic activities. The development of such machinery is accompanied by structural and functional adaptations to combat hypoxic ambient water of the swamps. The effectiveness of these structures depends on their progressive modifications.

Evolution of Bimodal System

A bimodal system may be defined as one in which an organ or organs of an animal at a given stage of its life history utilizes both water and air in its gas exchange mechanism. The particular organ(s) involved varies in different species.

During the late Devonian, the atmospheric oxygen was about one fifth of its present level, and drying of lagoons, swamps and freshwater bodies was very prevalent. During this period the lung seems to have evolved in the Dipnoi, Crossopterygii and Amphibia. The extinct fish *Cheirolepis*, which represents the possible ancestors of the group palaeoniscoidea possessed lungs as well as gills. The Palaeoniscoids later gave rise to the Chondrostean fishes represented by the modern forms *Polypterus* and *Polyodon*, and the Holostei represented by present day *Amia* and *Lepidosteus*. All these fishes became the masters of fresh as well as seawater during the Cenozoic era. The lungs became transformed into a specialized hydrostatic organ – the swimbladder (Fange, 1976). During the Tertiary and Quaternary periods, the level of atmospheric oxygen fell considerably 0.1 per cent PAL (Present Atmospheric Level), affecting the oxygen content of the water. Under these conditions the gills were unable to sustain the oxygen requirements of these fishes, especially in the freshwaters of rivers and swamps. Lungs were no longer available for aerial respiration as they had become modified into the specialized swim bladder. As such the advanced groups of teleostean fishes, represented by the modern forms of *Anabas*, *Colisa* (=*Trichogaster*), *Channa* (=*Ophicephalus*), *Clarias*, *Heteropneustes*, *Monopterus* (=*Amphipnous*), etc. survived, because they developed other types of accessory respiratory organs (Munshi and Hughes, 1992). Many of these air-breathing organs are essentially modifications of the gills (Munshi, 1980, 1985). The swimbladder in these fishes is either absent or very much reduced (Moitra and Munshi, 1997).

The linings of the accessory respiratory organs are essentially mucoid in nature, while lungs have a surfactant lining of phospholipids (Hughes, 1965; Pattle, 1976; Hughes and Weibel, 1978).

In dual breathers the bimodal gas exchange machinery (gills-skin and air-breathing organs) chiefly function for the uptake of oxygen while carbon dioxide is easily lost into the water via the gills or skin. In hypoxic swampy conditions there is a chance of transbranchial loss of oxygen from the gills when they are ventilated by comparatively more hypoxic ambient water. Under such adverse ecological conditions the dual breathers stop gill ventilation and depend on direct air-breathing. Such respiratory manipulation results in the accumulation of carbon dioxide in the fish body and a serious problem of acid-base balance develops. The air-breathing fishes regulate the perfusion and ventilatory mechanism to combat such situations.

It is interesting to discuss the findings of scientists on the impact of modifications of water and air-breathing organs on their relative efficiency in oxygen and carbon dioxide transport through water and air, respectively.

Cardiovascular Responses

In bimodal-breathing vertebrates the cardiovascular system is greatly modified to accommodate the combined activities of the gills, skin and air-breathing organs. Differentiation of 'pulmonary' (suprabranchial) and systemic circulation in air-breathing teleosts (except *H. fossilis* and *C. batrachus*) is the first step to completely separate the two systems in higher vertebrates. In air-breathing teleosts the total cardiac ouput first enters the branchial circulation and the post-branchial partially oxygenated blood from the first and second pair of efferent branchial arteries perfuses air-breathing organs for gaseous exchange and then returns to the systemic venous circulation (jugular vein). On the other hand, the third and fourth pairs of efferent branchial arteries shunt blood to the systemic arterial circulation through comparatively reduced gill resistance. Perfusion of blood to air-breathing organs results in the extravascular resistance pathway. Such vascular resistance may be an adaptation to lower the velocity of blood in the vascular papillae for maxium oxygen loading. Vascular papillae of *Anabas, Channa, Monopterus* are modified to slow down the blood flow for higher oxygen uptake. On the other hand, the third and the fourth pairs of branchial arches shunt blood to the systemic circulation through reduced gill resistance and large sized blood vessels. Because of the degree of development of the four pairs of gills, the gill microcirculation shows wide interspecific variation. The symposium was a platform to discuss points relating to the modification in the cardiovascular system of air-breathing fishes with special reference to the reorganization of perfusion distribution, axial flow separation and spatial shunts (Olson *et al.*, 1994; Munshi, 1992; Munshi *et al.*, 1994).

Acid-Base Balance

The maintenance of relatively constant pH values in the body fluids at a given temperature is one of the important tasks of the regulatory systems for homeostasis in animals. Since the effectiveness of most enzyme systems is governed by an optimum pH value of the medium, variation in the pH may reduce enzyme activity. There is a balance between production and elimination of H^+ and OH^- ions during normal steady-state conditions. In fishes, the buffer values of blood and intracellular compartments (Olson, 1992) are generally much smaller than in higher vertebrates (Sinha and Munshi, 1980). Fishes are also handicapped in the respiratory compensation because of physical limitations or the energetic problems with long-term hyperventilation of the viscous respiratory medium, water. As such, gill ventilation is little affected during acid-base disturbance (Dejours, 1973). Because of the limitation of gills in the adjustment of the buffering mechanism, the excretory systems take care of steady-state pH adjustment in fishes (Heisler, 1984). When air-breathing fishes terminate aquatic oxygen uptake through gills and skin because of hypoxic ambient water, the transition to exclusively air-breathing causes considerable increase in blood plasma carbon dioxide. The retention of CO_2 in the blood leads to increased levels of HCO_3, CO_2 and lowered pH in the blood and tissues (Howell, 1970). With the development of hypercapnia during air-breathing, bicarbonate is shifted to the intracellular space in order to protect the intracellular pH value (Heisler, 1984). The available information on the mechanism of preferential intracellular pH protection in air-breathing fishes is limited and thus generated detailed discussion in the symposium on Water/Air Transition in Biology (Heisler, 1982).

Environmental Pollution

The abiotic factors of the environment have shown a gradual change in the past few decades. This is mainly due to deforestation and other major environment changes as large-scale industrialization, urbanization, operations like mining, land leveling blasting, constructions of dams for power generation and irrigation. These cause water, soil and air-pollution. In agriculture, to boost production fertilizers, pesticides and herbicides are extensively used which result in pollution of water bodies, and create a danger to aquatic animals (Munshi and Singh, 1971; Roy *et al.*, 1986; Dutta, 1996). Increase in the acidity of the water, higher free CO_2, lower DO, (Dissolved Oxygen) and change in concentration of many dissolved cations and anions have posed a serious threat to aquatic animals by interacting adversely with them (Sinha and Munshi, 1981; Munshi and Singh, 1992). The functioning of the ecosystem may be pushed to a critical stage when it may not be possible to restore homeostasis through ordinary, non-emergency adjustment processes and therefore structural and functional changes may occur in the organisms. Changes in the structure of vital organs like, gills, skin, liver, kidney, gonads are evident in the non-target animals due to different types of environmental pollutants. The endocrine system is helpful in the coordination of the physiological and biochemical processes to regain homeostasis. The toxicants also alter the endocrinal coordination.

It is our primary duty to monitor environmental pollution and suggest methods for its control.

Mechanism of Detoxification

The effects of chemical and physical alterations of the environment are manifested at the organismal level by causing death or by impairing vital functions. Selye (1976) described the response of an individual as a succession of physiological and biochemical reactions and divided it into three main parts : (a) alarm stage – characterized by fight or fight response, (b) resistance stage – offering long-term protection against the toxicant, and (c) the exhaustion stage – depletion of energy store in the previous two stages followed by death. During these stages the organisms try to eliminate the toxicants outside the body and form a barrier against their entries. The organisms try to trap the toxicant and then expel it outside either in a membrane-bound envelope or in a watery medium either as urine or ambient water. But before that the toxicants are inactivated by the process of oxidation, methylation, conjugation etc., which require energy and enzymes. The tripeptide glutathione is associated with the detoxification of environmental xenobiotics. But there is lack of information regarding its role in aquatic organisms. Mitochondria-rich cells are primarily thought to be connected with ion-regulation, but in recent studies the increased number in the case of heavy metals, acid and pesticide pollution suggests their role in the detoxification process. Increased lymphoid spaces are also observed in the gill and skin epithelium, which play protective roles and help in detoxification by accumulating the entered toxicants and expelling them out with the help of some energy pumps, specially by the mitochondria-rich cells (Dutta *et al.*, 1996). Macrophages also help in detoxification by a process of pinocytosis and phagocytosis. This gives rise to secondary lysosome in which digestion and breakdown of engulfed material occurs (Munshi *et al.*, 1990).

Environmental regulation is a complex science and a challenge to mankind. The responsibility now lies not only with the politicians who are at the helm of affairs but also with the biologists.

The valuable papers by scientists from different countries will go a long way to take this difficult problem.

Chapter 9

Structure of the Heart of *Amphipnous cuchia* (Ham.) Amphipnoidae Pieces

Introduction

A review of the past literature shows that most of the workers in the past such as Gegenbaur (1891), Senior (1918) and Parson (1929) have described and discussed the structural peculiarities of conus arteriosus in elasmobranch and its disappearance in teleosts. Goodrich, (1930) in his well known treatise the "Structure and development of vertebrate" has summarized the relevant literature on this subject.

General accounts of teleostean hearts are available from the works of Danforth, (1912), Mott, 1950, Awati and Bal, 1934 and Karandikar and Thakur, (1954), Prakash, (1953) studied the heart of *H. fossilis* with special reference to its conducting system. A comprehensive and detailed account of the structure of the teleostean heart is given by Singh, (1960), who has described in detail the structure of heart of eight species belonging to six different families of freshwater teleosts. Liem, (1961) has studied the structure of the heart of *Fluta alba* a fish of the family Symbranchidae (order – Symbranchiformes). Saxena and Baxshi, (1965) studied the heart of *Orienus plagiostomus*. The more recent studies on the functional morphology of heart of fishes have been made by Johansen and Hanson, (1968), Randall, (1968) and Saxsena, (1970). Effect of salt solution and hormone like substances on heart beat have been studied by Hyde, 1908, Seyama and Inisawa, 1907, Huntsman, 1931 and Singh, 1971. Intrapericardial and intracardiac pressure and events of cardiac cycles in *Mustelus* have been studied, Sudae, 1905, a,b).

In this chapter the structure of the heart of *Amphipnous cuchia* (Ham) an air-breathing eel like fish of India has been studied. It belongs to the family Amphipnoidae, placed in order Symbranchiformes.

Amphipnous cuchia was first described by Taylon, (1830) who gave an elaborate account of its anatomy and bionomics. The fish has remarkable habit of distending their respiratory air sacs for gaseous exchange. It is a sluggish fish but its movements sometimes are very quick and resemble those of a typical eel. According to Hora, 1934, it is found in holes and crevices in the muddy banks of marshes and slow-running rivers and is particularly abundant in bhils of deltaic districts of lower Bengal. *A. cuchia* is often found wriggling about in wet grasses several yards away from any piece of water.

The gills of *Amphipnous* are greatly reduced and it is generally believed that it has lost practically all its power of aquatic respiration. Even in aquarium containing aerated water. *A. cuchia* distends its air chambers by gulping in air at the surface (Hora, 1935, Munshi and Singh, 1968).

Materials and Methods

Specimens were collected from the ponds and swamps of Alamnagar, Dist. Saharsha, Bihar and were kept in aquaria of P.G. Department of Zoology, Bhagalpur University. This study is based on 35 fishes.

After opening the abdomen and ligaturing of ventral aorta and important veins of an anestherized fish, 9-10 per cent formalin was injected through the hepatic vein by means of a syringe. The specimens were then immersed in 9-10 per cent formalin for a few days for hardening. The heart was then carefully removed for dissection and freehand sectioning.

Characteristics of the regression lines of the logarithm of heart weight, heart length, to log body weight and total length were calculated by the method of least square using facit and Moscal electronic calculator.

Observation

The elongated heart is situated far posterior to the cleithral symphysis (Table 9.1) in the pericardial chamber. It consists mainly of four chambers as in other teleosts, the sinus venosus, auricle, ventricle and bulbus arteriosus (Figure 9.1).

Sinus Venosus

The sinus venosus is reduced and is not quite discernible from the ductus cuvier. It is a thin walled squarish chamber situated on the dorsal aspect of the auricle. When the sinus venosus is opened from dorsal side a quite distinct median elevation is visible which partly divides the sinus venosus into two halves. The slit like sinu-auricular aperture is situated just in front of the median elevation and it has got no valves. The sinus venosus receives blood from the right and left anterior cardinal veins, hepatic vein and right posterior cardinal vein. There are no valves at the opening of these veins into the sinus venosus (Figures 9.1 and 9.2).

Auricle

The auricle is an extremely thin walled voluminous chamber with a smooth external surface. It envelopes a greater part of the bulbus arteriosus and some part of the ventricle (Figures 9.1–9.3). The inner surface of auricle is provided with a few muscular ridges distributed in the anterior and posterior region, where it joins with the bulbus arteriosus. Below the sinu-auricular aperture on the posterioventral wall of the auricle is an auriculo-ventricular aperture guarded by valves. The auriculo-ventricular valves are four in number (Figure 9.6). They are situated on the dorsal and ventral side of the auricle. These pocket like auriculo-ventricular valves are unequal in size and placed around the opening. These valves have their concavity towards the ventricle and convexity facing the auricular chamber. This facilitates flow of blood from the auricle into the ventricle (Figures 9.4–9.6).

Mode of working of the auriculo-ventricular valves:– The blood from the auricle forces upon the auriculo-ventricular valves and separates them. Thus the blood passes into the ventricle and when the ventricle is filled with blood, the concavities of the valves also receive blood, and create a pressure upon them which meet together to shut off the auriculo-ventricular aperture. Thus no blood could re-enter the auricle (Figures 9.4 and 9.6).

Ventricle

The conical shaped ventricle is a thick walled chamber. Anteriorly it is partly covered by the auricle (Figures 9.1–9.3). The ventricle wall consists of an outer cortical layer and an inner spongy layer. The inner layer of the ventricle has a network of muscular bars called columneae carneae which increase the surface of the ventricle. Distinct chordae tendineae run from the ventricular wall to the auriculo-ventricular valves (Figures 9.4 and 9.8). The cavity of ventricle is small in comparison to bulbus arteriosus and is not regular. The ventricle opens into bulbus arteriosus by ventriculo-bulbar aperture guarded by two semilunar valves. The valves are laterally arranged. The concavity of the valve is turned towards the lumen of the bulbus arteriosus (Figures 9.4 and 9.7).

Mode of working of the ventriculo-bulbar valves- The blood from the ventricle passes into the bulbus arteriosus and with its filling the concavity of the semilunar valves get inflated which results in the closure of the bulbar aperture. In this way the backward flow of blood from the bulbus arteriosus to the ventricle is prevented (Figures 9.4 and 9.7).

Bulbus Arteriosus

The bulbus arteriosus is a pear shaped structure situated anteriorly to the ventricle. The auricle completely encloses the bulbus arteriosus except on its ventral aspect which remains only partly covered (Figures 9.1–9.3). It has moderately thick wall and its cavity is divided by several septate ridges encroaching into its lumen (Figures 9.4 and 9.5). Anteriorly the bulbus arteriosus opens into the ventral aorta.

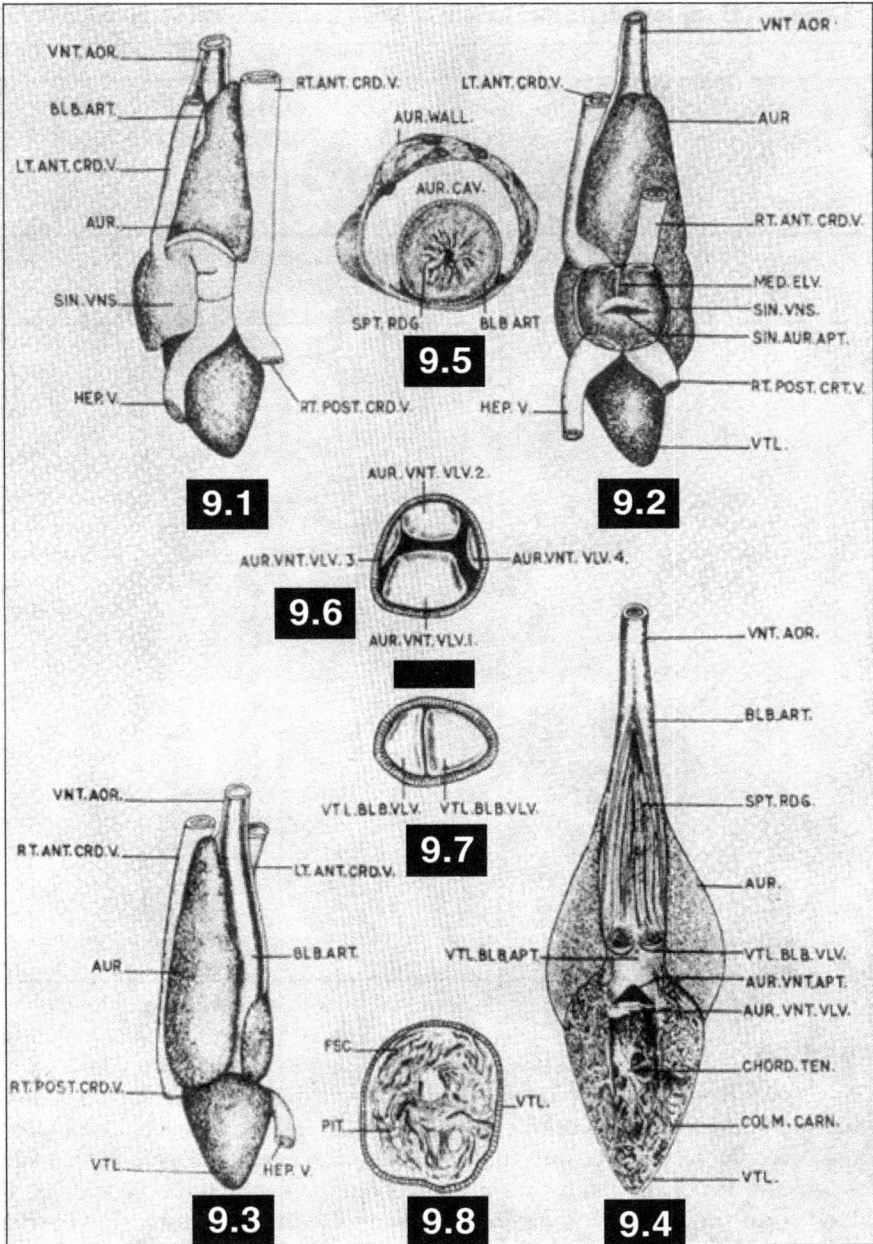

Figures 9.1–9.8

9.1: Dorsal view of the heart of *Amphipnous cuchia* x 2 Nat. Size; 9.2: Dorsally dissected heart of *A. cuchia* x 2 Nat. size; 9.3: Ventral view of the heart of *A. cuchia* x 2 Nat. Size; 9.4: Ventrally dissected heart of *A. cuchia* x 2 Nat. Size; 9.5: Free hand T.S. of auricle with bulbus arteriosus; 9.6: The auriculo-ventricular valves in surface view; 9.7: The ventriculo-bulbar valves in surface view; 9.8: Free hand T.S. of ventricle.

Table 9.1: Body length (L) and length of heart from cleithral symphysis (Y)

L	Log L	b Log L	A + b log L	Ý
1	0	0	-0.91239	0.1223
10	1	0.9502	0.0378	1.091
100	2	1.9004	0.988	9.728
1000	3	2.8506	1.9382	86.73

Body Length (L) cm	Length of Heart from Cleithral Symphysis (Y) cm
41.6	4.2
47.0	4.9
50.0	4.8
52.7	5.5
53.0	5.2
56.8	5.8
59.5	5.9
60.8	5.8
63.2	6.5
64.0	6.5
65.5	6.4
66.4	6.7
67.5	6.8
69.5	6.3
71.8	6.3
73.5	7.6
75.0	8.3
77.0	8.9

Circulation

The blood from the body is collected by right ductus cuvieri. Right anterior cardinal vein and Right posterior cardinal vein, single hepatic vein and anterior cardinal vein. These veins empty the blood into the sinus venosus from where it passes into the auricle through sinu-auricular aperture which has no valves. Then the blood from the auricle moves into the ventricle through auriculo-ventricular aperture which is guarded by two pairs of unequal sized pocket like valves. The ventricle pumps the blood into the bulbus arteriosus through ventriculo-bulbar aperture which is also guarded by two semilunar valves. From the bulbus arteriosus the blood passes into the ventral aorta.

Table 9.2: Body length (L) and heart length (X).

L	Log L	b Log L	A + b log L	X
1	0	0	-1.11453	0.07682
10	1	0.939024	-0.175506	1.6675
100	2	1.878048	0.763518	5.801
1000	3	2.817072	1.70254	50.41

Body Length (L)	Heart Length (X)
41.6	2.8
47.0	2.6
50.0	2.8
52.7	3.3
56.8	3.4
59.5	3.8
60.8	3.6
63.2	3.5
65.5	3.8
66.4	4.0
67.5	3.9
69.5	4.7
71.8	4.0
73.5	4.5
77.6	4.6

Statistical Observation

There is high degree of correlation between the heart and body weight (Table 9.3) as indicated by correlation coefficient (r=0.99).

The heart is one of the most important biophysical parameters. Its weight increases with the increase in body weight. When the heart weight is plotted against body weight on log/log coordinates for 20 fishes it gives a straight line (Figure 9.10) and the best fit was obtained with equation given below:

$$A = aW^b$$

where,

A: Heart weight

a: Intercept on the Y axis giving the value of heart weight of one gram fish

W: Weight of the fish

b: Regression coefficient

Therefore, the heart weight with respect to body weight may be represented by the equation:

$$A = 0.00062\ W^{L25}$$

Heart length (X) also is well correlated with body (L) (Table 9.2) (r = 0.93) and when plotted on log/log co-ordinates shows a slope of 0.93 (Figure 9.11) and the heart length for one cm fish is 0.0768 cm hence the length of heart $X = 0.0768\ L^{0.93}$.

Table 9.3: Body weight (W) and Heart weight (A)

L	Log L	b Log L	A + b log L	X
1	0	0	-3.21	0.00062
10	1	1.25	-1.95	0.011097
100	2	2.50	-0.71	0.1950
1000	3	3.75	0.54	3.468

Body Weight (G) (W)	Heart Weight (G) (A)
52.3	0.119
75.5	0.1232
94.0	0.1656
105.0	0.172
148.0	0.3432
160.0	0.343
174.5	0.370
189.0	0.452
208.0	0.499
252.0	0.505
261.0	0.632
267.0	0.672
291.0	0.790
338.0	0.858
366.0	1.159
405.0	1.176
408.0	1.215
459.0	1.379
469.0	1.576

When the data on the length of the heart from cleithral symphysis (y) was plotted against the body length (L) on double logarithmic grid gives a more or less straight line (Figure 9.9). However, the best fitting straight line was obtained by the method of least squares.

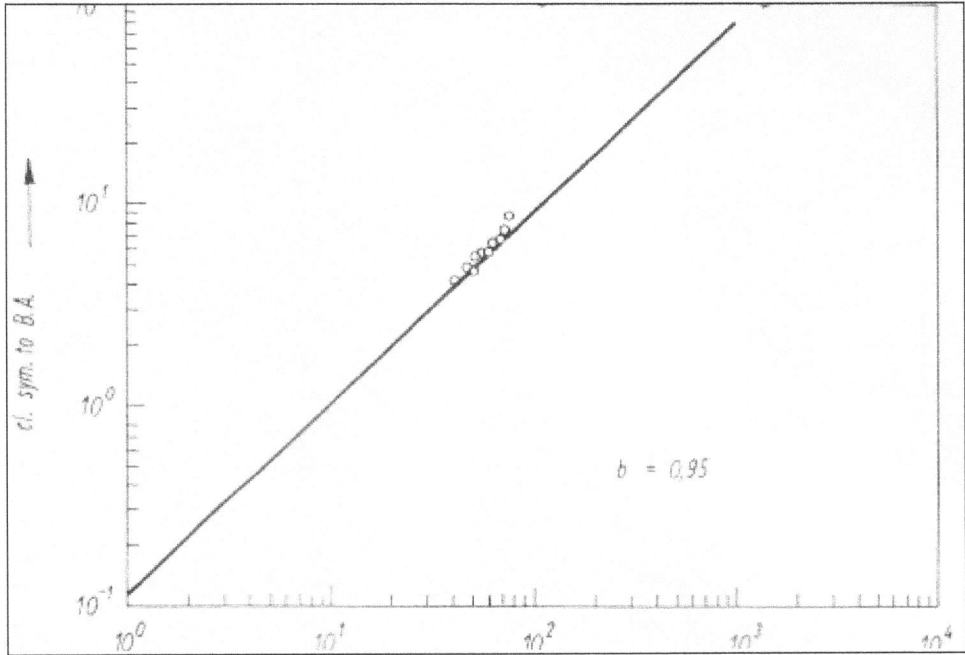

Figure 9.9: Log/Log Graph showing the relationship between the length of heart from cleithral symphysis and body length.

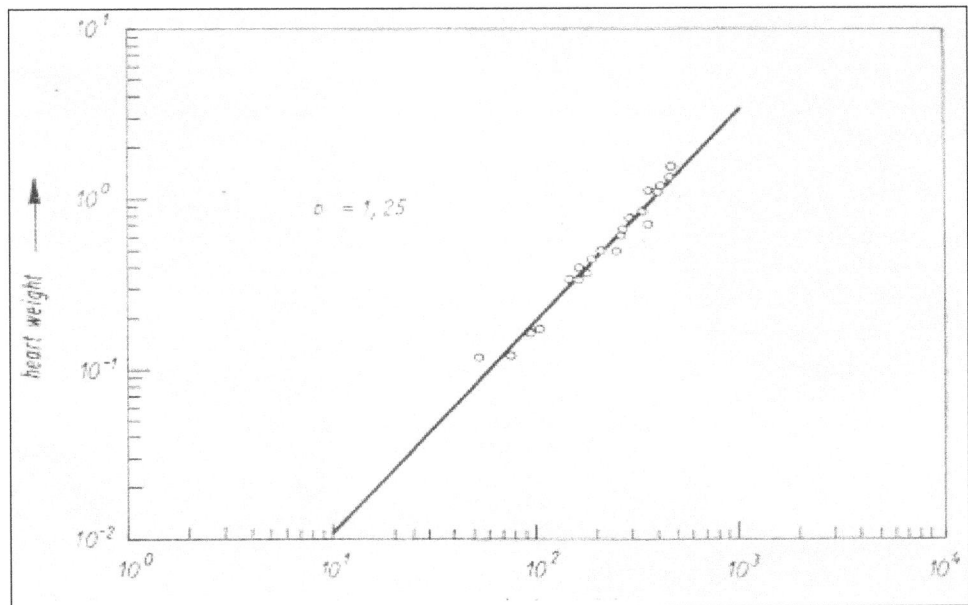

Figure 9.10: Log/Log Graph showing the relationship between heart weight and body weight.

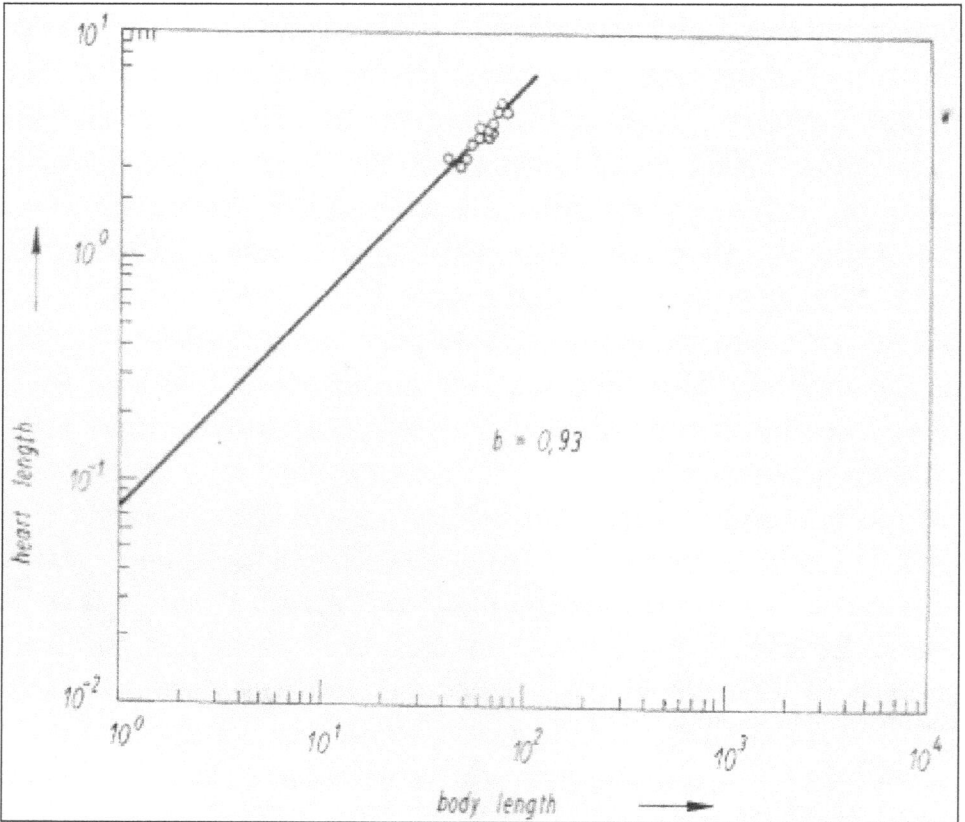

Figure 9.11: Log/Log Graph showing the relationship between the length of heart and body length.

The relationship between the heart length from cleithral symphysis and body length (Table 9.1) may be expressed by the equation

$$Y = aL^b$$

where

Y: Length of heart from cleithral symphysis.

Normally the position of heart in teleosts is near the gills. But in *A. cuchia* the heart is situated far behind. Wu and Liu, 1943 thought that "this condition has been apparently evolved in correspondence to the shape of body". This suggestion, however, did not get support from Morr, 1950 who showed that in Eel which has equally elongated body, the heart is not found in that position. It was suggested by Liem (1961) that the position of the heart might be involved in an hemodynamic role, Like *Fluta*, when *Amphipnous cuchia* is held head upwards, its heart remains filled with blood, whereas that of a true eel (Skramilk, 1935) will rapidly become empty when held in that position. It was reported by Munshi and Singh, (1968) and also found in the laboratory that *A. cuchia* is frequently found with its head upward making a sharp angle with the trunk to engulf air. In general according to Bennighoff, (1933) "there is definite tendency of the developing heart to grow equally in all directions, the ultimate shape of the teleostean heart thus depend on the immediate available space usually determined by the Cleithra and surrounding musculature".

There is some indication that the burrowing habit of this fish has something to do with the position of the heart. The space between the heart and cleithral symphysis is occupied by strong muscles which are used in burrowing. If the heart would have been situated near the cleithral symphysis the process of burrowing movement could have squeezing effect on the heart.

The sinus-venosus of *A. cuchia* is reduced much like those of other air-breathing fishes such as *Ophicephalus*, Singh, 1960 and *Fluta*, Liem, 1961. In purely water breathing fishes such as *Mystus, Wallago, Hilsa, Catla, Labeo, Cirrhina, Notopterus* and *Mastacembelus* the sinus venosus is comparatively better developed. Moreover, in *A. cuchia* the sinus venosus is partly divided into two chambers by a distinct median elevation. This type of elevation is not reported in Indian freshwater teleosts (Singh, 1960). Similar condition has been reported in *Fluta*, Liem, (1961) the sinu-auricular aperture has no valves as in *Ophicephalus* (= *Channa*), Singh, (1960) but it differs from *Fluta* which has a valve as reported by Liem, 1961.

The auricle is large and it encloses major part of the bulbus arteriosus and to some extent the ventricle also. The wall of the auricle is thin and its inner surface generally is devoid of muscular ridges, except its anterior and posterior part adjacent to bulbus arteriosus.

In most of the teleostean hearts studied the auricle is of spongy nature (Singh, 1960, Liem, 1961) but *A. cuchia* differs from them in having smooth walled auricle. The auriculo-ventricular aperture is guarded by valves. These pocket like valves are four in number and placed around the auriculo-ventricular aperture. Most of the workers such as Walter, (1928), Kingsley, (1926), Goodrich, (1930), Awarti and Bal, (1932), Wu and Liu, (1943), Mott, (1950) and Prakash, (1953) have reported the occurrence of only two auriculo-ventricular valves. Mitra and Ghosh, 1932 first reported the presence of a second pair of auriculo-ventricular valve in *Catla* and *Cirrhina*, Karandikar and Thakur, 1954 found them in a marine teleosts *Sciaenoides brunneus*, Singh, (1960) has described two pairs of semilunar or pocket like valves at the auriculo-ventricular aperture in most of the Indian freshwater teleosts. The second pair of auriculo-ventricular valves has not been reported in *Fluta* (Liem, 1961.

The conical shaped ventricle is thick walled having a small lumen as in most fishes studied. Its internal surface bears a large number of pits which increase the capacity of the ventricle. However, the lumen of the ventricle is larger than the bulbus arteriosus in which it opens. The ventricle opens through ventriculo-bulbar aperture into the bulbus arteriosus which is guarded by a pair of laterally situated semilunar valves. This condition has also been found in *Mastacembalus*, Singh, (1960). In *Catla catla*, a freshwater carp it is dorso-ventrally situated, Singh, (1961).

The bulbus arteriosus in *Amphipnous* is a tubular structure with no base. In *Mystus, Wallago, Catla* and *Mastacembelus* the bulbus arteriosus has at its base a swollen bulbus structure. Moreover, in this fish the bulbus arteriosus is entirely encircled by the auricle except on its posterior region. In this respect it differs from *Fluta*, internally as in *Ophicephalus* the bulbus arteriosus is divided by septate ridges.

The statistically computed data on the regression line of the heart weight against body weight, the heart length and the length of the heart from cleithral symphysis against the total body length reveals that (1) the heart weight increases with increasing body weight, (2) the heart length, and length of the heart from cleithral symphysis also increase with the increasing body length, (3) As the regression coefficient of the heart length and length of heart from cleithral symphysis against the total body length is less than one, the length specific heart length and length of heart from cleithral symphysis per unit body length decrease with the increasing body length. But the decrease is not very sharp because both have regression coefficient more than 0.9, (4) As the regression line of the heart weight against body weight is more than one, the heart weight per unit body weight increase with increasing body weight.

According to Skramlik, (1935) heart weight and body weight relationship for inactive and active fish are 0.15 per cent and 2.5 per cent body weight respectively. *Amphipnous cuchia* is an inactive fish and here we find that it is only 0.19 per cent body weight Kharakter Svyazimezhdm, (1972) claimed that the treatment of allometric equation to heart weight and body weight data is unnecessary. A table for the length of heart from cleithral symphysis against total body length is given by Liem, 1961 and he also found that length of heart from cleithral symphysis increase with the increasing total body length.

Microcirculation of Gills and Accessory Respiratory Organs of the Walking Catfish *Clarias batrachus*

The walking catfish, *Clarias batrachus*, a member of the sub-order Siluroidei, is native to much of southeast Asia where it is predominantly found in shallow water of ponds and swamps (Das, 1927). *Clarias*, like another siluroid, *Heteropneustes*, has evolved accessory respiratory organs in its branchial chamber that permit survival during periods when dissolved oxygen falls to levels of debilitating to solely aquatic breathing vertebrates. However, the structure of the air-breathing organs in these two catfish is strikingly different.

The ability to extract oxygen directly from the air has enabled *Clarias* to make brief sojourns on land, an activity mechanically assisted by considerable body flexibility and the ability to achieve a modicum of traction with their spiny pectoral fins. Although earlier reports indicated that *Clarias* could estivate under dehydrating conditions, similar to lungfish, the skin is incapable of retaining water loss and desiccation is a limiting factor to survival during drought and terrestrial excursions (Bruton, 1979). Nevertheless, during rainy periods, or in dew-laden grass, *Clarias* is able to travel considerable distance on land, an attribute that undoubtedly has accounted for much of the success of this fish in its rapid spread over southern Florida.

Development of accessory respiratory organs in air-breathing fish has, for hemodynamic and other physiological reasons, necessitated reorganization of the macrocirculation and such adaptations are often accompanied by changes in the vasculature of the gills themselves (Johansen, 1970; Satchell, 1976; Olson, 1994). Micro-circulatory modifications in gills and accessory respiratory organs of air-breathing fish are also becoming evident, although many of these are of unknown physiological significance (Olson, 1994). The cephalic vasculature of Clariidae has been examined by gross dissection and light microscopy (Nawar, 1955, Munshi, 1961) and, more recently, using transmission electron microscopy of vascular corrosion replica was performed in the present study to elaborate upon earlier studies and to clarify a number of points raised by them. The ultrastructure of the respiratory surface was also examined to corroborate the earlier findings of Lewis (1979,b).

Methods

Two groups of *Clarias batrachus* were used in this study. Fifteen *Clarias* were collected in North Bihar, India and maintained at the University of Bhagalpur in 500 liter aquaria. They were fed on alternate days with pieces of fish and chopped goat liver. They were maintained at the University of Notre Dame I 100 liter aquaria and fed commercial trout pellets (Purina, St Louis, MO). All fish were maintained in the laboratory for at least one month prior to use. Tissue samples were collected and corrosion replicas were prepared in the respective laboratories. Freeze drying and electron microscopy were performed at the University of Notre Dame which I visited several times for work.

Corrosion Replica

Ventral aortic cannulation and methyl methacrylate infusion procedures have been described previously (Olson, 1985). Fish (75± 10 g) were anesthetized in MS-222) ethyl-m-aminobenzoate) and the ventral aorta cannulated with polyethylene tubing (PE 50). The sinus venosus was severed and the fish was perfused for 15 min at constant pressure (20 ± 2 mmHg) with phosphate-buffered Ringer followed by the methyl methacrylate resin Mercox. Perfusion was terminated at the onset of polymerization. The carcass was the placed in warm (-50°C) water for 1 hour to ensure complete polymerization and macerated over the following week with alternate solutions of 20 per cent NaOH and 5 per cent HNO_3. When free of tissue, plastic vascular replicas were rinsed in distilled water and air dried. Select areas of vasculature replicas were removed with a hot wire and affixed to a aluminum specimen stub with either silver paste or double-stick tape. The interlamellar filamental vasculature was exposed by carefully breaking away the lamellae from a mounted filament using a fine-tipped probe. Mounted replicas were sputter coated with gold and examined with a JEOL T300 scanning electron microscope.

Gill and Air-breathing Organ Tissue

Clarias were anesthetized in MS-222 (ethyl-m-aminobenzoate; 1:10,000 wt: vol) and killed by cervical dislocation. Gills and air-breathing organs were exposed and rinsed first with a gentle stream of 2.5 per cent glutaraldehyde in 0.1 M phosphate buffer (pH 7.4:22°Courier List) for an additional 3 min. The tissues were then dissected from the head, placed in cold (4°Courier List) buffer.

Prior to SEM examination the tissues were rinsed three times (30 min each) with distilled water then dehydrated in a graded series of water; tert-butyl alcohol and freeze-dried using the method of Inque and Osatake (1988). Dried tissue was affixed to SEM specimen stubs with double-stick tape. The epithelium was removed from some of the samples by tangential sectioning with an acetone-cleaned razor. Samples were gold coated and examined with the SEM as described above.

Results

The following nomenclature is employed for orientation of gill and fan filamental vasculature. Basal, portion of filament closest to gill arch or tissue support peripheral, free end of filament; medial, plane, parallel to filament and passing through the afferent and efferent filamental arteries (*i.e.* center of filament), lateral away from medial plane; and afferent and efferent portion of filament near afferent and efferent filamental arteries, respectively.

Clarias has four pairs of well developed gills and paired enlarged recesses in the posterior buccopharyngeal region of the branchial cavity to accommodate additional respiratory organs; gill fans, arborescent (dendritic) organs, and a respiratory membrane that lies the suprabranchial chamber (Figures 9.12–9.14). Gills fans from the second and third gill arches extend into the suprabranchial chamber and divide it into an anterior and posterior recess (Figure 9.14). The arborescent organ from the second gill arch extends into the anterior recess and the fourth-arch arborescent organ fills the posterior recess.

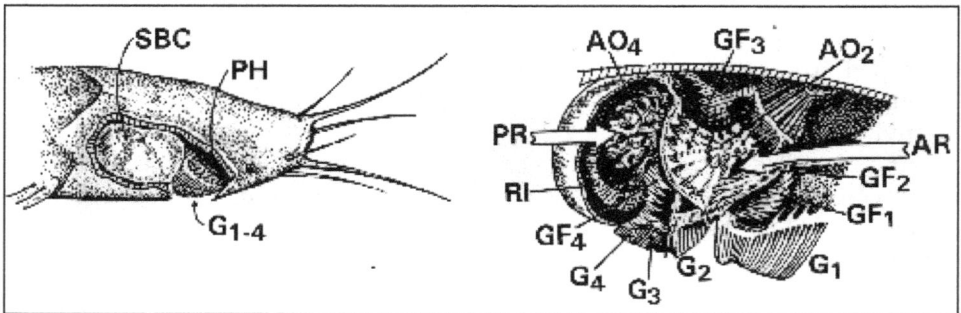

Figure 9.12: Right lateral view of *Clarias* showing the position of the suprabranchial chamber (SBC), pharynx (PH), and gills (G_{1-4}). Approximately = 1.5

Figure 9.13: Right lateral view of *Clarias* showing *in situ* relationship between four gill arches ($G_1 - G_2$), arborscent organs from the second (AO_2) and fourth (AO_4) gill arches and respiratory islet (RI) on the surface of the suprabranchial chamber. Large arrows indicate anterior (AB) and posterior (PR) receses Redrawn from Munshi, 1961. Approximately = 4.

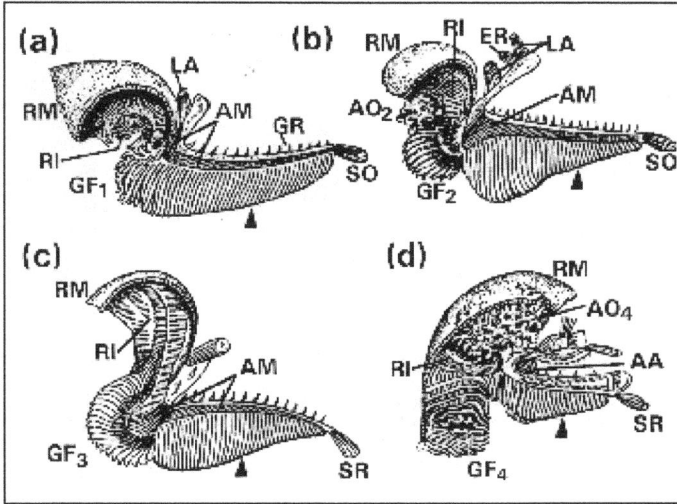

Figure 9.14: Drawing of gill arches 1-4 (a-d, respectively) and associated accessory respiratory organs, lateral view, anterior on right. The gill arch skeleton supports gill rakers (GR) and gill filaments of the outer (opercular) hemibranch of the first through third gill arches are fused to form gill fans (GF$_{1-4}$). Filaments forming the fourth fan (GF$_2$) are detached from the arch skeleton. Respiratory islet (RI) tissue is further extended from the arch onto the suprabranchial chamber as a respiratory membrane. A small arborescent organ (AO$_2$) is supported by the fourth arch. Abbreviations for muscles AA, abductor arcus branchialis, AM, abductor, ER, epiarculia rectus, LA, levator arcus branchialis, SO, subarculia oblique, SR, subarculia rectus. Redrawn from Munshi, 1961. Approximately x 6.

A bony skeleton in each gill arch supports two rows of long filaments (paired hemibranchs) on its ventral surface and gill rakers on the dorsal aspect. Toward the posterior end of the arch, at the point where the arch skeleton bends and travels dorsally, filaments of the aboral hemibranch (opercular side) become shortened and fuse with each other to form gill fans. The fans are moved by the filament abductor muscles. The most posterior filaments lose their attachment to the arch skeleton and continue into the respiratory membrane of the suprabranchial chambers. There are no filaments attached to the dorsal segment of the arch skeleton (epibranchial) in arches 2-4. The epibranchial segments of the second and fourth branchial arches bear the second and fourth arborescent (dendritic) organs, respectively. *Clarias* gill filaments are in general, similar to those found in water-breathing teleosts and the lamellae are well developed. Respiratory lamellae of air-breathing organs are smaller than those on gill tissue and are characteristically folded at right angles so that the outer (free) margin is parallel to the epithelial surface.

Vasculature

General Circulation

The overall vascular organization of the gills of individual gill arches, and respiratory organs and schematically illustrated. Afferent branchial arteries to the first gill arches arise directly from the ventral surface of the bulbus. The remaining branchial arteries are supplied from a short ventral aorta. The second pair of afferent branchial arteries originate from the anterior end of the ventral aorta, whereas the third and fourth pairs bifurcate from a common trunk, the latter arising from the dorsal wall of the ventral aorta. Afferent branchial arteries to the fourth gill arhes have the greatest diameter. As the afferent branchial artery enters the gill arch tissue it divides into a recurrent branch, supplying the anterior third of the arch, and concurrent branch that supplies the posterior arch fans, suprabranchial epithelium, and in the second and fourth arches, the arborescent organs. Dorsally, the respective efferent branchial arteries drain the gill and air-breathing organs into the dorsal aorta. The efferent branchial artery from the first arch enters the carotid labyrinth and a main branch turns caudad and unites with the efferent branchial from the second arch. Their common trunk, the epibranchial artery, then travels caudad toward the dorsal aorta and anastomoses with the contralateral epibranchial just prior to joining the confluence of the fourth-arch epibranchials. Another large vessel, draining the respiratory membrane of the suprabranchial chamber, feeds into the epibranchials. Efferent branchials from the third ad fourth arches remain separated, almost to the origin of the dorsal aorta. The smaller third-arch efferent branchial artery anastomoses with the ventral margin of the fourth. Gill and air-breathing organ circulations are in series with the systemic circulation.

Gill Filament

Filaments from the gill arch have three circulatory networks respiratory (arterio-arterial), nutrient (arteriovenous), and an interlamellar system (also arteriovenous). The respiratory vasculature is a simple set of parallel pathways across the respiratory lamellae that ultimately delivers blood to the dorsal aorta. Vascular connections to the nutrient and interlamellar systems are less obvious. These two pathways return blood directly to the heart by way of the branchial veins.

The respiratory pathway in the filament consists of an afferent filamental artery, afferent lamellar arterioles, lamellar sinus, afferent lamellar arterioles, an efferent filamental artery. Long afferent filament arteries have a gently taper and they lack ampullae or other luminal irregularities. Paired afferent lamellar arterioles arise from the mediolateral wall of the filamental artery and supply individual lamellae, although at the base of the filament the arterioles may feed several ipsilateral lamellae. Afferent lamellar arterioles travel nearly laterally to pass between, and then over, longitudinal (collateral) and connecting vessels of the interlamellar system. The arteriole then travels a short distance toward the efferent side and gives rise to the lamellar sinus.

The outer border of the lamellar sinus is slightly dilated and travels, uninterrupted, across the lamella as the outer marginal channel are organized into parallel rows thereby creating 3-6 linear vascular channels. Medial to these channels, the pillar cells appear randomly positioned and the lamellar sinusoid becomes irregular. The inner vascular margin of the lamella is irregular and there is no evidence of a thoroughfare pathway similar to the outer marginal channel. Lamellae in the basal region of the filament are nearly rectangular, but peripherally they become broader with an elongated "leading" edge on the efferent side that faces the water current. Short efferent lamellar arterioles drain the lamellae into the efferent filamental artery and from there blood flows into the afferent branchial artery.

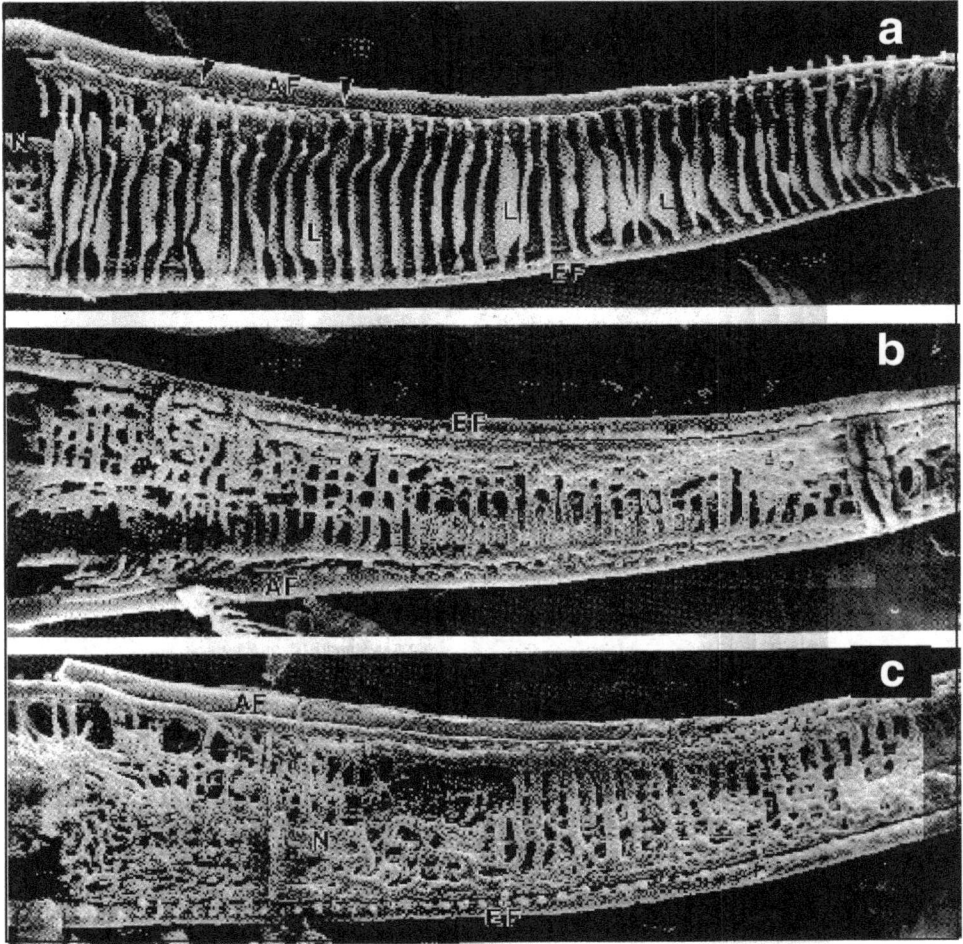

Figure 9.15: Corrosion replicas of gill filaments from the first gill arch showing three filamental vascular networks.

a) The arterioarterial (respiratory) pathway consists of the afferent filamental artery (AF), lamellae (L) and efferent filamental artery (EF). The interlamellar vasculars is visible between inner margins of adjacent lamellae in the body of the filament. Afferent lamellar arterioles pass between longitudinal segments of the interlamellar vasculative on route to the lamellae (arrowheads). b) Lamellae removed to reveal the interlamellar system between AF and EF, nutrient vessels were not filled in the filament. C) Lamellae removed to show interlamellar system of small bore nutrient capillaries that often travel parallel to the interlamellar vessels. Large nutrient arteries (N; a,c) run the length of the filament. Scanning electron micrograph (SEM), x 20 all figures.

Nutrient vessels in *Clarias* filaments originate primarily from small tortuous vessels, the latter arising from the base of the efferent filamental artery or the efferent branchial artery. Tortuous vessels anastomose to form a large nutrient artery that enters the base of the filament core. This artery travels the length of the filament, giving off numerous branches and smaller arterioles that eventually form a capillary network across the filament. Nutrient arteries and arterioles, much like afferent and efferent filamental arterioles, are readily identified in corrosion replicas by the longitudinal impressions of their endothelium. Collecting veins drain the capillary back to the base of the filament and into the branchial venous network.

The interlamellar system consists of a paired ladder like arrangement of comparatively large-bore vessels that travel the length of the filament body on both sides of the cartilaginous support rod. Interlamellar vessels traverse the filament medial to, and in the area between, the inner margins of the lamellae. The relationship between interlamellar vessels and lamellae is evident. Interlamellar vessels are connected along the long axis of the filament by collateral vessels, the latter are closely approximated to the afferent and efferent filamental arteries. These collateral vessels appear to drain the interlamellar fluid from the filament. In most instances the collateral vessels form a plexus that surrounds the afferent and efferent lamellar arterioles. Random anastomotic connections between adjacent interlamellar vessels are also evident from mid-filament toward the efferent collateral.

Figure 9.16: Corrosion replica of lamellae on a filament from the first arch. The outer marginal channel (arrowheads) forms a continuous vascular pathway around the free edge of the lamellar sinus. Pillar cells are arranged in rowes forming 3-6 linear vascular channels near the outer channel. Closer to the filament body the lamellar sinusoides become irregular and channelisation is lost. Arrow indicates direction of blood flow. SEM x 1,800.

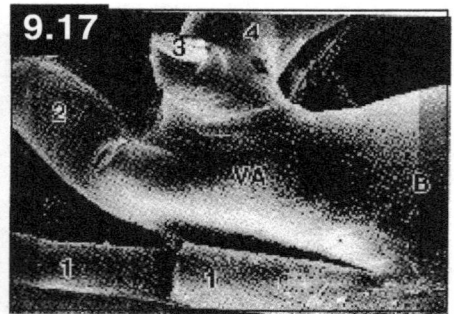

Figure 9.17: Origin of ventral aorta and afferent branchial arteries from bulbus arteriosus (B), Independent afferent branchial arteries to the first gill arches (1) arise directly from the ventral aspect of the bulbus. The ventral aorta (VA) gives rise to afferent branchial arteries to the second (2) and a short, large diameter common trunk arises dorsally and divides into afferent branchial arteries for the third (3) and fourth (4) arches, Corrosion replica. SEM x 10.

The interlamellar system is supplied by small feeder arterioles that originate from the medial border of the efferent filamental artery. As many as one feeder arteriole for each pair of efferent lamellar arterioles can be observed, although one feeder per two pair of lamellar arterioles may be more common. Feeder arterioles are not found on the afferent filamental artery basal to the origin of lamellar arterioles. Replicas of feeder arterioles have characteristic arterial like striations, for the first 20=40µm after their origin from the afferent branchial artery. The striated segment appears to have relatively high resistance and thus may serve to step down pressure from the afferent filamental artery to the interlamellar vessels. Distally, the striations are lost and the vessel lumen becomes irregular. As the striations disappear, the arterioles pass between a pair of collateral vessels adjacent to the efferent filamental artery (one of each pair of collaterals is connected to interlamellar systems on opposite sides of the filament). Feeder arterioles then travel a short distance and may branch numerous times before joining the interlamellar vessels. There is usually a slight sphi172

ncter like constriction on the feeder arterioles immediately prior to their anastomosis with the interlamellar vessels. The lumen of the interlamellar vessels is regular, with balloon-like dilations, suggestive of a low-pressure sinus. Endothelial cell impressions were not readily apparent on interlamellar vessel replicas. There was no evidence in this study of anastomoses between nutrient vessels and the interlamellar system in the filament.

Arborescent Organ, Gill Fan and Suprabranchial Chamber

Paired rows of lamellae covering the arborescent organs (Figure 9.12), gill fans (not shown), and suprabranchial epithelium (Figure 9.13) are supplied and drained by vessels that resemble gill filamental vessels, but have been modified to adapt to a flat respiratory epithelium. This vasculature is best studied in replicas of gill fans and the suprabranchial chamber epithelium where both lamellar (air) and tissues sides of the circulation can be readily examined. Our studies indicate that the arborescent organ vasculature is also similar; however, it was difficult to expose vessels beneath the lamellar surface without excessive destruction of the replicas.

Respiratory lamellae are supplied and drained by a sheet of parallel arteries that branch three to five successive times as they approach the terminal afferent and efferent arteries. Lamellae are organized into paired rows that form the respiratory islets. Each row of lamellae receives blood from a single afferent artery that travels beneath the outer border of the islet. Both lamellar rows are drained into a common efferent artery that travels beneath and between the lamellar rows. Thus each islet has two afferent arteries along the lateral borders and a single, central, efferent artery. This two-to-one, afferent-efferent vessel ratio is maintained from the main (first generation) distributing vessels all the way to the terminal arteries. Lamellae are perfused by third through fifth generation arteries.

Numerous anastomotic connections are found between adjacent terminal efferent. These anastomotic vessels form an extensive plexus interconnecting the efferent arteries along their entire length. The anastomotic vessels always travel on the tissue side of the afferent arteries and are not present on the lamellar side of either afferent or efferent arteries. Anastomoses between afferent arteries were not observed.

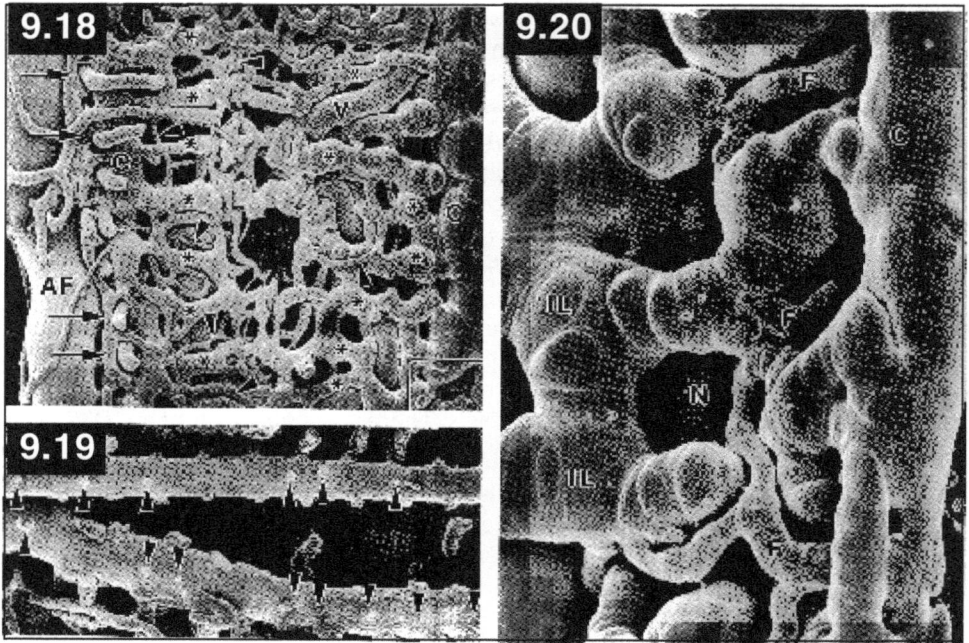

Figure 9.18: Corrosion replica showing relationship between interlamellar and nutrient systems in gill filament. Transverse interlamellar vessels (*) are drained at afferent and efferent borders by collateral vessels. Near the afferent filamental artery (AP) the collateral vessel forms a plexus around the afferent lamellar arterioles (arrows). Nutrient vessels (arrowheads) have a smaller and more uniform diameter than interlamellar vessels. Nutrients drain into a separate vein (V. Small vessel (box in lower right corner) is a feeder from the efferent filamental artery to the interlamellar system. SEM, x 155.

Figure 9.19: Replicas showing medial borders of two efferent filamental arteries and the origin of feeder arterioles (arrowheads) that supply the interlamellar system. Vascular stumps of efferent lamellar arterioles are evident on the lateral borders of the filamental arteries. SEM x 180.

Figure 9.20: Replica of efferent side of filament showing collateral vessel (c) and interlamellar system (IL). Feeder vessels (F) originate from the efferent filamental artery and have a striated arteriole-like appearance near their origin (*). They then pass between the paired C, lose the striated texture, and branch to supply several interlamellar vessels. Note irregular bubble-like distensions in IL wall, N, nutrient artery. SEM x 630.

Vascular replicas of both afferent and efferent arteries have a striated endothelial relief, indicative of high pressure vessels. In some replicas, the efferent artery replicas also contained numerous constrictions. It is not known if this is due to vasoconstrictory tonus or decreased intravascular pressure in postlamellar vessels during the methacrylate infusion procedure. Anastomotic connections between the efferent arteries have a similar striated appearance.

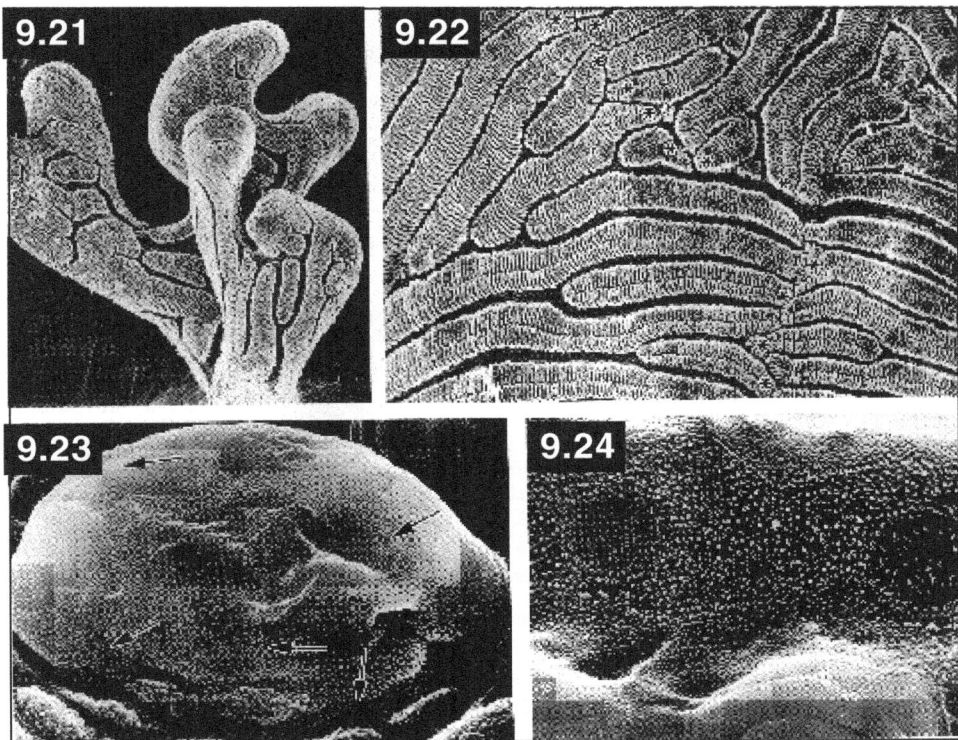

Figure 9.21: Replica of portion of arborescent organ vasculature. Rows of paired lamellae form respiratory islets that over the tissue. SEM, x 30.

Figure 9.22: Respiratory vasculature from suprabranchial chamber. Respiratory islets form long rows of paired lamellae. A "seam" (along *) indicates islets derived from different arches. SEM x 60.

Figure 9.23: Surface of respiratory lamellae from suprabranchial chamber. Microvillous epithelial cells appear partially covered with a mucus-like shroud. Cell junctions are paired microridges (arrows). Note crenate pavement cells at base of lamellae and on non-respiratory surface. SEM x 3,000

Figure 9.24: Interlamellar epithelium from arborescent organ has microvillous epithelium. Round innocyte-like cells with a microvillous borders are also present. SEM x 4,300.

Short afferent and efferent lamellar arterioles connect lamellae to their terminal arteries. Afferent lamellar arterioles are often curled and are somewhat longer than the efferent arterioles. Lamellae have a slightly enlarged outer marginal channel and two to three rows of parallel channels medial to the outer one. Medial to these, the lamellar sinus pathways become irregular and a distinct inner marginal channel is lacking. Islet lamellae are shorter and have fewer channels than those on the gill filaments. On several occasions, the lamellae appeared to split midway across the lamella. In this instance, a single afferent lamellar arteriole and initial lamellar sinus

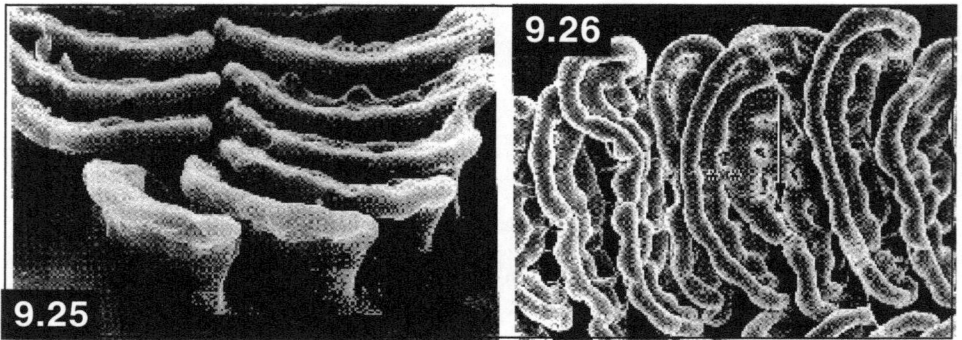

Figure 9.25: Respiratory lamellae at the end of an islet on arborescent organ. The outer border of air-breathing organ lamellae are characteristically folded parallel to the general organ surface. SEM x 770.

Figure 9.26: Replicas of respiratory lamellae on the suprabranchial chamber. Plane of micrograph is parallel to respiratory surface. Pillar cells are arranged in rows forming an outer marginal channel and several medial linear an outer marginal channel(**), thus splitting the efferent side of the lamella into two separate lamellae, drained by distinct efferent lamellar arterioles. Arrow indicate direction of blood flow. SEM x 650.

formed two sinusoides and efferent lamellar arterioles. One split lamella was observed in approximately every two or three islets.

A second circulatory network was also observed in the islets. These vessels were found on the lamellar side of the afferent and efferent arteries. This system resembled the interlamellar system of the gill filament with transverse interlamellar-like vessels connecting longitudinal vessels, the latter similar to the filament collateral vessels. Branches of the collateral-like vessels traveled between the arteries and connected this system to larger drainage vessels that, in turn, traveled on the tissue side of the arteries. Vasculature connections supplying this interlamellar-like system could not be identified.

Surface Architecture

Three general cell types, pavement, mucus, and ionocyte (chloride cell), were selectively distributed in specific areas of the epithelium. This pattern of cell distribution in gill filaments and accessory respiratory organs was found to be quite similar and specific cells were generally restricted to homologous locations on the various tissues. A fourth structure was also observed between the junctional complexes of cells.

The surface structure of pavement cells exhibited one of three general patterns. 1) whorled microridges, 2) short, irregular microridges with occasional microvilli, or 3) short microvilli. Pavement cell type was correlated with its location on the tissue.

Pavement cells with long, whorled microridges were found on the arch support skeleton and in the non-respiratory area bordering the suprabranchial chamber. These cells were round or oblong and 8-10µm in diameter. Frequently, a single microridge, as long as 40-50μm, would form a coiled spiral over the cell surface. Occasional branches in the microridges were also present.

Short microridges with occasional microvilli were common characteristics of pavement cells on the lamellar epithelium (*i.e.* along the afferent and efferent margins) of both arch and fan filaments and on the non-respiratory areas between the respiratory islets on the epithelium of the suprabranchial chamber (Figure 9.27a,b). Microridges at the periphery of the cell were usually longest, upto 10-15 μm and encircled the cell. Centrally located microridges were progressively shorter and randomly oriented. In many instances it was not possible to make a distinction between very short microridges and microvilli (Figure 9.27a,b). In some specimens, pavement epithelial cells in the non-respiratory areas between lamellae on the suprabranchial chamber had a crenate surface and the microridges were short, broad and densely packed. Cell junctions between microridges pavement cells appeared as long, paired ridges, with each cell contributing one ridge.

Pavement cells on all respiratory lamellae were covered with microvilli. Many pavement cells in the space between adjacent lamellae, *i.e.*, the interlamellar epithelium, were also microvillous. Pavement cell microvilli were less dense and had smaller diameters than microvilli often appeared to be partially shrouded in mucus or some similar extracellular matrix. Globular droplets were also observed on the surface of some cells in the interlamellar filamental epithelium (Figure 9.27e). Cell junctions between these cells appeared as paired microvilli or short microridges.

Mucus cells were randomly distributed over much of the epithelium (Figure 9.27). They were most prevalent on the afferent border of the gill filament and in the interlamellar epithelium. Diameter of the aperture of mucus cells on the interlamellar epithelium was the largest (2-6μm) > gill lamellae (1-3 μm) > non-respiratory area peripheral to suprabranchial epithelium (0.5-1.5μm). non-respiratory area of suprabranchial epithelium and arborescent organs (0.5-2 μm). Mucus cells were very seldom observed on respiratory lamellae of suprabranchial epithelium and arborescent organs.

Ionocytes were rare except on gill and fan filaments. A few ionocytes were observed on lamellae, interlamellar filamental epithelium, and afferent filamental epithelium, and afferent filamental borders, shears were most common on the efferent border of the filament (Figure 9.27). Surface diameters of ionocytes were usually greater than those of neighboring pavement cells. Ionocytes were densely covered with microvilli and occasional short microridges. Round cells resembling ionocytes were observed in the interlamellar epithelium of suprabranchial chamber and arborescent organ epithelia. These cells often appeared covered with mucus or glycocalys, had large diameter microvilli, and were bordered by a paired microvillous ring.

In many areas of gill and air-breathing tissue, the space between microridges or microvilli of adjacent pavement cells, or between pavement cells and ionocytes, was filled with fine microvilli or short microridges (Figures 9.27a,b,d,f). Usually only a single row of fine microvilli was present, although occasionally two roows were observed (Figure 9.27b).

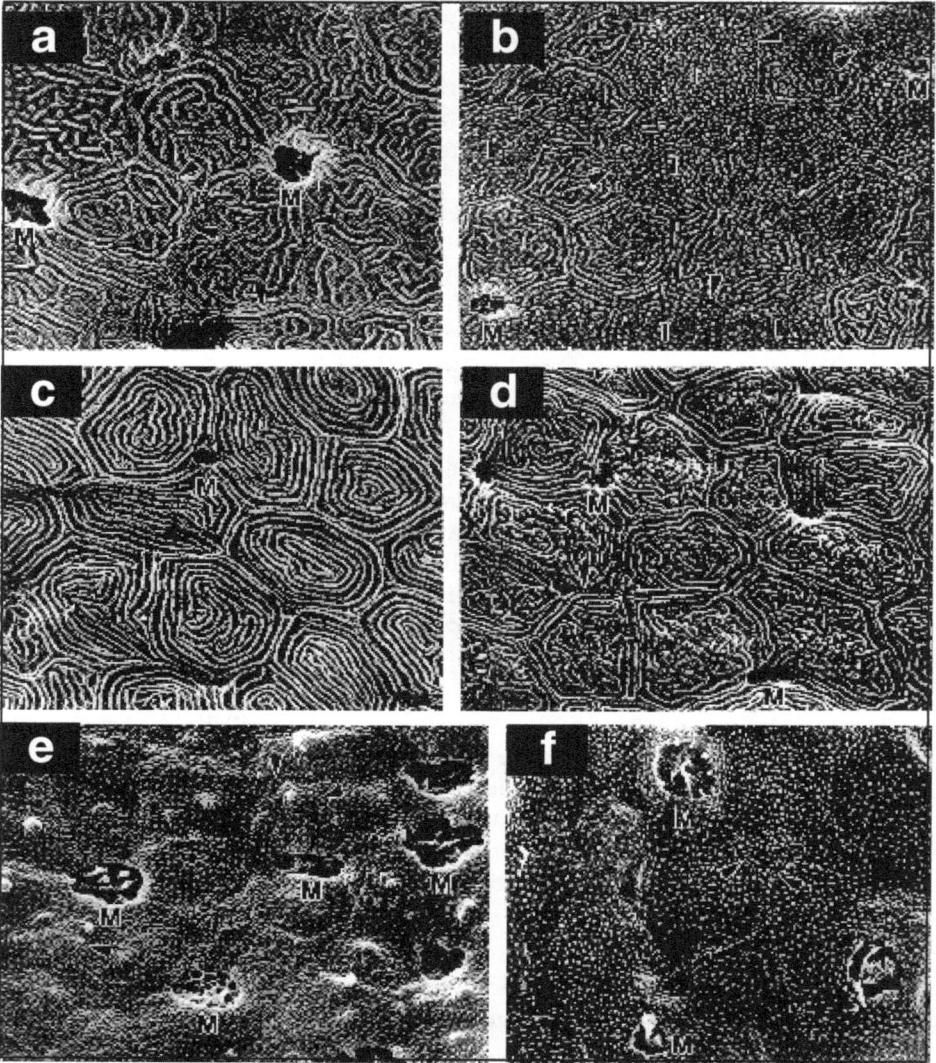

Figure 9.27(a-f): Surface feature of *Clarias* respiratory organ epithelia.

a: Afferent border of filament near tip; b: Efferent border of filament; c: Non-respiratory area peripheral to suprabranchial chamber; d: Afferent border of filament on fan; e: Interlamellar filament epithelium near afferent border of gill filament; f: Gill lamellar surface. Pavement cells are the most common cell type and their membranous projections are either long, whorled microridges (c), short, irregular microridges (a,b,d), or fine microvilli (e,f). Innocytes (I), characterized by dense microvilli are especially prevalent on the efferent border of filament (b). Mucus cell (M) density is greatest in the interlamellar epithelium (e). Note fine microvilli (arrowheads) in intercellular spaces between adjacent pavement cells or between pavement cells and innocytes. SEM a, x 4,300 b-f x 3,000.

Measurements were made from several microridges of the diameter of microvilli or width of microridges.Ionocyte microvilli were generally broadest (0.24 μm) followed by lamellar microvilli (–0.19 μm) and microridges (-0.15 μm). Microridges width appeared slightly greater in cells with shorter microridges. Fine microvilli in the intercellular spaces were around 0.12 μm in diameter. Microridges on crenate epithelial cells of ten exceeded 0.30 μm in width, although it was not clear that these were true microridges.

Discussion

Accessory respiratory organs (ARO) have evolved in a surprising number of fish and are found in varying degrees of complexity. Three physiological problems are encountered by the addition of ARO to a gill-systemic circulation : 1) forced partial unsaturation of mixed oxygenated and deoxygenated blood, 2) trans-branchial loss of oxygen into hypoxic water, and 3) hemodynamic consequences of additional vascular circuitry (Satchel, 1976; Graham, 1994). Although experimental confirmation is limited, there are three anatomical modifications of the macrocirculation that may affect the efficiency of ABO as respiratory organs: 1) organization of "conduit" vessels that connect the heart, gills, ARO, and systemic tissues, 2) axial flow separation of deoxygenated systemic venous blood and oxygenated ARO blood to transit through the heart; and 3) spatial and temporal shunts that allow selective perfusion of gill or ARO (Olson, 1994). The extent to which anatomical modifications resolve the physiological problems determines, to a large extent, the overall efficiency of the ABO in respiratory exchange.

ABO in Clariidae enable these fish to survive in hypoxic water (Bevan and Kramer, 1987; Das, 1927; Ghosh *et al.,* 1990, Jordan, 1976) and they support gas exchange during brief terrestrial sojourns (Das, 1927; Jordan, 1976) and during drought when the gills may be covered with viscus mud (Bruton, 1979). Superficially, Clarias ARO appear to be complex and highly specialized organs. However, at the tissue level, and certainly with respect to any sophistication of the of the circulatory anatomy, these structures are actually quite primitive.

Clarias ARO (gill fans, dendritic organs, and suprabranchial epithelium) are derived from gill tissue and they retain the filament-lamella structural and vascular relationships characteristic of gills with only a few, and generally minor, modifications. The macrocirculation is organized as if ARO tissues were simply additional gill filaments and, with the exception of multiple ventral aortae and a direct connection between the suprabranchial chamber and the dorsal aorta (see below),there is no other apparent specialization. Thus ARO in these fish appear to supplement branchial respiration rather than replace it. However, the ability of these fish to survive a variety of adverse environmental conditions clearly illustrates that the ARO provide a selective physiological advantage.

While there are few similarities in the gross features of the tree-like evaginated dendritic accessory respiratory organs of *Clarias* and the invaginated air sacs of *Heteropneustes,* striking comparisons can be found in the fine structure of the respiratory epithelium and in the organization of the macro- and micro-circulation. Both genera have multiple ventral aortas, direct vascular connections between the ARO and dorsal aorta and ARO retain a filament-lamella structure typical of gill tissues (Olson *et al.*, 1990). This suggests that a prototype air-breathing organ originated in a common silurid ancestor and that the grass structure of the ARO was modified after the genera diverged. These relationships are detailed in the appropriate sections below.

General Vascular Organisation

Gill and ARO circulations in *Clarias* are in parallel with each other and collectively they are in series with the systemic circulation (Nawar, 1955; Munshi, 1961, present study). Because the ARO are extensions of gill filaments, there has been little modification in the overall complexity of the conduit vessels other than to increase their length and spatial shunts appear to be absent. Anatomical modifications to enhance axial blood flow separations in the great veins and during transit through the heart are also lacking. Temporal shunts may occur 1) between different branchial arteries 2) between gills and ARO 3) between outflow pathways to the systemic circulation, or 4) within either gill filament or ARO lamellae. These may functionally replace spatial shunts, which are generally absent from *Clarias*, in modulating gas exchange processes.

The origin of afferent branchial arteries in *Clarias batrachus* more nearly resembles that found in *Heteropneustes fossilis* (Olson *et al.*, 1990) than that reported for *Clarias lasera* (Nawar, 1955). In *C. lazera* the fist pair of afferent branchials originate from a somewhat elongated segment of the ventral aorta, the latter extending from the anterior conical apex of the bulbus (Nawar, 1955). In both *C. batrachus* and H. fossilis (Olson *et al.*, 1990), afferent arteries to the first arches originate from the ventral aspect of the bulbus. In *H. fossilis* (Olson *et al.*, 1990), there is a short common trunk before the vessels bifurcate to form the afferent branchials, whereas separate afferent branchial arteries to the first arch arise directly from the bulbus wall in *C. batrachus*. The situation is reversed with respect to the origin of the second-arch afferent branchial arteries. Nawar (1955) observed that second-arch afferent branchials originated directly from the dorsal surface of the bulbus in *C. lazera*, whereas in both *C. batrachus* and *H. fossilis* (Olson *et al.*, 1990), there is a short common trunk before the vessels bifurcate to form the afferent branchials, whereas separate afferent branchial arteries to the first arch arise directly from the bulbus wall in *C. batrachus*. The situation is reversed with respect to the origin of the second-arch afferent branchial arteries. Nawar (1955) observed that second-arch afferent branchials originated directly from the dorsal surface of the bulbus in *C. lazera*, whereas in both *C. batrachus* (present study) and *H. fossilis* (Olson *et al.*, 1990) they originated from the anterior extension of the ventral aorta. In all three species, the common root of the third-fourth branchial arteries, the common root of the third-fourth branchial arteries arises from the mid-dorsal surface of the bulbus.

Olson *et al.* (1990) hypothesized that the ventral origin of the first afferent branchial arteries of *H. fossilis* might provide some degree of control of flow distribution between and anterior and posterior arches as it appeared that dilation of the latter would mechanically compress the former. A similar situation is also appearent in *C. batrachus*.

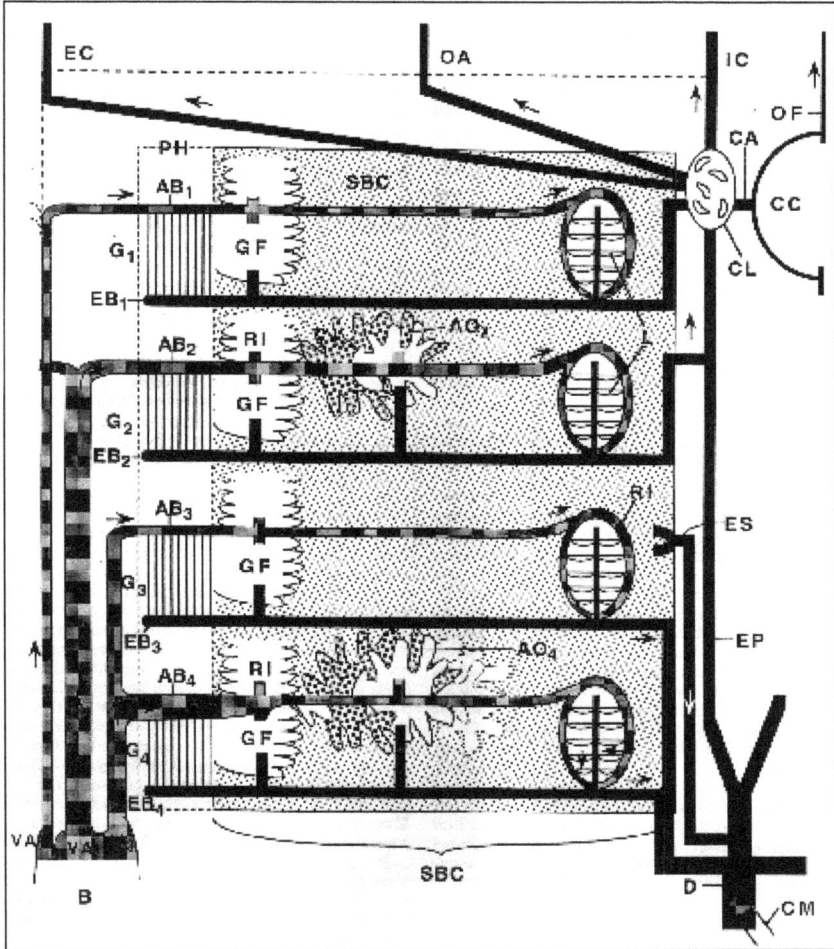

Figure 9.28: Schematic of central respiratory vasculature in *Clarias*.

Redistribution of flow between the arches, *i.e.* a temporal shunt, would be desirable under circumstances that necessitate increased aerial gas exchange because the posterior branchial arteries of *C. lazera* are so different. Clearly, the earlier observations of Nawar (1955) need to be confirmed with modern corrosion methods.

There is a direct connection in *Clarias* between the efferent arterial circulation of the suprabranchial chamber and the first-second arch epibranchial arteries. In *C. batrachus* this connection is slightly posterior to the anastomosis of the epibranchials from either side of the head, whereas it is anterior to the anastomosis in *C. lazera* (Nawar, 1955). Thus, unlike the gills fans and dendritic organs, whose efferent arteries drain into the efferent branchial artery of the gill arch, arterial efferent from the suprabranchial organs drains into the efferent branchial artery of the gill arch, arterial effluent from the suprabranchial organs drains directly, or nearly so, into the dorsal aorta. This connection may serve as a low resistance conduit from the respiratory islets of the suprabranchial chamber. This connection may take on additional significance when the fish is in hypoxic water or viscus mud. Presumably, under these circumstances, the most dorsal regions of an air-filled buccopharyngeal cavity will be less likely to become in contact with water or mud and respiration may be sustained across these surfaces. A direct connection between the respiratory islets of the suprabranchial chamber and the dorsal aorta would thereby serve as a physiological shunt enhancing delivery of oxygen to the dorsal aorta.

The efferent arterial drainge of accessory respiratory organs of *Clarias* also shows some homology to that of *H. fossilis* in spite of the structural differences in respiratory organs. In a previous study of *H. fossilis* we (Olson *et al.*, 1990) observed direct connections between the efferent air-sac artery and the dorsal aorta in the distal region of the air-sac. This is functionally identical to the above-mentioned low-resistance conduit vessels in *Clarias* and may similarly benefit *H. fossilis* respiration in hypoxic or muddy environments.

Gill Circulation

All four gill arches of *Clarias* are highly developed and appear well equipped for aquatic respiration, a fact borne out by the ability of the fish to survive indefinitely in normoxic water (Jordan, 1976; Munshi *et al.*, 1976). Lamellae are present on all filaments and their vascular morphology is generally indistinguishable from that found in water-breathing teleosts (Olson, 1991). A continuous, and comparatively large-bore, outer marginal channel undoubtedly serves as a preferential pathway for red cells. The linear arrays of pillar cells parallel to the outer marginal channel provides an additional three to five channels of flow across the lamella. Pillar cells medial to these channels are less ordered and flow becomes more sinusoidal. As in other fish (Olson, 1991) the loss of lamellar channelization occurs as the free lamella becomes embedded in filamental tissue. This random medial pathway, anatomically unsuitable for gas exchange but important in plasma hormone metabolism (Olson, 1991), may be less accessible to red cells and thereby perfused by plasma rich blood.

The absence of a continuous large-bore inner marginal channel in gill lamellae indicates that there are no special provisions for intralamellar shunting of red cells away from the outer borders of the lamella when the fish is in oxygen depleted water. In addition, there are no lamellar bypass shunts in the gill filament that might perform a similar function. If *Clarias* has any ability to prevent oxygen loss from the blood while in hypoxic water it will have to be accomplished either through gill vasoconstriction or dilation of ARO vessels.

Because respiratory islets appear to be only slightly modified gill filaments that are displaced from the gill arch, it seems unlikely that vascular innervation, or receptor populations, of the two tissues would be different enough to selectively regulate blood distribution between the gill filaments and ARO. However, we propose that temporal shunting of blood from gills to ARO can be accomplished through local hypoxic vasoconstriction, without the need for intricate control systems. It has been shown in water-breathing teleosts, such as *Oncorhynchus mikiss* and cod (*Gadus morhua*), that hypoxia produces a direct myogenetic vasoconstriction of gill vessels (Ristori and Laurent, 1977; Petterson and Johansen, 1982). Furthermore, this effect can be elicited by reducing PO_2 in either perfusate or ambient water (Petterson and Johansen, 1982). We propose that when *Clarias* is in hypoxic water, gill vascular resistance is increased, thereby decreasing blood flow through the gill filaments and reducing oxygen loss to the environment. However, vascular resistance will not increase in the respiratoy islets because the ARO are exposed to normoxic air in the suprabranchial chamber. Because gill and islet circulations are in parallel, an increase in branchial resistance will divert blood flow to the islets, thereby effecting a temporal, and physiologically beneficial shunt. Temporal shunts could also occur in areas of the ARO that are under ventilated or covered by hypoxic water or mud. A similar mechanism probably occurs in other bimodal breathing fish where spatial shunts are also present (Olson, 1994).

Vessels in the body of the gill filament arise from the post-lamellar circulation and therefore can not form a lamellar bypass. Recent evidence (reviewed in Olson, 1991) indicates that there are two distinct circulatory networks in the filament body of many teleosts and this is evident in *Clarias* gills. These two circulations have been tentatively labeled as the nutrient pathway and the interlamellar circulation (Olson, 1991), although their vascular connections and physiological functions are still incompletely understood. Corrosion replicas of *Clarias* gills shed further light on these vascular pathways.

Filamental nutrient vessels are narrow-bore, capillary-sized, vessels often found closely associated with interlamellar vessels. Nutrient vessels may be supplied by small arteries originating from the efferent filamental artery branchial artery (Olson, 1991). These arteries are formed from the condensation of numerous tortuous arterioles typical of the so-called secondary circulation (Vogel, 1985). Nutrient vessels in the catfish (*Ictalurus punctatus*) are drained from the filament by a series of veins that are distinct from the collateral vessels of the interlamellar system (Boland and Olson, 1979). A similar arrangement appears in *Clarias* gills. No anastomoses were evident between nutrient vessels and the interlamellar system in *Clarias* filaments.

The interlamellar system is quite likely a low pressure network and can be greatly distended by elevated perfusion pressure, as has been shown experimentally in the trout (Olson, 1983). Narrow diameter feeder vessels supplying the interlamellar system appear to be designed to step-down pressure from the efferent filamental artery. This process may occur in the first few μm along the feeder where the vascular replicas appear to undergo an abrupt transition from a striated, arterial-like endothelial cell relief to an amorphic surface similar to that of the interlamellar vasculature. Sphincter-like constrictions at the junction of the feeder vessel and the interlamellar vessels indicate that further regulation of perfusion pressure of flow is possible.

Interlamellar vessels form a lattice network that encircles both afferent and efferent lamellar arterioles near the origin of the arterioles from their respective filamental arteries. This arrangement is common in teleosts (Olson, 1991) and elasmobranches (Olson and Kent, 1980), but its significance is unknown. It has been suggested that distension of the interlamellar system may influence lamellar perfusion through its effects on transmural pressure in lamellar arterioles (Smith, 1977; Boland and Olson, 1979). It is not clear if the pressure generated in the interlamellar system is sufficient to affect the comparatively high pressure in the arterioles. Alternatively, vasoactive substances formed in the interlamellar vasculature may readily diffuse across the thin vessel wall and directly affect lamellar arteriolar resistance.

Vasculature of Accessory Respiratory Organs

Vascular organization of the ARO is virtually identical to that of the gill filament with only a few minor modifications. Lamellae that would be on opposite sides of a gill filament are now oriented in parallel rows to accommodate the planar orientation of the ARO epithelium. To accomplish this, the gill filament has been unfolded along the axis of the efferent filamental artery and thus two afferent filamental arteries are formed for each respiratory islet (Munshi, 1961). Air sac islets of *Heteropneustes* are similarly arranged (Olson *et al.*, 1990). Gill filamental arteries directly to lamellar arterioles, whereas their equivalent in the ARO bifurcate several times before terminating and may supply lamellae along much of their length. This branching accounts for considerable variation in the length of adjacent islets. Anastomotic connections between adjacent efferent arteries are also unique to islet vessels. Their function is not clear.

While the respiratory vasculature of islet tissue is as extensive as that found in the gill filament, the nutrient and interlamellar vasculature is considerably attenuated. Nutrient supply to islets appears to be provided by subepithelial vessels, where as the reduction in interlamellar vasculature may correlate with a general decrease in exposure of this epithelium to an aquatic environment. Although the function of the interlamellar circulation has not been resolved in any fish, it is thought to have a role in transbranchial exchange of water or electrolytes (Laurent, 1984; Payan *et al.*, 1984; Olson, 1991).

Islet lamellae are considerably smaller than those on the gill filament, yet the pattern of pillar cell channelization of lamellar blood flow in the outer margin is preserved. Gill lamellae are generally flat and nearly perpendicular to the filament surface. Islet lamellae are folded almost ninety degrees and it appears that the fold of one lamellae may support the free margin of its neighbour. This would prevent lamellae from collapsing onto the non-respiratory epithelium due to surface tension generated at the air-water interface of the ARO. Thus both lamellar surfaces remain available for gas exchange. Lamellar folds have also been observed in islet lamellae of *Heteropneustes* (Olson *et al.*, 1990) where they probably perform a similar function.

Surface Ultrastructure

Gill epithelium is comprised of three main cell types pavement, mucus, and ionocytes (Laurent and Dunel, 1980; Laurent, 1984). These cells are also found on gill and accessory respiratory epithelia of *Clarias* (Lewis, 1979a,b; Hughes and Munshi, 1973, 1979, present study), although there are several variations.

Pavement cells with long, whorled, microridges are epidermal cells characteristic of fish skin and non-respiratory surfaces (Olson, 1995) and are also observed some distance from the respiratory surfaces in *Clarias* (Lewis, 1979a,b; Hughes and Munshi, 1973, 1979). Closer to the respiratory surfaces, *i.e.* on the gill filament body and in the areas between respiratory islets, there is a reduction in microridges length and the whorled pattern becomes less apparent.

An interesting modification is found in the external membrane of pavement cells from both gill and accessory respiratory organ lamellae. Here, microridges are for the most part, absent and the epithelium is studded with numerous microvilli (Lewis, 1979a,b). A microvillous epithelium found in a variety of water-breathing teleosts (Olson, 1995) and Lewis (1979a,b) suggested its presence may be correlated with the unique mode of *Clarias* respiration. Interestingly, Hughes and Munshi (1979) observed that lamellae of *Heteropneustes fossilis* were also covered with microvilli, whereas Rajbanshi (1977) reported that this fish has microridges on its lamellar surface. The respiratory surfaces of other bimodal breathing fish are also modified; microridges and microvilli are completely absent from the smooth ARO respiratory epithelium of *Anabas testudineus* and *Monopterus cuchia* (Munshi *et al.*, 1986, 1990). While the transition from microridges to a microvillous or smooth respiratory epithelium may be correlated with some property of epithelial function, it is doubtful if this is solely an adaptation for bimodal breathing. Microvillous epithelia are also found on gill lamellae of obligate aquatic breathing catfish such as *Ictalurus punctata* and *I. melas* and gill lamellae of some pelagic fish, such as the bluefish *Pomatomus salarix* and Atlantic mackerel *Scomber scombrus* lack both microridges and microvilli (Olson, 1995). Although additional studies on other Siluriformes are needed, it may be that a microvillous epithelium evolved as a general feature of the Siluriformes and in that case may have served a more primitive function, unrelated to aerial respiration.

The fine microvilli found in the intercellular surface between epithelial cells may be from replacement cells that are about to emerge on epithelial surface, or they may be a different, these microvilli have not been previously described in other SEM studies of fish epithelia, including two previous studies of *Clarias* (Lewis, 1979a,b).

It is not known if the topography of crenate-like islet epithelial cells bordering the lamellae is an artifact or indicative of a specific physiological function. Similar cells have been observed on gill lamellae of the air-breathing mudskippers *Periopthalamus chrysospilos* and *Boleopthalamus boddaerti* and they have been proposed to enhance mucus retention while the gills are exposed to air (Low *et al.*, 1988).

However, if they serve a similar function in *Clarias* they would retain mucus at the base of the ARO lamellae, which was not observed, nor could they account for the mucus-like shroud over the lamellar surface.

Vasculature of the Head and Respiratory Organs in an Obligate Air-breathing Fish, the Swamp Eel *Monopterus* (=*Amphipnous*) *cuchia*

Methyl methacrylate vascular corrosion replicas were used to examine the macro-circulation in the head region and the micro-circulation of respiratory vessels in the air-breathing swamp eel *Monopterus cuchia*. Fixed respiratory tissue was also examined by SEM to verify capillary orientation. The respiratory and systemic circulation are only partially separated, presumably resulting in supply of mixed oxygenated and venous blood to the tissues. A long ventral aorta gives rise directly to the coronary and hypo-branchial arteries. Two large shunt vessels connect the ventral aorta to the dorsal aorta, whereas the remaining ventral aortic flow goes to the respiratory islets and gills. Only two pairs of vestigial gill arches remain, equivalent to the second and third arches, yet five pairs of aortic arches were identified. Most aortic arches supply the respiratory islets. Respiratory islet capillaries are tightly coiled spirals with only a fraction of their total length in contact with the respiratory epithelium. Valve-like endothelial cells delimit the capillary spirals and are unlike endothelial cells in other vertebrates. The gills are highly modified in that the lamellae are reduced to a single-channel capillary with a characteristic three dimensional zig zag pathway. There are no arterio-arterial lamellar shunts, although the afferent branchial artery supplying the gill arches also supplies respiratory islets distally. A modified interlamellar filamental vasculature is present in gill tissue but absent or greatly reduced in the respiratory islets. The macro and micro-circulatory systems of *M. cuchia* have been considerably modified presumably to accommodate aerial respiration. Some of these modifications involve retention of primitive vessel types, whereas others, especially in the micro-circulation, incorporate new architectural designs some of the whose functions are not readily apparent.

The swamp (mud) Eel *Monopterus (=Amphipnous) cuchia* is an obligatory air breathing fish found in the Indian subcontinent and Burma (Hora, 1935). It inhabits holes and crevices in the muddy banks of swamps, lakes, and slow running rivers containing fresh and brackish waters. During the dry summer months the fish escapes desiccation by burrowing into the mud (Hora, 1935). The main respiratory organs are the highly vascularized epithelium of air sacs and the buccopharynx. In addition, the skin also plays an important role in gas exchange (Mittal and Munshi, 1971). Although it is an obligatory air breather (Lombolt and Johansen, 1976), this fish retains the piscine mechanism of ventilating water through the buccopharynx and rudimentary gills on the second and third gill arches (Munshi and Singh,'68; Lomholt and Johansen, 1974).

A variety of anatomical modifications are found in air-breathing fish that serve to enhance oxygen extraction directly from the air (Johansen, 1970; Munshi, 1985). These are accompanied to various degrees by restructuring the vascular system. This is seen not only in the micro-circulatory anatomy of the gas exchange organs but also in regard to the general organization and relationships of conduit vessels that supply the gills, accessory respiratory organs and systemic tissues.

Recent application of methyl methacrylate vascular corrosion techniques in the analysis of vessels from the air-breathing fish *Anabas testudineus* (Munshi *et al.*, 1986; Olson *et al.*, 1986b) has illustrated the advantages of producing a three-dimensional replica of major vessels and capillaries. With this technique the relationships of previously described vessels can be verified, better visualized, and new pathways, especially at the capillary level, can be examined in detail with the scanning electron microscope (SEM). The present study employs this technique in analysis of conduit vessels and respiratory epithelia of *M. cuchia*. There are two general goals: 1) to provide a better understanding of the cardio-vascular adaptations that accompany air breathing and 2) to provide anatomical reference, which can be used in the further systematic studies of synbranchiforms (Rosen and Greenwood, 1976).

Materials and Methods

Animals

Live specimens of *M. cuchia* were purchased from fishermen near Purnea in North Bihar, India, and transported to the University of Bhagalpur. They were maintained in 1,000 litre plastic pools and fed chopped goat liver and live earthworms.

Corrosion Replicas

Methyl methacrylate corrosion replicas were prepared as described previously (Olson, '85). Briefly, a fish was anesthetized in MS 222 (ethyl-aminobenzoate) and the ventral aorta was cannulated. Phosphate-buffered Ringer solution, containing 100 USP/ml heparin, was infused at physiological pressure (30 mm Hg) until fluid flowing from the sinus venosus was devoid of red cells. Methyl methacrylate (Mercox) was mixed with catalyst and infused (also at 30 mm Hg). At the onset of polymerization the ventral aorta was clamped and the fish placed in 60°C tap water for 2 hr to ensure complete polymerization. The carcass was macerated in alternate rinses of 20 per cent NaOH, water, and 3 per cent HNO_3. When free of tissue, replicas were rinsed in distilled water and air dried. Fifteen specimens were examined.

The macrovasculature was examined under a stereomicroscope and photographed with a Polaroid MP-4 copy camera. Areas selected for further examination were severed from the preparation, using a hot wire, and mounted on an SEM stub with silver paste. They were then coated with carbon and gold and examined with a JEOL T 300 scanning electron microscope (SEM).

Air-Sac Tissue

Eight specimens of *M. cuchia* were anesthetized in MS 222. The mouth was irrigated with, tap water for 5-10 min and then with 2-5 per cent glutaraldehyde in 0.1 M. phosphate buffer (pH 7.4). The head was then severed and placed in cold (4°C) 2.5 per cent, glutaraldehyde for an additional half hour. The respiratory epithelium was then dissected from the head and fixed for an additional hour in 6 per cent glutaraldehyde, post fixed in 1 per cent OsO_4 (1 hr), dehydrated in a graded alcohol series (to 70 per cent), and stored at 4°C until further processing. Prior to examination, the tissue were dehydrated in 100 per cent alcohol and dried using the critical point method with CO_2 substitution. The epithelia were mounted, mucosal side, and examined with the SEM. In some preparations the surface was gently scraped with a scalped blade prior to coating in order to expose the respiratory vessels.

Results

The relative positions of the air-sac and rudimentary gill arches in the head of *M. cuchia* are shown in Figure 9.29. Schematic summarise of the arterial and venous pathways in the head region. References to distances between structure are made with respects to a mature, 50-60 cm fish.

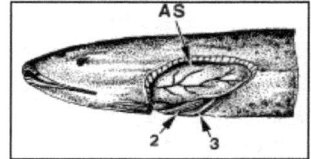

Figure 9.29: Diagram of head of *M. cuchia* showing relative position of air-sac (AS) and Second (2) and third (3) gill arches.

Macrocirculation

Arterial Vessels

The heart is situated far behind the head, around 10-11 cm from the tip of the snout. At the point of transition from the bulbus arteriosus to the ventral aorta, the vessel makes an abrupt ventral turn. A second bend, almost a fold in the vessel, redirects the aorta along its anterior course. A single coronary artery arises from the ventral surface of the ventral aorta in the region of the bend and after dividing into two main branches turns caudad to supply anteriorly, forming a second oblong, or ellipsoid, vascular loop. The ventral aortic branches do not reunite after their initial bifurcation.

The first vessels to arise from the hyo-mandibular trunk are the first arch branchial arches. These vessels run latero-posteriorly, then dorsally, to become the main air sac artery. The first gill arch does not bear gill filaments.

Distal to the origin of the first afferent branchial arteries, the hyo-mandibular trunks give rise to the comparatively small hyoidean arteries. The main vessels continue anteriorly as the mandibular arteries. Hyoidean arteries ramify over the hyoidean region of the pharynx; the latter bulges out in the intact fish to give the overall impression of a sac-like structure in front of the main respiratory air-sac. Hyoidean arteries supply the respiratory islets in this area. The mandibular arteries then turn dorso-laterally and divide into the mandibular and maxillary arteries. Small glossal arteries also arise from the mandibular artery at the point of flexure of the latter. Glossal arteries supply the respiratory islets in the anterior floor of the buccopharynx.

The dorsal aorta extends as a single vessel from the anastomosis of the shunt vessels into the tail of the fish. A small vessel arises from the right aortic wall immediately distal to the origin of the dorsal aorta. This vessel then branches and the posterior branch crosses over to the left side of the aorta this branch is broken in. Both branches travel anteriorly medial and slightly dorsal to the anterior cardinal veins, as paired lateral dorsal aortae. The coeliaco-mesenteric artery also arises from the

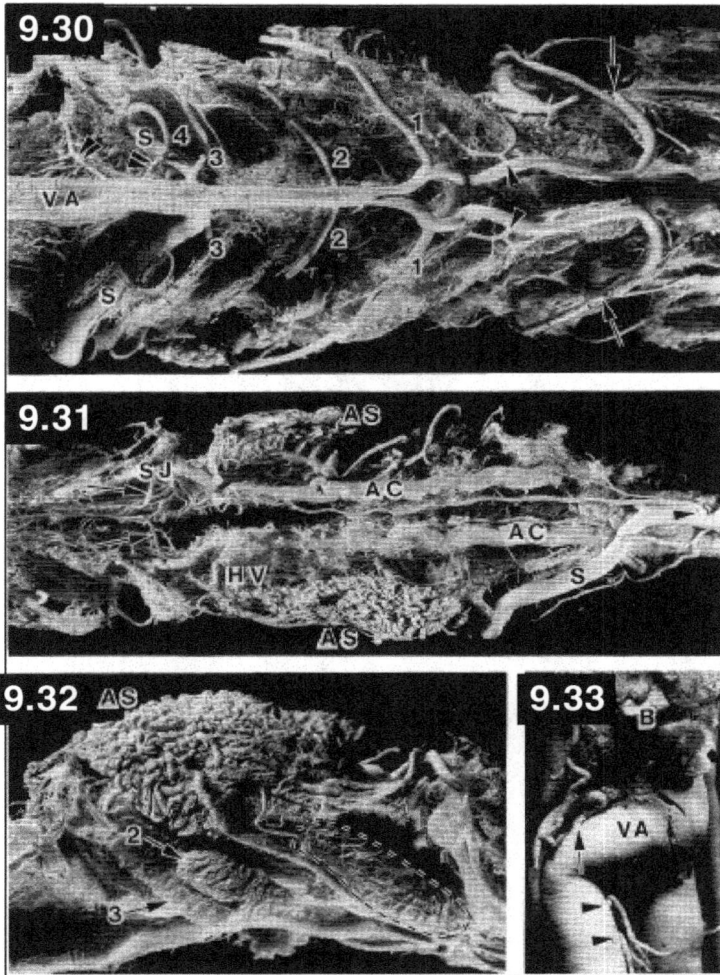

Figure 9.30: Ventral view of major arterial vessels in the head anterior in right, Arabic numerals indicate gill arch vessels or its equivalent, 8, shunt vessels, VA, ventral aorta, arrowheads, origin of hyoidean artery, arrow, bifurcation forming mandibular and maxillary arteries, double arrowheads, hypopharyngeal a X 3.

Figure 9.31 : Major arteries and veins in dorsal head; dorsal view, anterior to left. Right air sac has been partially removed to show air sac vein. AS, air sac, AC, anterior cardinal v; arrowhead, origin of lateral dorsal aortae (left lateral dorsal aorta has been broken), arrows, carotid a, X 2.

Figure 9.32: Lateral view of respiratory vasculature in head, anterior to right. Arrows point to second (2) and third (3) gill arches. First branchial artery supplies air sac (AS). Area enclosed within dashed line indication the hyoidean "air-sac". Note hyoidean artery on ventro-lateral surface of hyoidean air sac. X 2.5

Figure 9.33: Origin of ventral aorta (VA) from bubbus arteriosus (B). Arrow, origin of coronary artery, arrowheads, origin of hypobranchial arteries. Right lateral view, anterior loward bottom of page X 5.

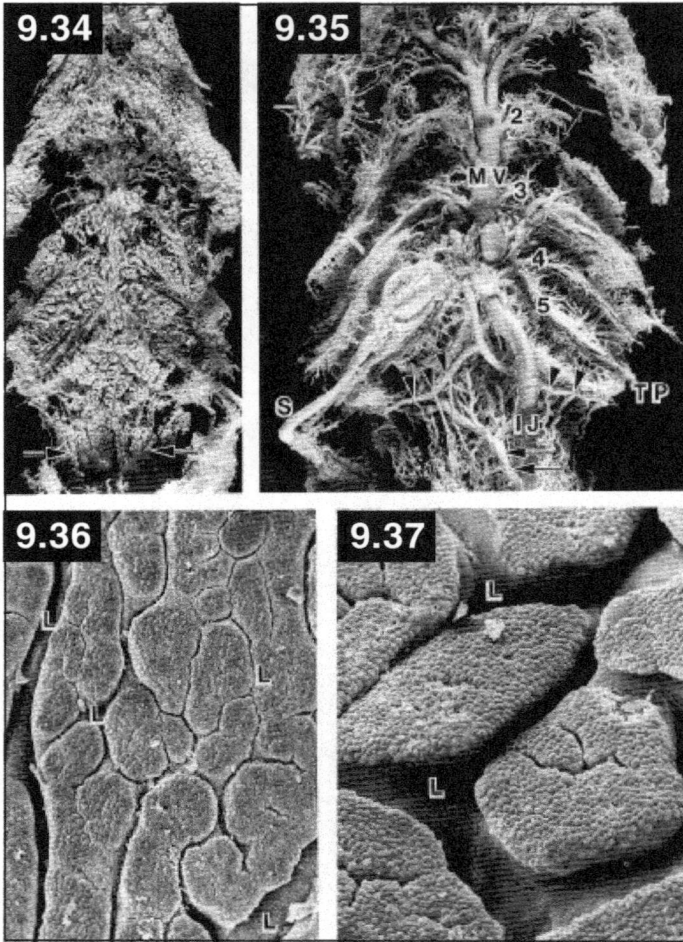

Figure 9.34: Dorsal view of floor of buccopharyngeal cavity; anterior to top of page. The vascular replica shown the density of respiratory vessels. Note respiratory islet vessels deep in the hypopharyngeal region (arrow) X 2.

Figure 9.35: Venous system in floor of buccopharyngeal tissue, ventral view, anterior to top of page. Arterial system has been removed (compare with Figure 9.30, which is same replica with arteries present). MV, median vein in ventral buccopharyngeal chamber. Arabic numbers (2-5) indicate venous drainage from corresponding gill arches or their equivalent. LJ, inferior jugular v; TP, tooth pad vessels from fifth ceratobranchial arch, 8, shunt vessel. Branches of the hypobranchial artery (arrowheads) course parallel to the corresponding veins that ultimately drain into the hypobranchial vein (arrow). X 3.5

Figure 9.36: Respiratory epithelium of air sac. Narrow and wide lanes (L.) of non-respiratory epithelium delineate the respiratory islets. X 136.

Figure 9.37: Respiratory epithelium of buccopharynx. Pebbled surface of respiratory islets results from underlying respiratory capillaries L., lanes of nonrespiratory epithelium. X 200.

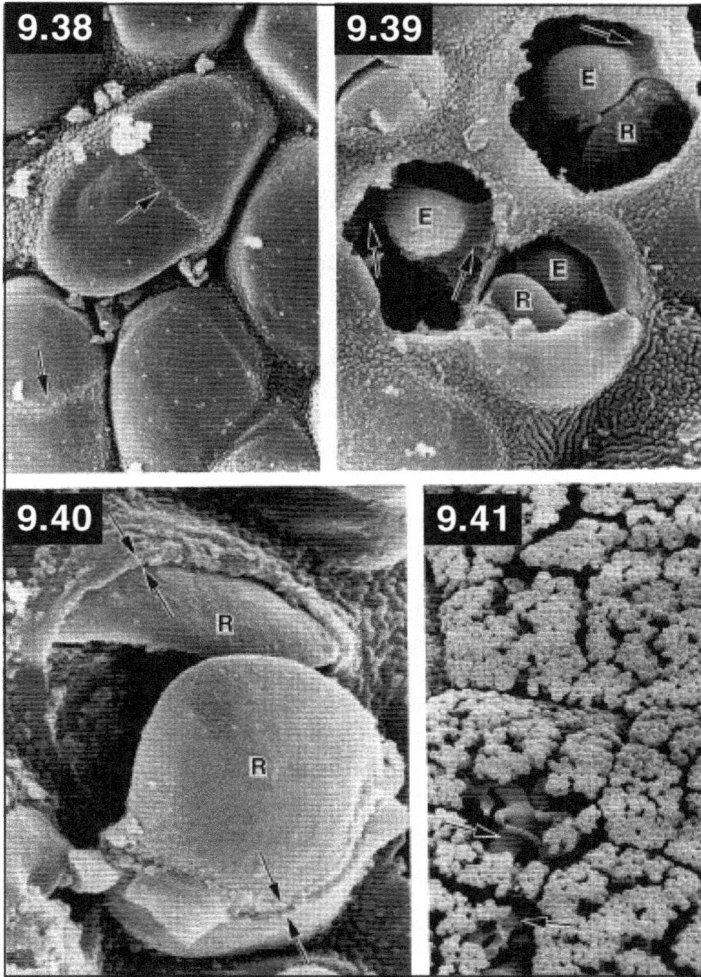

Figure 9.38: Higher magnification of respiratory epithelium. Smooth epithelial cells lie over respiratory capillaries, whereas, microvilla cells are seen in intercapillary areas. Note microvilla junction between smooth cells (arrows). Some debris also adhere to surface X 4300.

Figure 9.39: Smooth epithelial cells removed to expose respiratory capillaries. Valve-like endothelial cells (E) span the capillary forming hairpin loops. These cells are bulbus in their center and attach to, and form the capillary wall with cytoplasmic extension (arrow). Red blood cells (R) are forced over the endothelial cells and pass close to the epithelium. Note microridged and microvillous epithelial cells in lower right corner. X 4,200

Figure 9.40: Endothelial "valves" force red blood cells (R) to take centrifugal path at capillary loop, insuring minimal diffusion distance between red cell and air (arrow). Smooth epithelial cells partially removed. X 9,800.

Figure 9.41: Corrosion replica of respiratory islet capillaries, mucosal (epithelial) side. Several distributing vessels are viable beneath capillaries (arrow). X 160.

ventro-lateral aspect of the dorsal aorta in this region. Other systemic arteries arise periodically along the length of the dorsal aorta.

Venous Vessels

The major venous drainage in the head region is from the respiratory islets. The major veins are schematically illustrated. The superior jugular vein drains blood from the bucco-pharyngeal respiratory epithelium via mandibular, maxillary (not shown), hyoidean, and air sac veins and the inferior vena cava. Identification of many of these connections resulted in destruction of the replicas; hence micrographs are not presented. Central to its anastomosis with the air sac vein, the left superior jugular vein joins with the left hypoglossal vein and then drains into the anterior cardinal vein. The right hypoglossal vein joins the anterior cardinal vein close to the heart. Thus there is a distinct hypoglossal venous system in *M. cuchia*. Respiratory islets in the glossopharyngeal region are drained via progressively larger vessels into the hypoglossal veins and from there into the superior jugular veins.

Respiratory islets in the mandibular region and floor of the buccopharyngeal chamber drain into a large median vein centrally situated in the floor of the buccopharynx. The median vein also drains small veins from anterior portions of the mandible and veins from branchial arches 2,3,4 and 5. The fifth branchial vein also drains the tooth pad of the ceratobranchial arch. The median vein then joins the left inferior jugular vein. The hypopharyngeal vein drains the hypopharyngeal respiratory islets via long veins that course parallel to arterial branches from the hypopharyngeal artery. This vein also drains into the left inferior jugular vein. The left inferior jugular course posteriorly some distance before anastomosing with the anterior cardinal vein. The anterior cardinal veins empty into the ductus Cuvier.

Microcirculation

Respiratory Islets

The respiratory islets consist of clusters of raised gas exchange units. The basic pattern is similar in the air sac and buccopharyngeal epithelium. Between the islets are non-respiratory "lanes". The surface of each islet is covered with several hundred rounded epithelial cells that at higher magnification are shown to be associated with underlying capillaries. The epithelium immediately over the capillary is smooth, whereas adjacent epithelium has characteristic microridges or microvilli. Frequently the junctions between two smooth epithelial cells also contain fine microvilli. Removal of the smooth epithelial cells exposes the capillary loops. Each loop is centered around a "valve like" endothelial cell. This type of endothelial cell is characteristically bulbous in its center and extends across the capillary lumen. It is attached to the lateral margins by cytoplasmic extensions and flanges remainiscent of pillar cell flanges. The endothelial valves appear to force red blood cells to come into close contact with the endothelium on the respiratory side of the capillary, thus minimizing oxygen diffusion distance. When viewed from the mucosal (epithelial) side, corrosion replicas

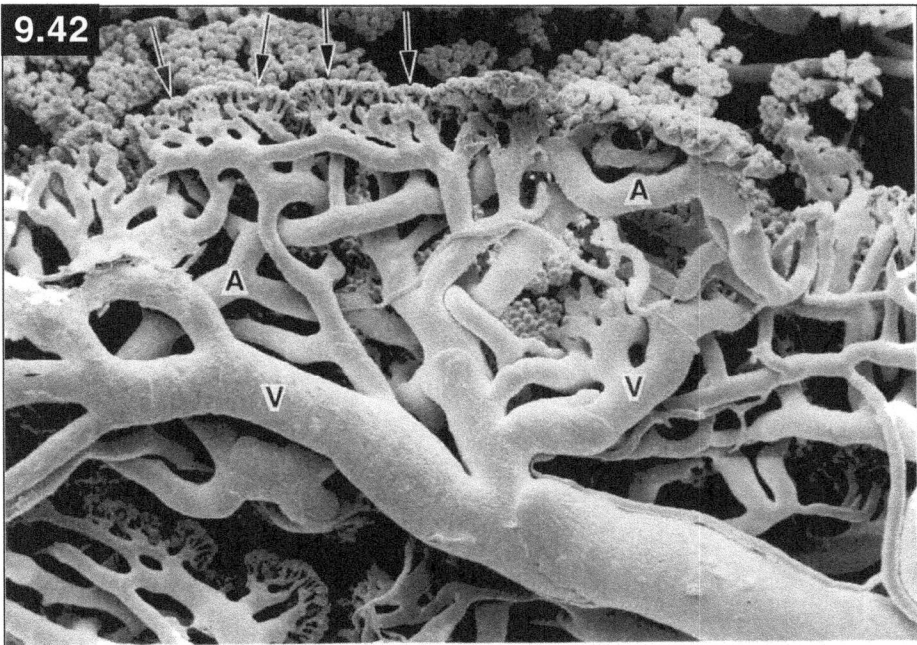

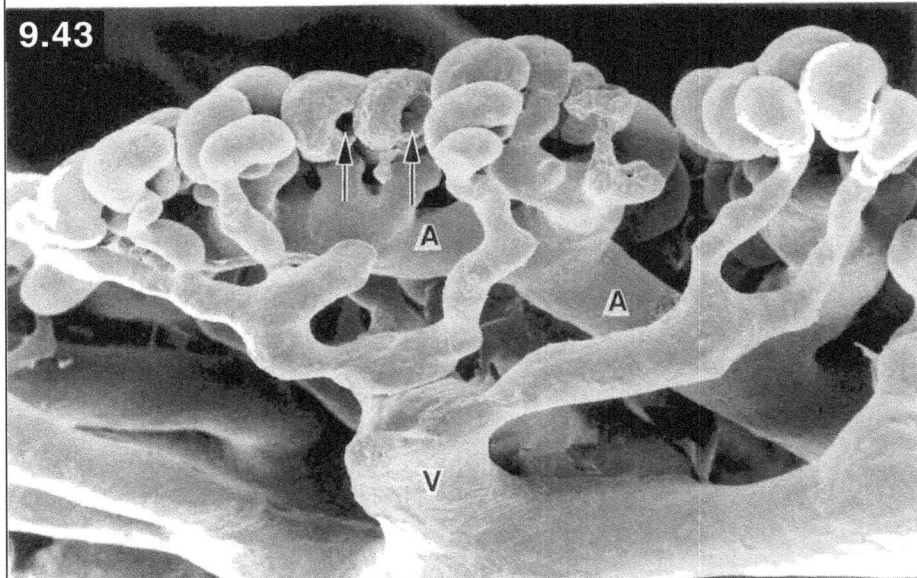

Figure 9.42: Vascular replica of respiratory epithelium seen in transverse view. Large artery (A) and veins (V) interwine to form a vascular lattice supplying the respiratory capillaries (arrow) x 140.

Figure 9.43: Respiratory capillaries are supplied by a central artery (a) and spiral radially to be drained by a peripheral vein (v). Note axial holes in capillaries produced by endothelial "valves" (arrow) x 1,020.

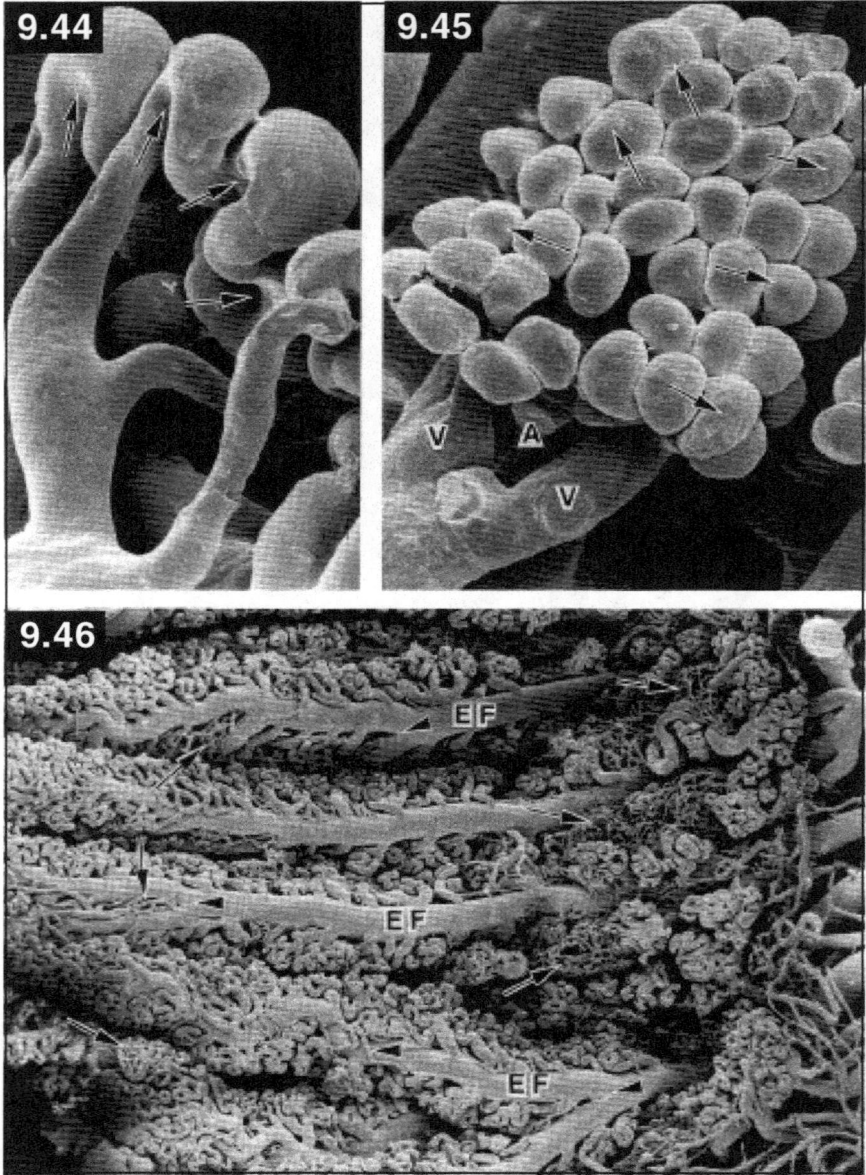

Figure 9.44: Spiral nature of respiratory capillaries. Arrows indicate position of endothelial "valve" x 1,950.

Figure 9.45: Surface appearance of respiratory rosette vasculature from pharyngeal epithelium. The central artery (A) supplies respiratory capillaries which radiate outward (arrows) toward peripheral veins (V). x 1,050.

Figure 9.46: Vasculature of anterior-ventral portion of second gill arch EF, efferent filamental arteries, arrowheads, anastomoses of afferent filamental arteries, arrows, non-respiratory venous-like vasculature. X 45.

Figure 9.47: Vasculature of the posterior-dorsal position of second gill arch (of Figure 9.46). Note extensive web of respiratory venous like capillaries (arrow). x 60.

Figure 9.48: Relationship between afferent (AB) and efferent (EB) branchial arteries in second of gill arch. The AB enters the arch and passes beneath the EB. The EB is formed from efferent filamental arteries (arrows) x 30.

Figure 9.49: Side view of pairs of second arch gill filaments viewed perpendicular to plane of arch and afferent (AB) and efferent (EB) branchial arteries. Long arrows, direction of flow in afferent filamental arteries, short arrows, direction of flow in efferent filamental arteries. Note small venous like vessel along afferent filamental artery (arrowhead). X 70.

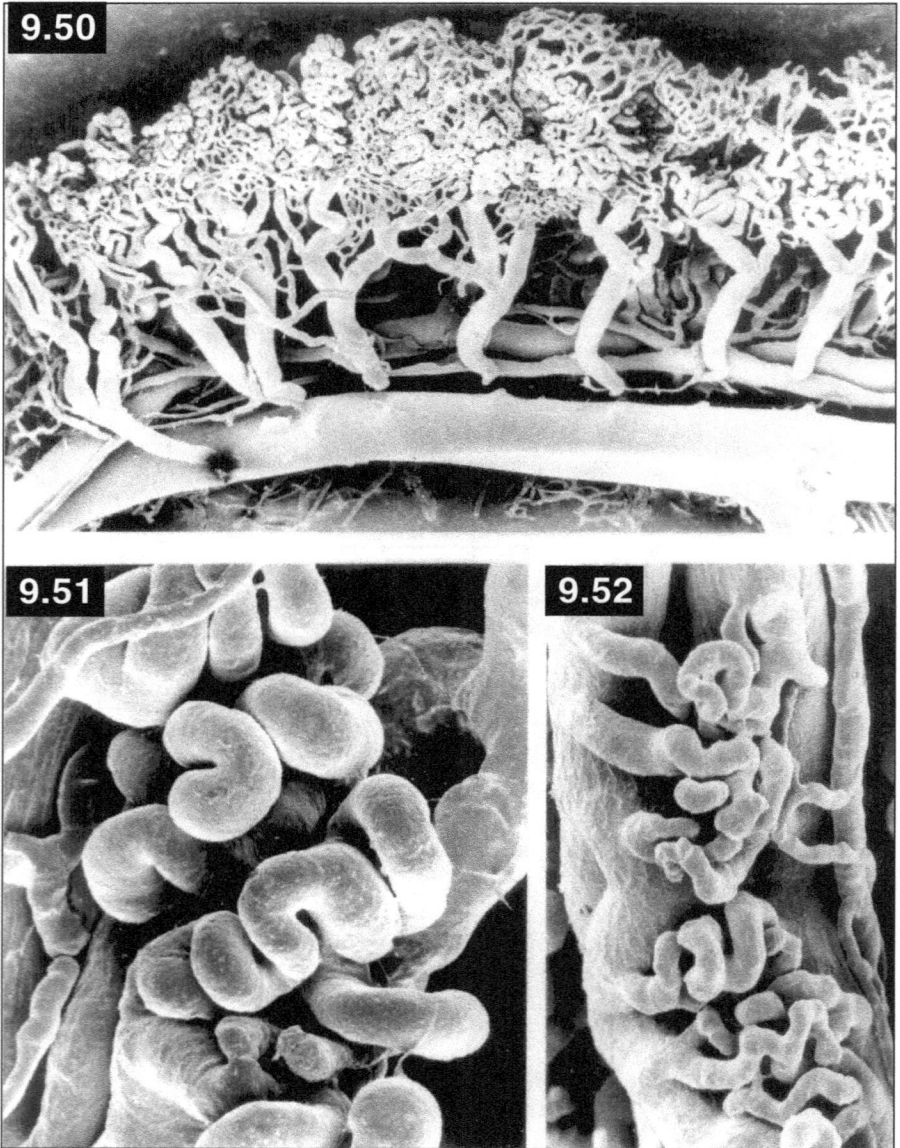

Figure 9.50: Capillary like lamellae of second arch filament. Blood flows from afferent filamental artery (left) through capillaries to afferent filamental artery (right). Note small venous-like vessel along afferent filamental artery (arrow-heads). X 300.

Figure 9.51: Higher magnification of lamellar capillaries from second arch filament. Solid lines indicate peripheral (outer marginal) loops of one capillary that apposed to epithelium. X 640).

Figure 9.52: Vessels of thirds arch filaments. Afferent branchial (AN) artery lies beneath efferent branchial (EB) artery. Note branching of afferent filamental arteries (arrowheads) and extensive venous like network (arrows). X 45.

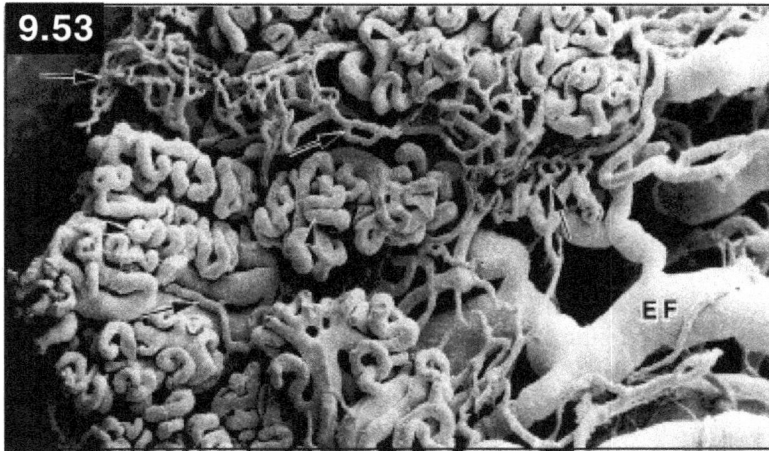

Figure 9.53: Higher magnification of third arch filaments, EF, efferent filamentary artery, arrowheads lamellar capillaries arrows, venous-like web. X 160

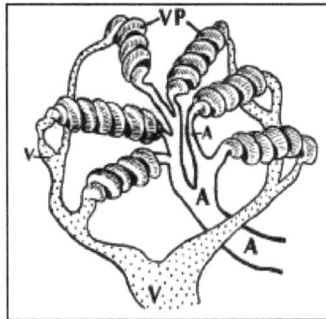

Figure 9.54: Diagram of typical vascular rosette in respiratory epithelium. One or several arteries (A, median arterioles) deliver blood to the center of the rosette, spiral capillaries of the vascular papillae radiate circumferentially and drain into peripheral veins (V).

of the respiratory vasculature have a pebbled appearance similar to the intact epithelium. Replicas viewed perpendicular to the plane of the epithelium reveal the intricate interweaving of arteriolar and venus segments that ultimately supply the respiratory capillaries. Each vascular unit, or rosette, within the islet consists of a somewhat centrally placed afferent arteriole that gives rise to between several dozen spiral capillaries (papillae) that typically radiate outward from the artery. The capillaries are drained by a venous plexus that runs circumferentially around the capillaries.

Gill

Filaments or filamental-like structures are only found on the second and third gill arches. Filaments in the second gill arch are for the most part covered with tissue and are not directly exposed to water currents. The shape of the filament and its associated vasculature varies somewhat with the relative position of the filament on the arch. Filaments on the anterioventral portion of the arch are longer with straight afferent and efferent filamental arteries. Afferent filamental arteries arise at regular intervals from the afferent branchial artery and supply blood to a single, or occasionally two filaments. Filaments may also branch peripherally, in which case each branch is then supplied by a bifurcation of the afferent filamental artery.

Lamellar sinusoids typical of most water breathing fish are absent from filaments of the second gill arch. Instead, blood flows from the afferent to the efferent filamental artery via single channel capillaries or several parallel capillaries supplied by a branching vessel; the latter being the equivalent of the lamellar arteriole. Multiple capillary drainage by single arterioles is more prevalent in the basal filamental areas. Large shunt-type anastomoses between afferent and efferent filamental arteries were not observed. Unlike the capillaries of the respiratory islets, gill capillaries do not spiral but take a three-dimensional, zig-zag course that is best illustrated. Thus only a short portion of the length of the capillary is apposed to the epithelium. This position is roughly equivalent to the outer marginal channel of gill lamellae from typical water-breathing teleosts. Efferent filamental arteries drain one or several filaments, or filament branches, into an efferent branchial artery. Afferent and efferent branchial arteries of gill arches two and three communicate with each other only via the lamellar capillaries. No large-bore shunts between afferent and efferent branchial arteries were observed in either second or third arches. In the posteriodorsal aspect of the arch the filaments are less organized into straight structures. The afferent and efferent vessels usually follow a more tortuous couse and branching along the length of the filament is more common. However, the relationship between afferent, efferent, and lamellar capillaries and the general zig-zag capillary pattern is unchanged.

A second vascular network is also found associated with the second and third gill arches. This system forms a web of small venous-like vessels around the filaments and occassionally between the lamellar capillaries. This system, or one very similar to it, is especially prevalent in the basal areas of the filaments around the efferent filamental arteries of the second arch. It covers most of the filaments of the third arch. The vasculature that feeds and drains these web-like vessels could not be identified.

The vasculature of the third arch is essentially identical to that of the second arch with respect to pathways and vessel types involved. Filaments on the third arch are smaller, however, and the filamental vessels tend to be more irregular than those of the second arch. Multiple branches of the efferent filamental artery are uncommon.

Discussion

It is obvious from this study and preceding ones (Wu and Liu, 1943; Liem, 1961; Munshi and Singh, 1968; Rosen and Greewood, 1976; Mishra *et al.*, 1978) that the major vessels in *Monopterus cuchia* and related species depart from the conventional pattern observed in other teleosts. Primitive aortic arches have been retained to perfuse the anterior region of the head, providing nutritive flow in some instances and perfusing the respiratory organs in others. The coronary and hypobranchial arteries are also fed directly by the ventral aorta, rather than being supplied by post-branchial (or post-gas exchange organ) vessels as they are in typical water-breathing fish. The concept of an arterio-arterial vasculature, *i.e.* ventral aorta gill dorsal aorta, which is the rule in typical water-breathing fish, is also eschewed in favour of an arterio-venous circuit through the accessory respiratory organs. Only in the rudimentary second and third gill arches is an arterio-arterial pathway found and here (second arch epibranchial artery), it is used to perfuse the brain via the carotid arteries.

As pointed out by Lombolt and Johansen (1976) the vascular arrangement in *M. cuchia* is not the most efficient design for oxygen delivery. Oxygenated venous blood returning from the respiratory islets mixes with systemic venous blood, resulting in partially oxygen-saturated blood returning to the ventricle. Other air-breathing teleosts such as *Channa* (Ishimatsu *et al.*, 1983) and *Anabas* (Olson *et al.*, 1986b) may be able to avoid, or at least minimize, this problem by decreasing the degree of mixing of oxygenated and deoxygenated venous blood in the veins and heart through a special arrangement of the venous return pathways. They also appear to be side to separate ventricular outflow by means of multiple ventral aortae (Ishimatsu *et al.*, 1983; Olson *et al.*, 1986b). The gills of these air-breathing fish are also modified so that some areas serve as large-bore bypass vessels (Olson *et al.*, 1986b). However, the venous system of *M. cuchia* does not have such specialization, and even if it did, the inordinately long ventral aorta (Wu and Liu, 1943; Liem, 1961; Mishra *et al.*, 1978) would promote mixing of arterial blood. Other factors such as hematological enhancements (increased hematocrit, hemoglobin concentration and oxygen carrying capacity; Lombolt and Johansen, 1976) and varied use of cutaneous respiration and intermittent aquatic ventilation (Lombolt and Johansen,1974; Graham *et al.*, 1987) may counter some of these constraints. It appears, however, that the vasculature arrangement of *M. cuchia* may be the limiting factor in maximal gas exchange efficiency. Even though spatial blood flow shunting is not highly efficient in *M. cuchia*, temporal shunting may assist in the oxygen delivery process at least during a portion of the ventilatory cycle. Lombolt and Johansen (1976) have shown that resistance in the shunt vessels is high but constant. Following a breath, ventral and dorsal aortic pressures progressively rise, which has been interpreted by Lombolt and Johansen (1976) to indicate that vasoactive changes occur in the air-breathing organs. This could temporally shift flow between systemic and respiratory organs, enhancing gas extraction from the air space after a breath and supporting tissue, oxygen delivery (by progressively increasing systemic flow) during breath holding. Clearly measurements of cardiac, output, or preferably organ perfusion, are needed to evaluate the efficiency (or presence) of temporal shunting.

Microcirculation

Respiratory Islets

The vasculature of the respiratory islets has several unique structural modifications. First, unlike the respiratory capillaries (lamellae) of water breathing fish or even the true gill vessels of air-breathing fish (including *M. cuchia*), the vascular pathway of the respiratory islets is an arterio-venous pathway (Munshi *et al.,* 1989). In typical gills of air-breathing fish and respiratory gills of air-breathing fish, the respiratory pathway is arterio-arterial (Laurent, 1984; Olson *et al.,* 1986b). The significance of these arteriovenous connections, from an evolutionary perspective, is unclear. However, from a hemodynamic aspect, the purpose of this arrangement appears to be to rapidly lower islet blood pressure to central venous pressures. Flow from the respiratory islets returns directly, to the heart. At the heart there is a common drainage of islet veins with systemic veins. Venous blood pressure in the respiratory islets must be similar to that of the systemic veins to ensure uniform drainage from each circuit and to prevent retrograde flow from one pathway back into the major veins of the other.

The second unique aspect of the respiratory islet circulation is the peculiar spiral nature of the islet capillaries. These capillaries, which may be derived from, or at least are analogous to, the outer marginal channels of gill lamellae (Munshi *et al.,* 1989) are to our knowledge, unique to *M. cuchia* and perhaps related species. The spiral nature of these vessels, with only a fraction of their length apparently close enough to the respiratory surface to facilitate gas ex-change, most probably subserves other functions as well.

Munshi *et al.* (1989) have suggested several functions for the spiral vessels and their related endothelial cell valves : 1) the increased resistance resulting from the unusual length of the capillary may increase the residence time of red cells in the capillaries and thereby maximize gas equilibrium 2) as more blood resides in the air-breathing organ (and coupled with number 1 above) this increases the amount of oxygen loaded per unit of time; 3) the "valves" force the red cells to the epithelial surface of the capillary wall and thus minimize diffusion distance between the air and red cells; and 4) the capillary loops cause a bulging of the epithelial surface above them and thereby increase the epithelial surface area of the otherwise flat, single sided exchange surface.

From our own observations, the need to lower islet venous pressure could certainly be achieved by the increased capillary length, although it might be argued that arterial resistance vessels would better serve this purpose. The longer spiral vessels might also enhance mixing and thus promote better overall oxygen saturation, however straight vessels with periodic endothelial irregularities such as those found in *Anabas testudineus* (Munshi *et al.,* 1986) would serve as well without the added resistance. Perhaps the advantage of the spiral vessels rests not in a hemodynamic attribute but an anatomical one, that of an increased surface area of the vascular endothelium.

Recent studies have shown that the fish gill, like the mammalian lung, is important in regulating circulating hormones such as catecholamines and angiotension (Nekvasil and Olson, 1986; Olson *et al.,* 1986a). The metabolic efficiency of the gill in hormone regulation is due in part to two anatomical characteristic. First, it is the only organ to receive the entire cardiac output and second, it has an extensive endothelial (pillar) cell surface area. The respiration of *M. cuchia* do not have similar characteristics, *i.e.*, the respiratory islets are in parallel with systemic circulation and the gills are reduced or absent altogether. However, the islets undoubtedly receive a large fraction of the cardiac output, perhaps nearly all during periods of the ventilatory cycle (see above). The coupled with the expanded surface area of islet capillaries may be sufficient for relatively efficient control of plasma hormones. Interesting in this regard, however, is the recent observations that the respiratory epithelium of *M. cuchia* unlike that found in other air-breathing teleosts, does not contain angiotension-converting enzyme (Olson *et al.*, 1987). This may indicate either that the islet vessels are not involved in angiotensin formation or that there are other differences in the renin-angiotension cascade in *M. cuchia*. The role of these vessels in metabolism of other hormones, *e.g.* catecholamines, prostaglandins, etc. remains to be determined.

Gills

Although the vestigial gills in *M. cuchia* are morphologically atypical, they appear to retain both lamellar and interlamellar (filamental) vasculature networks common in water – breathing teleosts. Presumably the gills contribute somewhat to the gas exchange process (Munshi *et al.*, 1989). Gill pathway has been greatly reduced to a single channel that zig-zigs in three dimensions as it proceed from the afferent to efferent filamental. Why the gill capillaries follow this pattern rather than the spiral pattern of islet capillaries is unknown, but it is probable that this function similarity between the two is interesting. We could not find any evidence of lamellar bypass shunts in the gills.

An interlamellar vasculature (viz., central sinus or venous sinus) similar to that found in typical water – breathing teleosts (Holand and Olson, 1979; Olson, 1989; Laurent, 1984) and function of gills of air-breathing fish such as *Amia calva* (Olson, 1981), *Anabas testudineus* (Olson *et al.*, 1986b) and *Heteropneustes fossilis* (Olson *et al.*, 1990) is not present in the gills of *M. cuchia*. The venous-like web observed in the present study is probably analogous to the interlamellar system and may represent modified interlamellar vessel. It appears likely that as the sheet-like lamellar and lamellar sinusoids became modified to single-channel, irregular capillaries in *M. cuchia* the anatomical framework upon which the interlamellar system was designed was also lost and the latter became less ordered as well. In fact, the interlamellar system in *M. cuchia* is very similar to the interlamellar system of elasmobranches. It further illustrates the highly conserved nature of the interlamellar vasculature among fish and suggests that some homeostatic function is unique to these vessels. It is interesting in this regard that in three species of air-breathing fish studied to date, *A. testudineus* (Olson *et al.*, 1986b; Munshi *et al.*, 1986), *H. fossilis* (Olson *et al.*, 1990) and *M.c uchia*. The common modality associated with retention of the interlamellar system is that the the tissues in which these vessels are found are in contact with an aqueous but not an aerial environment. Thus the inference can be made that the interlamellar system is involved in one or several blood-water exchange processes. The apposition of the interlamellar vessels with gill ionocytes (chloride cell) has previously been suggested to indicate that these vessels are somehow involved in chloride cell function.

Histochemistry and Functional Organization of the Heart and Aortae of an Obligatory Air-Breathing Mud-Eel, *Monopterus (=Amphipnous) cuchia* (Ham., 1822) (Synbranchiformes, Amphipnoidae)

The heart of *Monopterus (=Amphipnous) cuchia* is four chambered. The sinus venosus and the atrium give weak reactions for lipids and do not show significant SDH activity. The ventricle and the longitudinal ridges of bulbus arteriosus, however, give strong reactions for these. The inner layer of bulbus arteriosus is thick and mainly consists of elastic fibres.

The walls of the dorsal and the ventral aortae are typically three layered and show significant differences in their lipid distribution and SDH activity. The tunica adventitia is dominated by collagen fibres. The tunica media is mainly composed of elastic fibres and is thicker in ventral aorta than in dorsal aorta. From the tunica intima small valvules project into the lumen and are comparatively more numerous in the dorsal aorta. They may play an important role in the regulation of blood flow. The difference in the SDH activity and the distribution of lipids and elastic fibres in the various chambers of heart and the wall of the aortae reflects the differences in their metabolic activities and elastic modulus (N. Mishra, J. Ojha, J.S.D. Munshi and Mittal, A.K., 1978).

Introduction

Although the structural organization of heart, aortae and blood vessels of cyclostomes (Augustinsson *et al.*, 1956) selachians (Lauder, 1964; Johansen and Hanson, 1967; Satchell, 1971) and teleosts, both air-breathing and water-breathing (Dornesco and Santa, 1963; Randall, 1968; Light and Harris, 1973; Munshi and Mishra, 1974; Priede, 1975, 1976; Satchell, 1976) have drawn the attention of researchers in recent years for the histochemical localization of various chemical constituents in these organs that play an important role in understanding their physiology have not yet been studied on an adequate scale.

The present investigation has been designed to examine the histochemical localization and distribution of the structural proteins, lipids and succinic dehydrogenase (SDH) in the heart, the dorsal aorta and the ventral aorta of *Monopterus cuchia* (formerly *Amphipnous cuchia*, Rosen and Greenwood, 1976).

M. cuchia, native to India, Bangala desh and Burma, is uniquely adapted for air-breathing and lives a truly amphibious life. This fish depends mostly on a pair of specialized air-sacs for its gas exchange. This fish, although principally an air-breather, has retained the piscine mechanism of ventilating the rudimentary gills (Lomholt and Johansen, 1974).

Materials and Methods

Live specimens of *M. cuchia* (approx. 70 cms in length) were collected from the ponds and swamps of Alamnagar, Saharsa, in Bihar. Fishes were anaesthetized using MS222 (Sandoz) and then the heart, the ventral aorta and the dorsal aorta were dissected out, cut into small pieces (approx. 5 mm in length), washed in Ringer's saline solution and finally fixed in Bouins fluid, Zenker's fluid and Helly's fluids. After fixation and proper post-fixation treatment the materials were dehydrated in ethanol, embedded in paraffin wax and were cut at 6 μm. Paraffin section were stained with Ehrlich's haematoxylin and eosin for the study of general organization of heart and the aortae.

Mallory's triple stain (Jones, 1950), Mallory's P.T.A.H. (Pearse, 1968), Verhoeff's elastic stain (Lillie, 1954), Orcinol New-Fuchsin stain (Fullmer and Lillie, 1956), and Weigert's elastic stain (Jones, 1950) techniques were used for the demonstration of collagen and elastin.

For enzyme histochemistry, fresh frozen and for lipid histochemistry, 10 per cent neutral formalin fixed tissues were cut at 15-30 μm using an American optical cryostat at - 20°C.

Nitro-BT techniques (Nachlas *et al.*, 1957) with 20 min incubation in the substrate at 37°C was used for the localization of SDH activity. Sections incubated without substrate were used as control.

Sudan black B method was employed for the demonstration of lipids (Casselman, 1959). Few sections were also treated with absolute acetone for 1 hr. at room temperature for the extraction of lipids as controls.

Observations

Heart

The heart of *Monopterus cuchia* (approx. 70 cm in length) is situated far posterior (6.3 cms) to the cleithral symphysis in the pericardium. It is elongated (approx. 3.8 cms) in outline and consists of four chambers – the sinus venosus, the atrium, the ventricle and the bulbus arteriosus. The wall of each heart chamber has two distinct layers – the outer and the inner.

Sinus Venosus

The sinus venosus is a thin walled structure situated dorsal to the atrium. The inner layer of its wall is approx. 73 μm thick and formed of collagen and a few muscle fibres. The outer layer is thin (approx. 36 μm thick) and is mainly composed of collagen fibres.

Atrium

The atrium is situated ventral to the sinus venosus. The histochemical nature and the histological organization of the wall of the atrium is similar to that to sinus venosus.

Ventricle

The conical shaped ventricle is a thick walled chamber situated on the posterior aspect of the heart and partly covered anteriorly by the atrium. The inner layer of the wall of ventricle is comparatively thick (approx. 860 μm), spongy in nature and is formed of irregularly arranged cardiac muscle fibres. This layer is thrown into

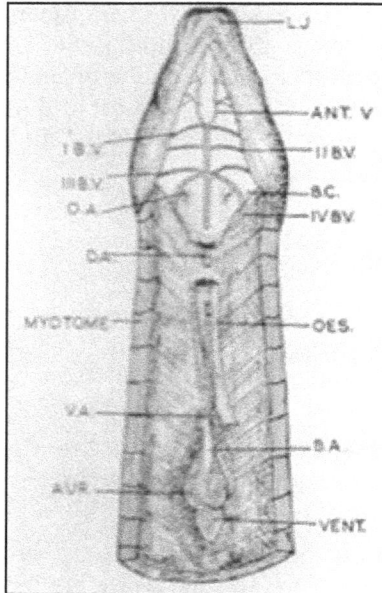

Figure 9.55: A diagram of the blood vessels *Monopterus cuchia.*

Abbreviation: Ant. V: Anterior vessel; AUR: Atrium; BA: *Bulbus arteriosus***; BC: Buccal caviy; BV: Branchial vessel; DA: Dorsal aorta; LJ: Lower jaw; OA: Oesophageal artery; OES: Oesophagus; VA: Ventral aorta; VENT: Ventricle.**

numerous irregular projections the trabeculae that protrude into the lumen of the ventricle. This layer is richly infiltrated with lipids and shows high succinic dehydrogenase activity. The outer layer (approx. 72μm thick) is made up of connective tissue which is mainly collagenous in nature.

Bulbus Arteriosus

The bulbus arteriosus is a thick-walled pear-shaped structure situated anterior to the ventricle and covered by artrium on its dorsal side. The inner layer of its wall is comparatively thick (approx. 369 μm) and mainly composed of elastic fibres interspersed with few muscle fibres and give moderate reactions for lipids. Numerous longitudinal ridges (average height 651 μm and width 147 μm) arise from this layer and project into the lumen of the bulbus arteriosus. These ridges are mainly composed of collagen fibres intermingled with few elastic fibres. The thin peripheral margins of these, however give strong reactions for lipid and SDH.

The outer layer of the wall is approximately 10 per cent μm thick and is mainly composed of collagen fibres intermingled with few elastic fibres.

Ventral Aorta

The ventral aorta (approx. average inner diameter 872 μm) emerges from the bulbus aorta and gives off four pairs of branchial vessels during its course. The wall

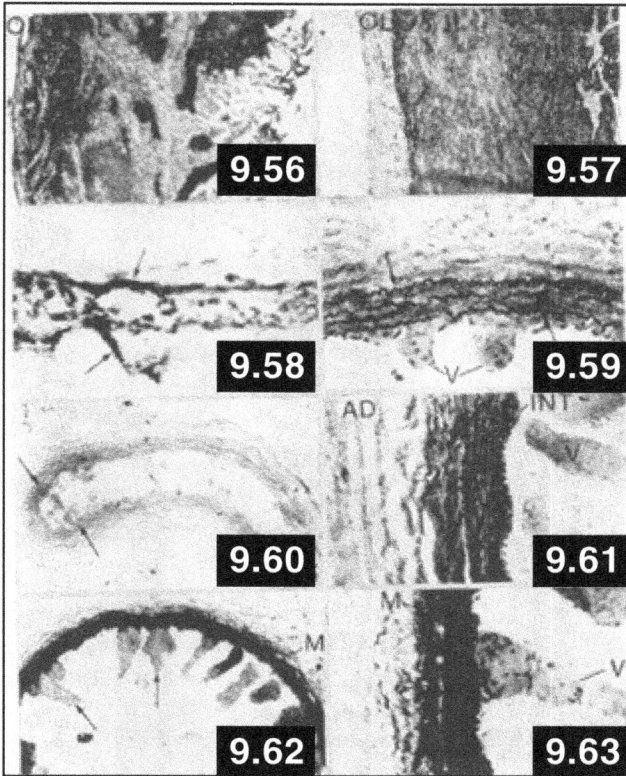

Figure 9.56: A part of the cross section of the ventricle showing its outer layer (OL) and the trabeculae (arrows) protruding from the inner layer into the lumen (Ehrlich's hematoxylin cosin) x 60

Figure 9.57: Part of the cross sections of the bulbus aorta showing the outer layer (IL) and the longitudinal ridges (arrows). (Ehrlich's hematoxylin cosin) x 70.

Figure 9.58: Part of the horizontal section of the bulbus aorta showing strong SDH area (arrows) in the longitudinal ridges. (Nitro-BT techinique for SDH) x 100.

Figure 9.59: Part of the cross section of the ventral aorta showing the presence of elastic fibres (arrows) in the media. Valvules (v) are are also media. Valvules (V) are also seen (Verhoeff) elastic stains x 300.

of the ventral aorta comprises three layers – the tunica adventitia, the tunica media and the tunica intima. The tunica adventitia is the outermost layer (approx.93 μm thick) and composed of collagen fibres interspersed with elastic fibres. The tunica media is the middle layer (approx. 60μm thick) and mainly composed of elastic fibres intercalated with few muscle fibres. It gives moderate reaction for SDH and lipids. The tunica intima is the inner most layer (approx. 12 μm thick) and is composed of endothelial cells arranged in a layer. Small (approx. average height 48 μm and width 24 μm) somewhat conical valvules arise from this layer and project into the lumen of row aorta. Each valvule contains lipid and shows moderate SDH activity.

Dorsal Aorta

The dorsal aorta (approx. average inner diameter 824 μm) is formed by the union of the fourth pair of branchial vessels near the 12th vertebra and gives rise to the systemic circulation. Its wall is composed of three distinct layers – an outer tunica adventitia (approx. 78 μm thick), middle tunica media (approx. 32 μm thick) and inner tunica intima (approx.9μm). Tunica adventitia is formed of collagen and the tunica media dominated by elastin. The tunica intima is composed of a single layer of endothelial cells. The valvules are conspicuously (approx. average height 119 μm and width 44 μm) developed and protrude and occupy a greater area of the lumen of the dorsal aorta.

Discussion

Like the systemic (=branchial) heart of fishes, the heart of *Monopterus cuchia* also consists of four chambers.

The presence of collagenous wall and the absence of SDH activity and low lipid content in the sinus venosus and atrium indicate their low level of metabolism. These findings suggest that probably the sinus venosus and atrium of this mud-eel play little or no role in driving blood through the heart and corroborate the earlier finding of Randall (1970) that in teleosts the rate of arterial filling is determined by venous pressure. However, the muscle fibres in atrium may initiate the contraction wave which in turn may play some role in the ventricular filling.

Intense intraventricular pressure is achieved by the contraction of the ventricle which helps the blood to flow to the different parts of the body. High level of metabolic activity is indicated by strong SDH reaction and the presence of lipid in the muscular layer of the ventricle. It is therefore suggested that for long lasting contraction rhythm, the cardiac muscles of the ventricle get energy by the breakdown of the lipid aerobically. Its outer collagenous layer may help in withstanding the tensions during its contraction (systole) and relaxation (diastole) cycle.

The dominance of the elastic fibres in the thicker inner layer of the bulbus arteriosus reflects its low elastic modulus. This property helps in withstanding the blood pressure generated by the ventricle during its systolic cycle. Because the thicker inner wall expands even with slight increase of pressure around it and therefore may help to reduce the force of blood in the ventral aorta. Its outer collagenous layer has comparatively higher elastic modulus because the elastic modulus of collagen is more than the elastin (Roach and Burton, 1957). This higher elastic modulus of the outer layer may act as a limiting factor for the excessive expansions of the bulbus arteriosus reveals its low metabolic state and therefore it may be suggested that this organ does not help in the propulation of blood and corroborates the earlier ridges show strong reactions for SDH and may be the active site for the initiation of contraction waves that pass through the ridges and perhaps regulate the velocity of blood flow in the ventral aorta.

In this fish also the ventral aorta is a trilaminate structure as mentioned earlier by Dornesco and santa (1963) in *Cyprinus carpio*. However, they have not studied the nature of the structural proteins that determine the elstic modulus of fish arteries (Satchell, 1971). Roach and Burton, 1957 have found out that elastin has much lower elastic modulus (30 N/cm^2/100 per cent elongation) than collagen (1 × 10^4 N/cm^2/ 100 per cent elongation). The ventral aorta seems to have a low value of elastic modulus, as it has high proportion of elastin in its wall (Satchell, 1971). Lander (1964) performed quantitative analysis of collagen and elastin in the arterial wall of five species of sharks and found that the ventral aorta contains elastin 31 per cent and collagen 46 per cent. The histochemical investigations of the ventral aorta of this fish gave an idea of the functional capacity of adventitia and the media separately. It seems that adventitia having more collagen will have higher elastic modulus than media which have rather a high percentage of elastin. The higher and lower elastic moduli of outer and the inner layers respectively, perhaps are well adapted for withstanding considerable vaso-constrictor tension in the ventral aorta of this fish gave an idea of the functional capacity of adventitia and the media separately. It seems that adventitia having more collagen will have higher elastic modules than media which have rather a high percentage of elastin. The higher and lower elastic moduli of outer and the inner layers respectively, perhaps are well adapted for withstanding considerable vaso-constrictor tension in the ventral aorta. Recently, Lomholt T. and Johansen 1976 have reported higher ventral aortic blood pressure (60 mm Hg systolic value). The intermediate SDH activity and moderate lipid content in media and valvules reveal their active role in the pulsatile of the blood vessel.

The dorsal aorta presents a similar picture, as it has also the intimal layer and as such it is a trilaminar structure and corroborates the earlier findings of Dornesco and Santa (1963) on *Cyprinus carpio*. The histochemistry of the structural proteins of the dorsal aorta reveals that the adventitia with collagen fibres will have more elastic modulus than the media which is dominated by elastin. The dorsal aorta seems to be more rigid in comparison to the ventral aorta, as the former has got a thicker adventitia layer. The media of the dorsal aorta is thinner than that of ventral aorta but contains more elastic tissue per unit area than that of the ventral aorta. Lauder, 1964 found out that the wall of the dorsal aorta contains 9 per cent elastin and 69 per cent collagen, in contrast to the ventral aorta which has about 31 per cent elastin and 46 per cent collagen. Satchell, 1971 pointed out that if media is stripped off from the rest of the wall and analysed separately, perhaps it will show a higher proportion of elastin than the whole lamina. This view of Satchell gets support from our histochemical findings that the thinner media is dominated by elastin. Comparative thinner elastic layer of the dorsal aorta may be correlated with lower vaso-constrictor tension due to lesser dorsal aortic blood pressure. This finding gets support from the work of Lomholt and Johansen, 1976 who have reported the considerable drop of blood pressure in dorsal aorta (40 mm Hg. Systolic value). This drop of blood pressure was correlated with the considerable vascular resistance in the shunt connecting this vessel. The presence of many valvules in the dorsal aorta is unique and they perhaps help to regulate blood flow in this blood vessel. The absence of SDH activity from the wall and the valvules of the dorsal aorta reveals their low metabolic activity in contrast to the ventral aorta and reflects the non-pulsatile nature of the dorsal aorta.

The presence of valvules in the ventral aorta and dorsal aorta of *M. cuchia* is very interesting. Dornesco and Santa, 1963 have also observed similar type of structures in the ventral aorta of the carp, *Cyprinus carpio* where the endothelium is quite often raised by the heaps of Plehn cells which were thought to be glandular in nature. He also noticed once the Plehn cells in dorsal aorta. It seems that the cells of tunica intima of ventral and dorsal aorta proliferate to give rise to these valvules which are physiologically very active in ventral aorta and inactive in dorsal aorta as indicated by their SDH activity. Further experimentation is however necessitated to ascertain their role in these blood vessels.

Histochemistry and functional organization of the heart, the ventral aorta and the dorsal aorta of an air-breathing fish, *Monopterus cuchia* have been described.

Presence of lipid and strong succinic dehydrogenase (SDH) reactions in ventricle and the longitudinal ridges of the bulbus aorta reflect their high metabolic activity during cardiac cycle.

Low elastic modulus of the bulbus arteriosus is indicated by the presence of well developed internal layer dominated by elastic fibres.

The ventral aorta has three typical layers. Adventitia is thicker and contains collagen as well as elastin, the media contain elastin and a few smooth muscle fibres. The tunica intima is thin and made up of a single layer of endothelial cells. Few small valvules are seen to emerge from the internal layer and project into the lumen. Valvules indicate their moderate metabolic activity. Thicker media dominated by elastic fibres reflects lower elastic modulus of the wall of ventral aorta.

The dorsal aorta has also trilaminar structure. The thinner media is rich in elastic fibres and thicker adventitia is dominated by collagen and therefore the wall of the dorsal aorta shows more elastic modulus than the ventral aorta which may be due to lower blood pressure in it. Many valvules in the dorsal aorta may help to regulate blood circulation in the dorsal aorta. Absence of SDH activity in the dorsal aorta indicates its lower metabolic activities.

Chapter 10

Fine Structure of the Respiratory Organs of the Climbing Perch
Anabas testudineus
(Pisces: Anabantidae)

An electron microscopic study has been made of the three respiratory organs of climbing perch. The gill structure is similar to that of the other teleosts but the thickness of the water/blood barrier is much greater, being as great as 20 μm in some specimens. The increased thickness is due to a multilayered epithelium which is thinner (3.5-7 μm) over the marginal channel of the secondary lamellae. The other two main layers, basement membrane and piller cell flange, are relatively thin (about 1μm).

The pillar cells have a typical structure, but in certain regions they are contiguous with one another and line well-defined blood channels. Some of the columns of basement membrane material in such regions may be common to adjacent pillar cells.

The air-breathing organs are (a) the lining of the *suprabranchial chambers*, and (b) the *labyrinthine plates* attached to the dorsal region of branchial arches. Electron microscopy showed that their structure is well adapted for gas exchange, the air/blood barriers being only 0.12-0.3 μm, comprising an epithelial layer, basement membrane, and thin capillary endothelium. The many parallel blood channels of the respiratory islets of both organs are separated by pillar-like structures which differ from the pillar cells of the secondary lamellae. Thus the hypothesis that the air-breathing organs represent modified gills is not supported by this study.

The fine structure of the non-respiratory region of the air-breathing organs is similar to that of the skin, and includes chemoreceptor-like cells. Evidence concerning the possible homology of pillar cells with plain muscle cells is discussed (Hughes, G.M. and Munshi, J.S.D., 1973)

Introduction

Gills are the typical respiratory organs of fish in their usual habitat of well-aerated water. However, some freshwater habitats contain low concentrations of oxygen and are liable to dry up periodically (Carter and Beadle, 1931). For fish living in such environments, aquatic respiration is sometimes supplemented by accessory air-breathing organs formed by modification of parts of the alimentary canal. Diverticula of the anterior regions of the gut form the lungs of Dipnoi and other groups close to the ancestry of tetrapods. Among teleosts, species of many different families inhabit surroundings in which aquatic respiration may be difficult, and may show convergences in their adaptations.

The climbing perch, *Anabas*, is a well-known example, in which a part of the pharynx has become modified to form a suprabranchial chamber. The structure and evolution of these accessory organs has attracted the attention of many workers including Zograff (1888), Henninger (1907), Rauther (1910), Das (1927), Marlier (1938) and Liem (1963). The hypothesis that the respiratory islets represent modified gill lamellae (Rauther, 1910; Carter, 1957; Norman, 1963; Munshi, 1968), was based upon the apparent presence of pillar cells which are such a characteristic feature of the secondary lamellae of fish gills. Recent electron microscope studies of *Anabas* air-breathing organs (Hughes and Munshi, 1968), showed that the "pillars" separating the capillaries in these organs are modified epithelial cells. This chapter gives a more detailed account of the results of these investigations.

Materials and Methods

Specimens of *Anabas* were collected near Calcutta and transported by air from India in plastic bags containing oxygen and a limited amount of water. The fish were kept in aquaria (25°C) at Bristol University and used for microscopy and in physiological experiments.

After stunning, small pieces of gill filament, labyrinthine plate, and the respiratory membrane lining the suprabranchial chamber, were fixed either in (1) 2.5 per cent osmium tetroxide buffered to pH 7.2 with 0.1 M phosphate buffer (Milloning, 1961), or (2) in 2.5 per cent glutaraldehyde in 0.1M sodium cacodylate buffer followed by post-fixation for 1 h in 2 per cent osmium tetroxide buffered with 0.1 M sodium cacodylate. Specimens were dehydrated in alcohol and embedded in Araldite. Sections were cut on an LKB ultramicrotome. They were mounted on carbon-coated grids, and stained with uranyl acetate followed by lead citrate (Reynolds, 1963), and viewed with an AEI 6G electron microscope.

Result

Gaseous exchange in *Anabas testudineus* takes place through three distinct sites. (Figure 10.1).

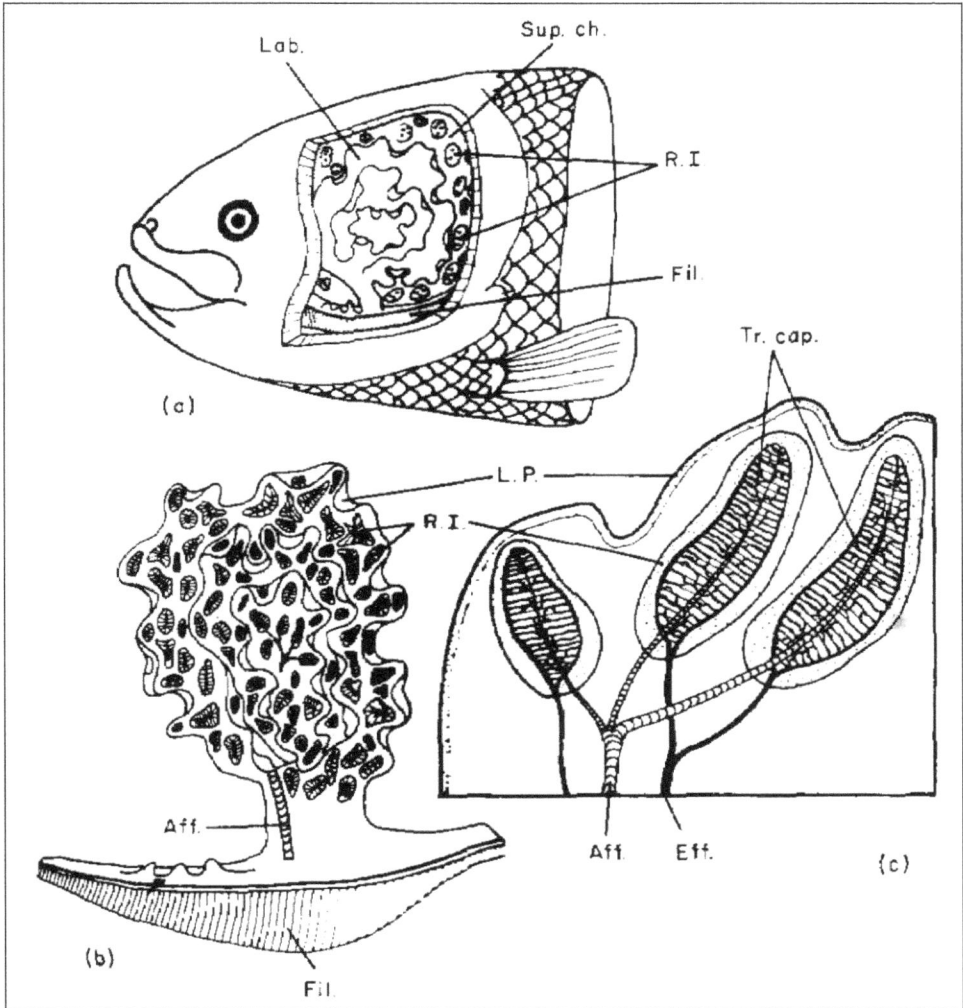

Figure 10.1: (a) Site view of *Anabas testudineus* with part of lateral wall removed to show the left suprabranchial chamber (Sup. Ch) with the labyrinthine organs (Lah) and its relation with the branchial arches to which are attached the gill filaments (Fil.). None the presence of respiratory islets (R.I) on the lining of the suprabranchial chamber. X 2; (b) Diagram of the entire labyrinthine organ borne by the first gill arch showing the respiratory islets (R.I.) on the labyrinthine plates (L.P.) and the root of the afferent branchial vessel (Aff.) that supplies blood to the respiratory islets. X 4; (c) Surface view of a labyrinthine plate showing details of the vascular supply to these respiratory islets. Relationships of the afferent (Aff.), efferent (Eff.) and transverse capillaries (Tr.cap.) are indicated. X 50

1. The secondary lamellae of the gill filaments,

2. The labyrinthine plates, and

3. The respiratory membrane lining the suprabranchial chambers.

The secondary lamellae are the sites of exchanges between water and blood, whereas the labyrinthine plates and respiratory membrane allow gaseous exchange between blood and the air contained in the suprabranchial chamber. This chamber is full of gas and is quite separate from the branchial chamber which contains water. All three organs have a large surface area (Hughes, Dube and Munshi, 1973) and provide regions where there is a small diffusion distance between blood and the external environment. Gills rely on the water for their support but the labyrinthine plates are stiffened and this enables them to maintain their position in air.

Secondary Gill Lamellae

Anabas gills are essentially the same as in most teleosts, consisting of gill filaments attached to the branchial arches, each filament having its surface area greatly increased by secondary lamellae. The secondary lamellae are short, flattened structures about 30 μm thick; microscopic examination showed that each consists of a middle vascular layer with epithelial coverage on both sides. The epithelium is multi-layered and varies from 5-18 μm in thickness, except around the free edge of the secondary lamella where it is only two-layered and 3-6 μm thick. In addition to the epithelial layers, a basement membrane and pillar cell flange layer can be clearly seen between the epithelium and the blood space.

Epithelial Layers

In the epithelial layers several cell types can be indentified:

1. Cuboidal cells

2. Mitochondria-rich (chloride) cells

3. Elongate columnar type cells with mitochondria and small or large vacuoles

4. Amoebocytes, with vacuoles having dark osmiophilic margins, and other types of lymphocytes.

5. Mucous cells.

The external cuboidal epithelial cells bear microvilli on their outer surface [Plate 10.1(a)]. Sometimes adjacent microvilli fuse at their tips and appear to enclose spaces. There are many mitochondria-rich (chloride) cells [Plate 10.1(a)], several of which can be seen on the outer epithelial surface. The nucleus is basal in position, and has a prominent nucleolus. Mitochondria are usually rod-shaped with well developed oblique cristae. Rough surfaced vesicles and vacuoles have been observed in these cells which also contain widely scattered dense granules, presumably of glycogen, in their cytoplasm. The third type of cell only differs from chloride cells in the presence of many vacuoles, but they may have the same origin. The mitochondria of these cells are usually arranged along the edge of the vacuoles [Plate 10.1(b)]. The fourth type of cell is represented by amoebocytes and lymphocytes. Amoebocytes are very common

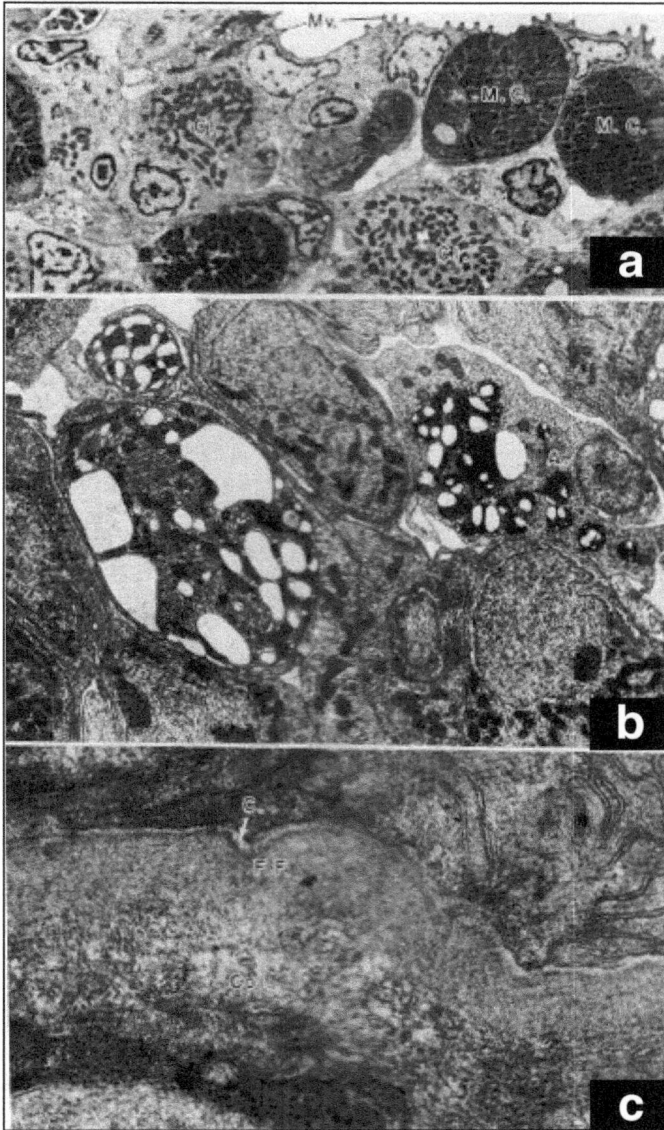

Plate 10.1

(a) Structure of the epithelium from the interlamellar (crypt) region of the primary gill lamella showing microvilli (Mv.), mucous cell (M.C.) and mitochondria-rich chloride cell (Cl) X 3400; (b) Structure of the epithelium of the interlamellar regions showing type of chloride cells with small and large vesicles, and with mitochondria concentrated near their periphery. The amoebocytes (A) have many vesicles with osmiophilic surfaces. x 9000; (c) Structure of the marginal blood channel of the secondary lamella near its origin from the gill filament. Notice the two layers of the basal lamina (clear (c) and fine fibrous (F.F) which together with the inner coarse fibrous (Col.) zone constitute the basement membrane. X 32000.

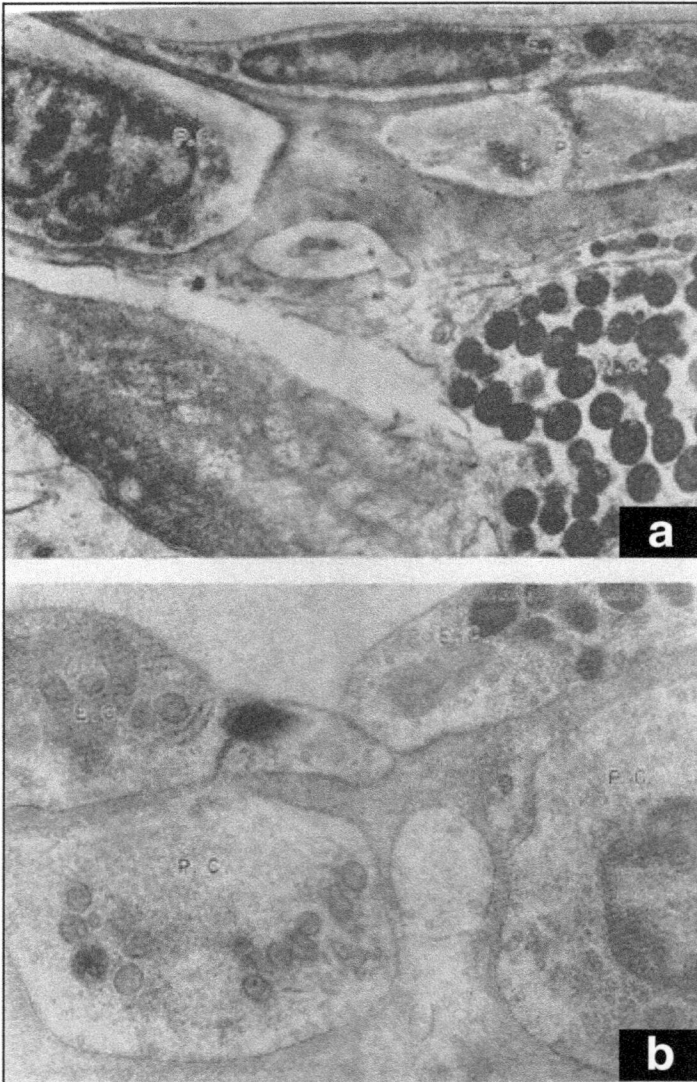

Plate 10.2
(a) Structure of an arteriole of the gill filament showing endothelial cells (E.C.), pericyte cells (P.C.) embedded in the basement membrane, and the dark osmiophilic granules of a Mast cell (M.C.). X 13,500; (b) More highly magnified view of the arteriole wall showing endothelial cells (E.C.), with characteristic osmiophilic granules, pericyte cells (P.C.) with micropinocytotic vesicles. Note the presence of fine fibrillae in the cytoplasm of the pericyte cells and their fibrous membranous coat. X 20,000.

components of the epithelium and are also found in the subepithelial connective tissue of the primary gill lamellae or gill filaments. These cells (7-9 μm in length) are irregular in shape with pseudopodia-like processes. The cytoplasm is characterized by the presence of many vacuoles surrounded by electrondense osmiophilic material [Plate 10.1(b)].

The fifth types of cell in the epithelium is the typical globular mucous cell which occurs on the secondary lamellae as well as in the crypts between them [Plate 10.1(a)]. In some specimens cysts of protozoan parasites were found embedded in the epithelium.

Basement Membrane

The basement membrane is fairly thick in the second, third and fourth pair of gills but is relatively thinner in the first gill. As seen in transverse sections of the secondary lamellae, it appears amorphous in nature, but in some sections of the marginal channel near its origin, all three layers – clear, fine and coarse fibrous layers, are seen [Plate 10.1(c)].

Vascular Structure of the Gills of *Anabas*

Arterioles of gill filament. These form the vessels through which blood enters the secondary lamellae and are characterized by an endothelial lining, with their characterstic osmiophilic granules, surrounded by a thick coat of connective tissue in which several distinctive types of cells are found [Plate 10.2(a)]. These cells will be described here as pericyte cells, which are characterized by prominent nuclei and lightly-staining cytoplasm in which fibrils are present, and their appearance suggests that they may be contractile. Mitochondria and vesicular bodies also occur in the cytoplasm and many micro-pinocytotic vesicles are attached to the plasma membrane [Plate 10.2(b)]. The pericyte cells are completely enveloped by basement membrane material and their characteristic structure is very similar to that of smooth muscle cells. A few Mast cells with osmiophilic bodies and amoebocytes are found in the connective tissue surrounding the blood vessels [Plate 10.2(a)].

Structure of blood channels within the secondary lamellae. In filaments of the 2nd and 3rd arches, pillar cells form a complete wall between two neighbouring blood channels of the secondary lamellae [Plate 10.3(c)]. There are 4-6 typical columns running through each pillar cell [Plate 10.3(a), (d)].

Each of the columns is formed of fibrillar material continuous with that of the basement membrane and, as in typical secondary lamellae, they are enclosed in deep infoldings of the plasma membrane [Plate 10.3(b)]. The gap between the appressed membranes is about 0.1 μm and there are occasional desmosomes.

Where the pillar cells are in direct contact with one another, with no intervening blood space, the columns run between the plasma membrane of adjacent cells. Thus two pillar cells may have common columns between them [Plate 10.3(c)].

In the secondary lamellae of the first gill, however, the pillar cells often remain separate from one another, as in a typical teleost gill. Each cell has a cylindrical cell body about 5 μm in height and 2-3 μm in diameter and contains a large centrally placed nucleus and a few mitochondria and multi-vesicular bodies [Plate 10.3(b)]. The flangers of these cells enclose the blood spaces and are thicker (0.4 µm) than the basement membrane (0.25 µm. Flanges of adjacent pillar cells are in contact with one another along irregularly shaped boundaries; some interdigitation is present but desmosomes are not often seen.

Pillar cells from the 3rd gill differ from those of the other gills, being shorter (4.3-4.7 µm and thicker (2.74-3.5 µm) structures that remain completely embedded in the basement membrane [Plate 10.4(a)], which covers the cell bodies; it varies in thickness from 0.1-0.7 µm, but at certain points is considerably thicker (1.2 µm). Thickening occurs particularly where the columns originate, the columns also being relatively thick (0.35 µm) Nuclei are bilobed and a well defined Golgi region is sometime recognizable. Elongated mitochondria with a few ill-defined cristae are distributed within the cytoplasm which also contains a few multivesicular bodies and a large number of smooth membraned vesicles.

The marginal channel around the free edge of a secondary lamella. In addition to the pillar cell flanges, this channel is also lined by several endothelial cells. The latter have elongate nuclei and many darkly staining osmiophilic bodies in their cytoplasm. The outer wall of this channel near the base of the secondary lamella is made up of a thick coat of basement membrane in which a layer of pericyte cells may be discerned [Plate 10.4(b)]. The epithelial layers overlying this channel are much thinner than those covering the main surface of the secondary lamella.

Structure of the Labyrinthine Plates

Studies with the light microscope have revealed that each plate is supported by thin laminar bone lying within soft tissues and bounded externally by a respiratory membrane in which vascular and non-vascular areas are present (Figure 10.1). Two types of connective tissue cells can be distinguished under the electron microscope, (a) spindle-shaped cells with long processes and (b) polygonal cells, intermingled with much fibrous material [Plate 10.5(c)].

The basement membrane lying between the epithelium and the connective tissue matrix of the labyrinthine tissue is very thin and is composed of electron-dense material [Plate 10.5(b)]. The thickness of the epithelial layers covering the non-vascular and vascular areas (islets) is quite different.

Non-Vascular Area

In this region the epithelium is multi-layerd and varies in thickness from 10-12 µm [Plate 10.5(a)]. The cells lying next to the basement membrane form a firm layer with desmosomes sometimes occurring at cell junctions [Plate 10.5(b)]. Bundles of fibrillar material (tonofibrillae) are present in the cytoplasm [Plate 10.5(b)]. The epithelial covering of the labyrinthine plates contains many lymphocytes and amoebocytes and even some well defined lymph spaces [Plate 10.5(a)]. The outer-most cell have electron dense cytoplasm with many ill-defined vacuoles [Plate 10.5(a)], Microvilli are found on the outer surface of some of these epithelial cells. Flask-shaped goblet cells are also present in the non-vascular areas.

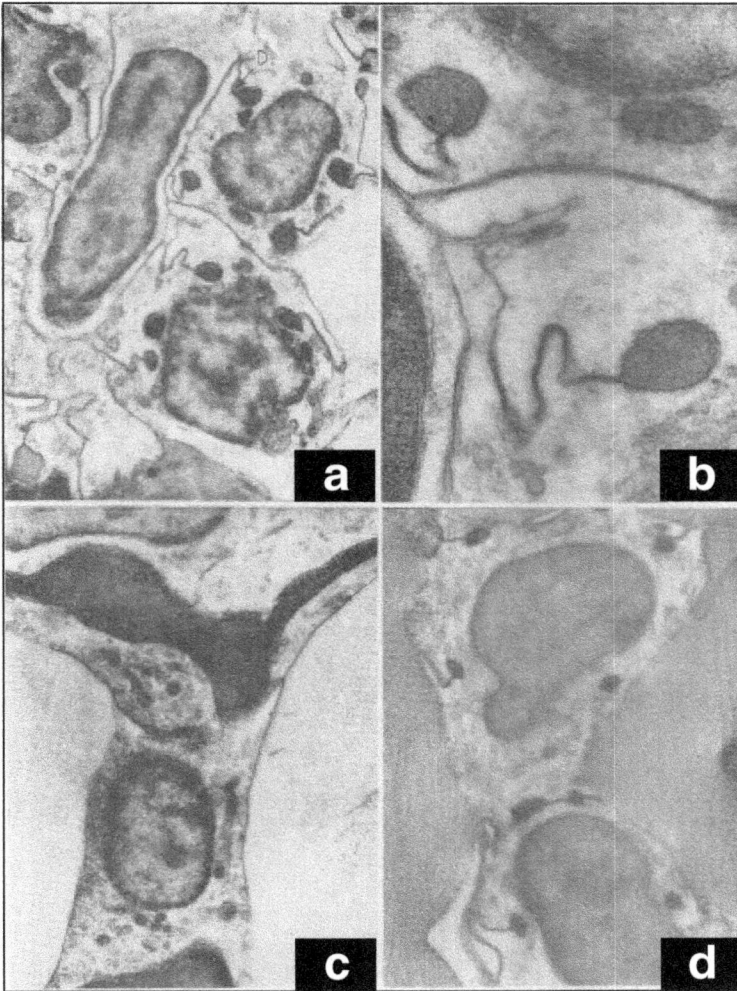

Plate 10.3

(a) Horizontal section passing through the vascular layer of a secondary lamella, showing pillar cells with their columns and desmosomes(D). X 14,000; (b) Part of (a) at higher magnification to show the columns contained within infoldings of the plasma membrane of the pillar cells.X 52,500; (c) Transverse section of a secondary lamella from the first gill arch showing structure of a typical pillar cell. X 14,000; (d) Horizontal section passing through the vascular layer of a secondary gill lamella showing two pillar cells, having common columns situated at the surface between their plasma membranes. X 17,500.

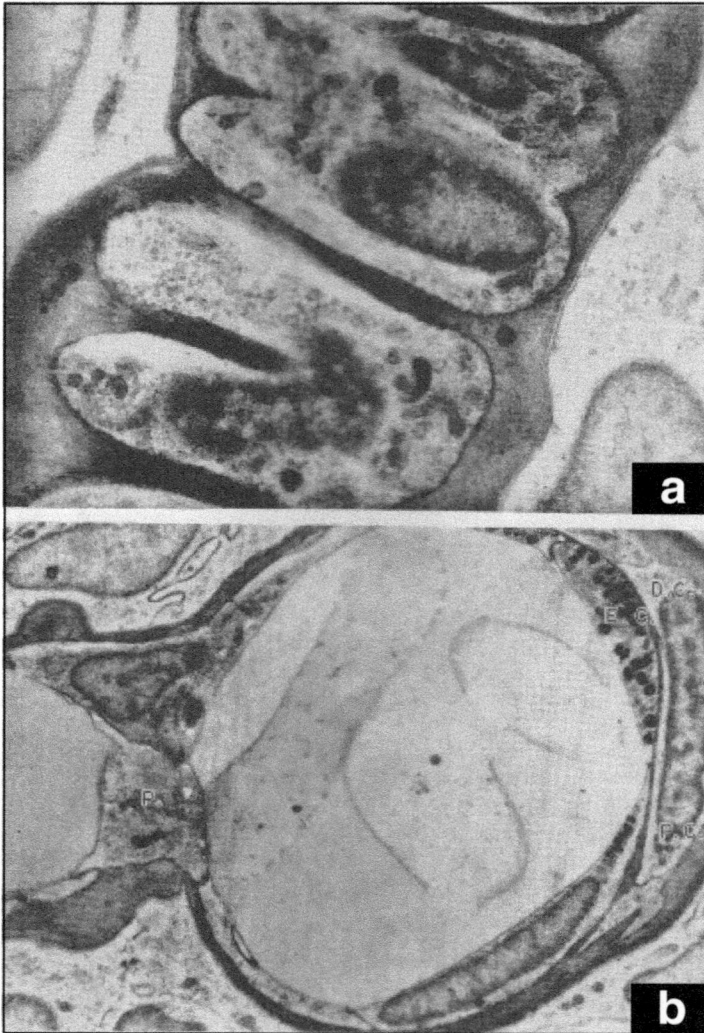

Plate 10.4

(a) Transverse section of a secondary lamella from the third gill arch showing the pillar cell structure. X 12,000; (b) Horizontal section of a primary gill lamella passing through the base of a secondary lamella and showing the structure of the marginal blood channel. Note the osmiophilic granules of the endothelial cell (E.C.), pericyte cells embedded in the basement membrane, and pillar cells. X 6000.

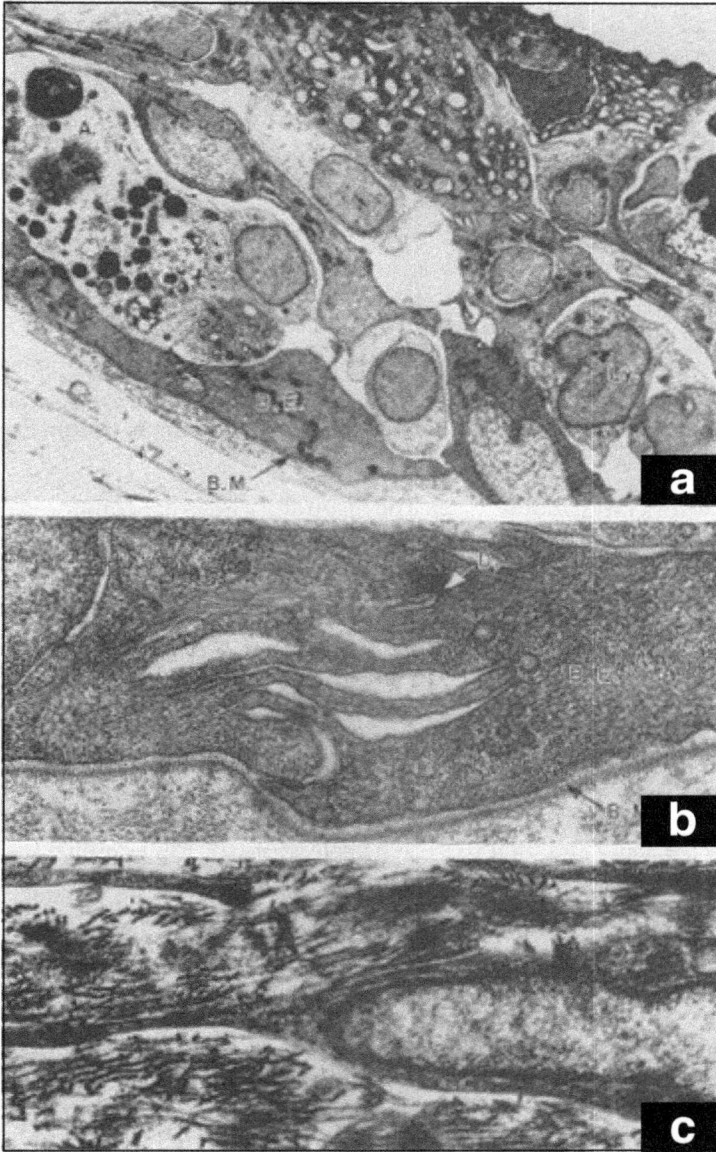

Plate 10.5

(a) Labyrinthine plate. Vertical section passing through a non-vascular region showing darkly-stained outer epithelial cells containing vacuoles, amoebocytes 9A), Lymphocytes (LY) in the lymphoid spaces, basal epithelial cells (B.E.), and basement membrane (B.M.) X 6000; (b) Similar section to (a), at higher magnification, to show the structure of the basal epithelial cells. Notice desmosomes (D) and tonofibrillae in the cytoplasm of these cells X 32,000; (c) Labyrinthine plate. Vertical section showing fibroblasts and coarse fibrillae of the connective tissue matrix. X 30,000.

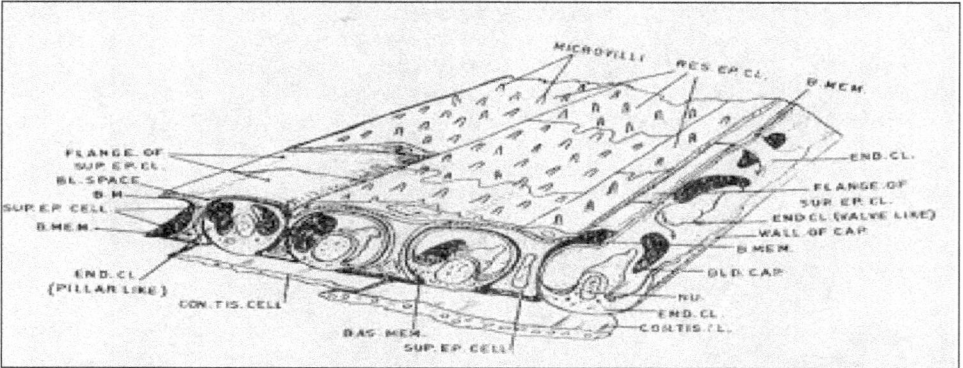

Figure 10.2: A 3-dimensional diagram to show the ultrastructure of a respiratory islet of *Anabas testudineus* as seen under the electron microscope. (Hughes and Munshi, 1973a).

Vascular area (=respiratory islets). The respiratory islets are composed of a series of parallel blood capillaries [Figure 10.1(c)], each channel being made up of a single row of endothelial cells. Their prominent cell bodies project into the lumen of the capillary while their extensions thin out and fuse to enclose the lumen [Plate 10.6(a)]. Many rod-shaped mitochondria, smooth-walled vesicles and a few multi-vesicular bodies are found in the cytoplasm. The endothelial lining of each blood capillary is covered externally by a thin basement membrane which continues round the sides of the endothelial cells, but is not present at their bases [Plate 10.6(a)]. These parallel capillaries are separated from each other, not by pillar cells but by epithelial cells which may be either single or several rolled into a single bundle [Plate 10.6(a)]. These tucked-in supporting epithelial cells give off long flanges that cover the adjacent capillaries externally and their tips often project into the inter-capillary spaces [Plate 10.6(b), (d)]. A basement membrane is present on the inner surface of the supporting epithelial cells and completely encloses the cell bodies.

Thus electron microscopy has shown that the supporting pillars are formed by epithelial cells together with a bordering basement membrane. Cell bodies of the respiratory epithelium invariably lie over the supporting cells and thin out over the external capillary surfaces. Thus each capillary of the respiratory islet is covered externally by a single layer of respiratory epithelial cells as seen in [Plate 10.6(c)] and sometimes also with flanges of the supporting epithelial cells [Plate 10.6(b), (d)]. Occasionally well-defined lymph spaces occur between the epithelium and blood lacuna and may contain lymphocytes and amoebocytes [Plate 10.6(d)].

Chemoreceptor-like Cells

In the non-vascular part of the labyrinthine plates certain specialized cells are found. They are conspicuously large in size, about 8 μm in length and stand out clearly in the electron-dense epithelial cover of the labyrinthine plate [Plate 10.7(a)]. Each cell is spindle-shaped, having a short, slender process at its apex. The nucleus is basal in position and a supranuclear Golgi zone is clearly visible. The granular

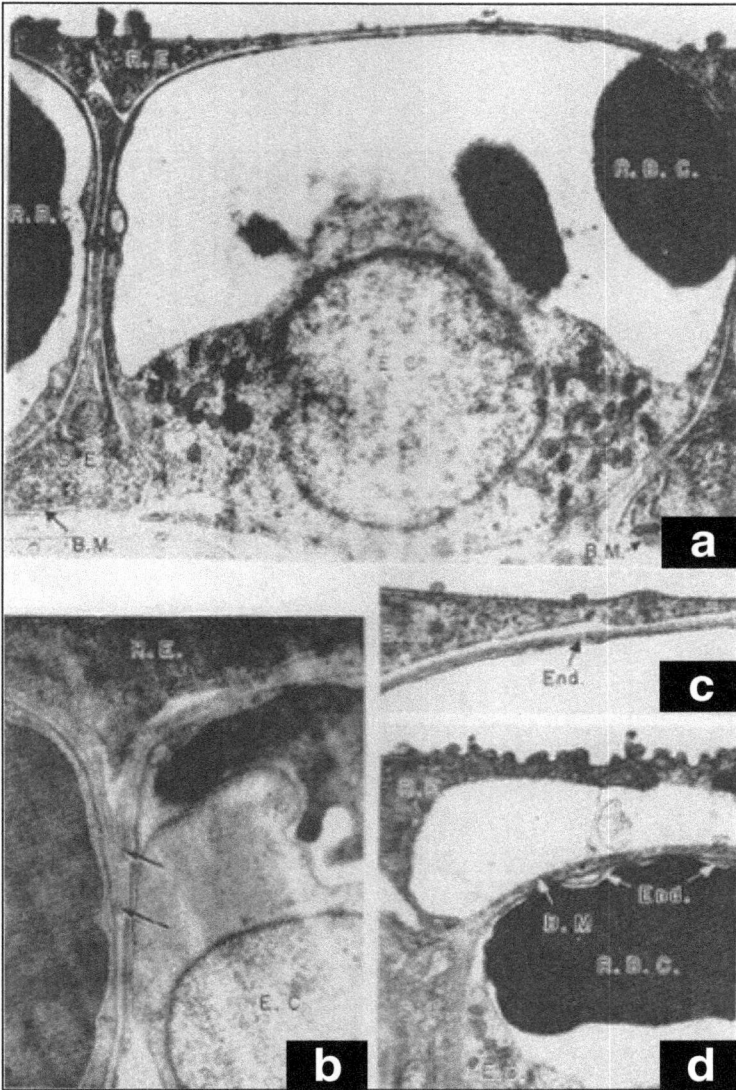

Plate 10.6

(a) Vertical section of labyrinthine plate passing through the respiratory islets (*i.e.* vascular regions), showing the structure of the blood capillaries and supporting inter-capillary epithelial cells (S.E.) and the respiratory epithelial cells (R.E.). Note structure of the conspicuously large endothelial cells (E.C.). The basement membrane is very thin and composed of electron-dense material (B.M.). X 16,000; (b) Respiratory islet of the labyrinthine plate. Vertical section showing the tips (arrows) of the supporting epithelial cells extending into the inter-capillary space. Note fibrillar material in the tongue-like processes of the endothelial cell. X 15,000; (c) Part of (a) highly magnified to show the nature of the air/blood barrier. Note the outer thin respiratory epithelial cell (S.E.), and the basement membrane (B.M.) and endothelium (End.). X 12,000.

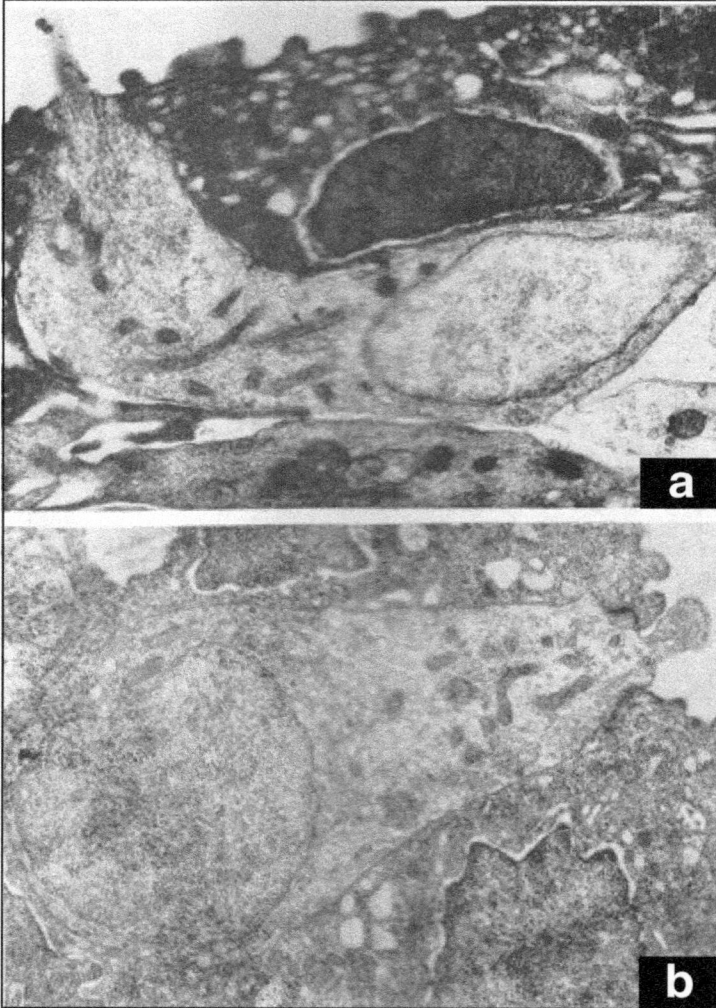

Plate 10.7

(a) Vertical section of a labyrinthine plate showing the general structure of a chemoreceptor. Note the presence of a single project microvillous. Golgi body (G). X 15,000; (b) Structure of a chemoreceptor from the respiratory membrane of the suprabranchial chamber. X 15,000.

cytoplasm contains few mitochondria. In some sections, the basal region of the cell is drawn out and may represent the axonal process. This cell is probably a chemoreceptor and is similar to those described by Whitear (1965, 1971) from the skin of fishes.

Respiratory Membrane of the Suprabranchial Cavity

The fine structure of respiratory islets from the lining of the suprabranchial cavity is essentially similar to that of the labyrinthine plates. These islets consist of many parallel endothelial channels separated by epithelial supporting cells, and a thin basement membrane (0.05 µm) (Figure 10.2). Horizontal sections of these islets show that the wall between two adjacent channels is formed from two flanges of endothelial cells, two layers of basement membrane, and the epithelial supporting cells [Plate 10.8(a), (b)]. The total distance separating adjacent capillaries varies from 0.3-0.7 µm. In vertical sections [Plate 10.9(a)] the relationship between respiratory epithelial cells, supporting pillar-like epithelial cells and the blood capillaries can be distinguished. Endothelial cells have large spherical nuclei (3-6 µm) and very characteristic tongue-like processes 1.0-2.0 µm in length which may act as small valves [Plate 10.8(a), (c)]. These processes project in what is probably the direction of blood flow [Plate 10.8(a)]. The electronmicrographs suggest that the valves can bend at different angles as indicated by the wrinkles in their plasma membrane [Plate 10.8(c), (d)]. A structural basis for contractility may be the fibrillar material present in their cytoplasm [Plates 10.6(b) and 10.8(c)]. These valves modify significantly the nature of the endothelial channels for they seem to divide each channel into a system of ring-like spaces at regular intervalves [Plates 10.8(c) abd 10.9(c)]. More detailed inspection of injected whole mount preparations under the light microscope [Plate 10.10] showed that the appearance of ring-like spaces is probably because of the presence of the valve-like processes. At higher magnification the ring-like spaces are somewhat arrow-shaped [Plate 10.10(b)], presumably due to the directional action of the valves and the pockets formed on both sides of the valves [Plate 10.11(b)].

The air-blood pathway is composed of a double layer of epithelial cells (Single layer of respiratory epithelium and flange of supporting cells = 0.18 µm, a basement membrane (0.05 µm) and thin extensions of endothelial cells (0.01 µm). The total thickness measures 0.2-0.3 µm. The average of the total thickness measurements was 0.21 µm.

Chemoreceptor-like cells. Specialized cells are found in the non-vascular part of the lining of the suprabranchial chamber which have the characteristics of integumental chemoreceptors [Plate 10.7(b)]. These sensory cells have their slender apical parts projecting freely from the surface. The prominent round nucleus is situated in the basal region of the cell body; the Golgi zone is supranuclear in position, and consists of several parallel arranged flattened vesicles. Mitochondria and rough-surfaced vesicles with ribosomes are found in the cytoplasm.

Structure of Vascular Buds and Formation of New Capillaries

The development of new blood channels has been observed at the tips of the primary gill lamellae of the 3rd arch. At the tips of these gill filaments primary afferent and efferent vessels have transverse connections. It is from the wall of these transverse

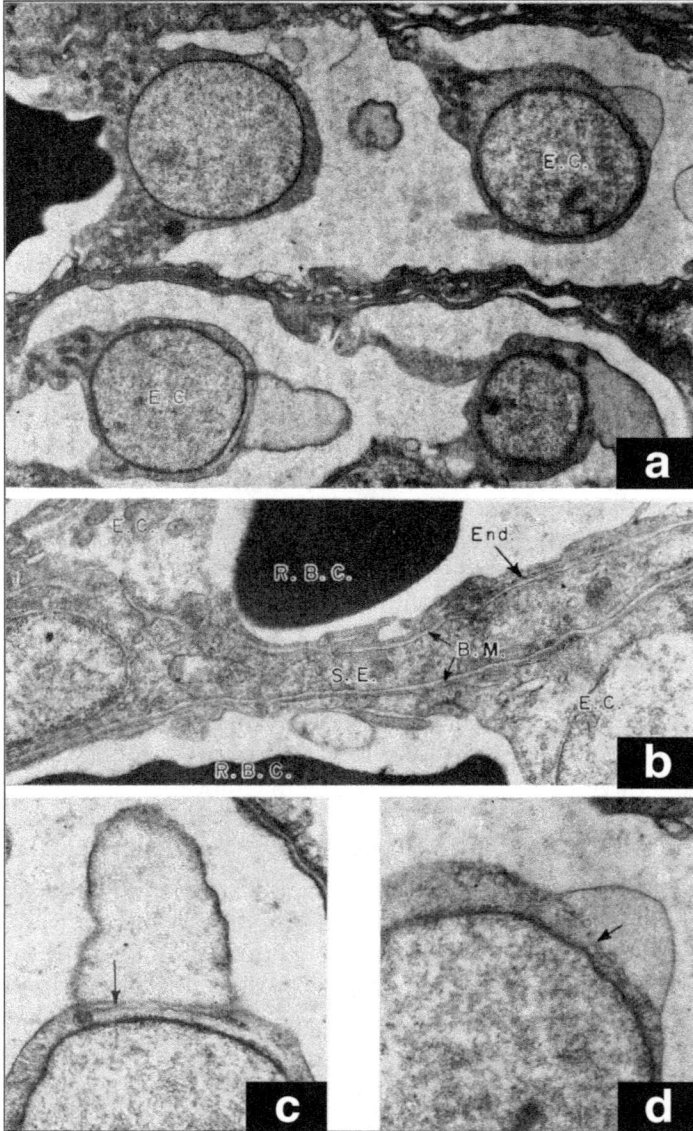

Plate 10.8

(a) Respiratory membrane of the suprabranchial cavity. Horizontal section showing the structure of two adjacent blood capillaries. Note the characteristic structure of the endothelial cells (E.C.) and their tongue like processes. X 8000; (b) Enlarged view of (a) to show structure of the inter-capillary region. Note the position of the supporting epithelial cell (S.E.), covered on both sides by the basement membrane (B.M.), and endothelium (End.) X 18,000; (c) and (d) Enlarged views of endothelial cells from Plate VIII(a) with their tongue-like processes at different stages of projection. Note the presence of wrinkles and pores (arrows) in the contacted plasma membrane of the cells and fine fibrillae in the tongue-like processes. X 16,000.

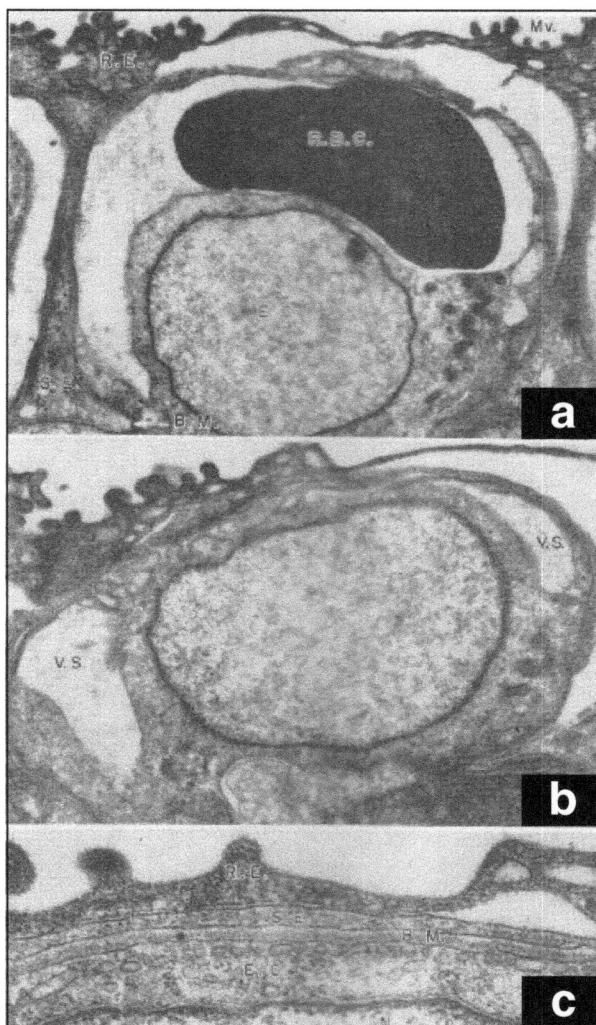

Plate 10.9

(a) Respiratory islet of suprabranchial chambers. Transverse section of blood capillary showing the respiratory epithelium (R.E.), supporting the pillar-like epithelial cell (S.E.), basement membrane (B.M) and an endothelial cell (E.C.). Note the distribution of microvilli (Mv.) on the respiratory epithelium and mitochondria in the cytoplasm of the endothelial cells. X 15,000; (b) Similar section to above but passing through a ring-like blood capillary of the respiratory membrane (compare with Plate X). Note vascular spaces (V.S.) on both sides of the endothelial cell, and the zigzag line of the basement membrane on the outer surface of the endothelial cell. X 20,000; (c) Region of the ring-capillary at higher magnification where the endothelial cell body meets the basement membrane. Note all three layers, outermost respiratory epithelium (R.E.), supporting epithelial cells (S.E.), basement membrane (B.M.), and cytoplasm of the endothelial cell (E.C.). The micropinocytotic vesicles in the cytoplasm of the endothelial cells are also apparent. X 50,000.

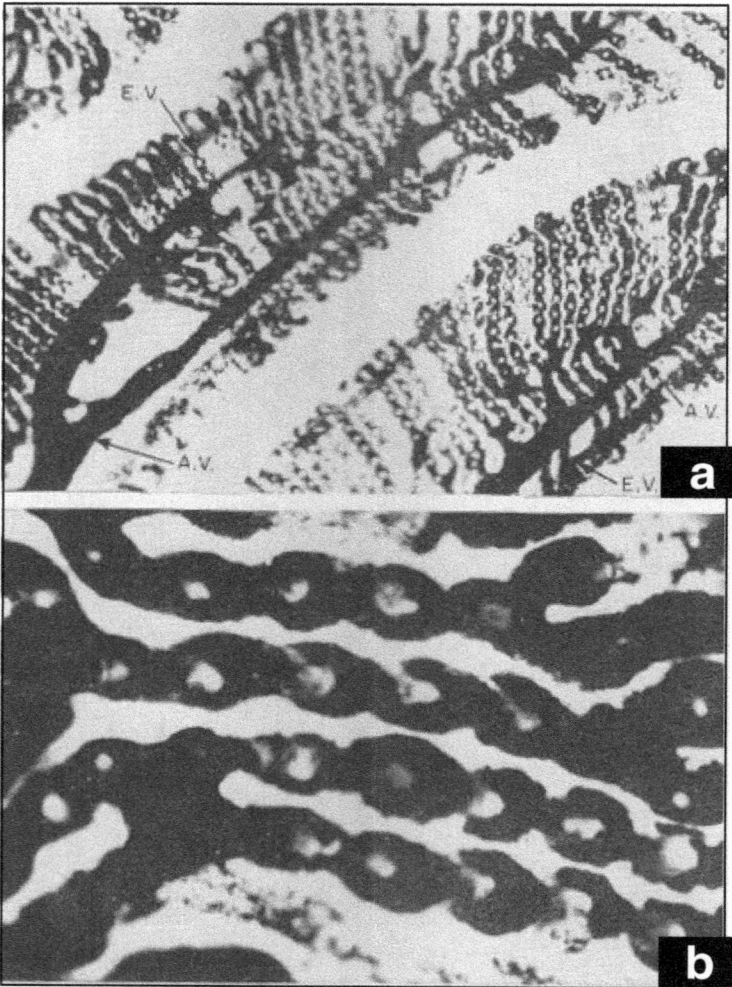

Plate 10.10

Whole mount preparations of respiratory membrane from *Anabas testudineus.*

(a) India ink injected preparation showing the pattern of the blood capillaries. Note the beaded and rin-like appearance of the capillary.

A.V. – Afferent vessel; E.V. – Efferent vessel) x 400.

(b) More magnified view of a part of the same preparation as Plate X(a) to show pocket-shaped capillary spaces. The white spaces within the injected capillary show the position of the endothelial cell bodies. X 1000.

vessels that vascular buds develop. A vascular bud is essentially composed of a group of embryonic mesenchymal cells completely surrounded by a basement membrane [Plate 10.11(a)]. These mesenchymal cells are characterized by having large electron-dense nuclei. In some vascular buds the cell outlines are not clear and many fibres run along their margins. Otherwise the embryonic cells are arranged in a linear order [Plate 10.11(b)] and in some cases ill-defined spaces can be detected.

A vascular bud at an advanced stage of development takes the appearance of a typical middle layer of a secondary lamella. The surrounding membrane is fibrous in nature and gives off some column-like structures that interconnect the basement membranes of opposite sides. Some of the mesenchymal cells take up the position of pillar cells whereas other remain as pericyte cells. Other mesenchymal cells give rise to endothelial cells with typical elongate nuclei and dark osmiophilic bodies in the cytoplasm. There is a great deal of connective tissue between the pericyte cells.

Discussion

Light microscope studies (Munshi, 1968) suggested that several changes have occurred in the nature of the gills of *Anabas* that might be associated with their air-breathing habits. Investigations using the electron microscope have confirmed that the water/blood pathway is thicker than in other teleosts although it is composed of the same three layers; the epithelial layer (5-18 μm), the basement membrane (0.6 μm) and the flange of the pillar cells (0.6 μm), the total thickness being 6-20 μm (Hughes and Munshi, 1968). In some other specimens the thickness may be as low as 5 μm and these are probably of a different variety. The most effective surface for oxygen transfer in *Anabas* gills seems to be the free edge or margin of the secondary lamellae where the total water/blood distance is only 3.5-7 μm. Even this distance is greater than that of many fish such as *Pollack* where it is only about 3 μm (Hughes and Grimstone, 1965), but in other fish, distances about 6 μm are common (Hughes and Wright, 1970). As most CO_2 is released from the gills (Hughes and Singh, 1970). It appears that the thickness of the water/blood barrier scarcely impedes CO_2 transfer, presumably because of its high solubility in water. The increased thickness in *Anabas* is mainly due to the multilayered epithelium, an adaptation which might restrict the collaps of secondary lamellae when the fish is out of water.

There are many mitochondria-rich cells, some of which may develop large vacuoles. These vacuoles are membrane-bounded and may represent enlarged parts of smooth endoplasmic reticulum or invaginations of the plasma membrane.

It must be remembered that *Anabas* is usually found in confined fresh or brackish waters of shallow ponds overgrown with vegetation (Hora, 1935). Therefore the presence of ion-regulating cells in large numbers would be understandable. These mitochondria rich cells differ, however, in structure from the so-called chloride cells of other teleosts which are generally considered to be concerned with ion-regulation. The chloride cells of *Fundulus heteroclitus* have an apical cavity and an extensive system of tubular elements of the agranular endoplasmic reticulum (Philpot and Copeland, 1963), and so there are certainly some similarities. Amoebocytes of the epithelium contain lysosomes and vacuoles and probably serve to protect the fish

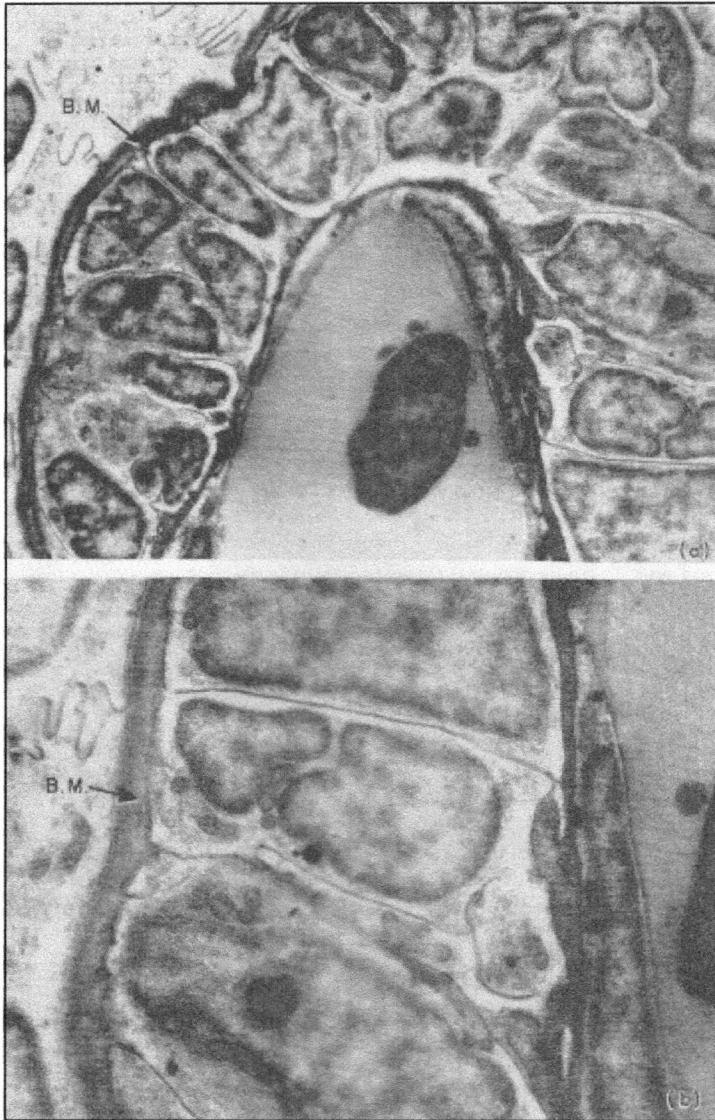

Plate 10.11

(a) Section through a developing secondary lamella bud at the tip of a primary gill filament. Note the nest of embryonic mesenchymal cells covered by thick basement membrane. X 7000; (b) Enlarged view of a part of Plate 10.11(a) to illustrate linear arrangement of the mesenchymal cells. Note columns arising from the basement membrane and running between the mesenchymal cells. X 14000.

from infection by protozoan parasites and bacteria, which are frequent in the stagnant and polluted waters that they inhabit.

The pillar cell columns sometimes remain at the surface rather than being contained in infoldings of the plasma membrane. This occurs especially where there are blocks of contiguous pillar cells, and may have been the original position. Infolding could have evolved later and might serve to reduce the danger of any thromboplastic effect that the collagen in the columns might have (Hughes and Weibel, 1972).

Some observations made on the developing secondary lamellae of *Anabas* bring out several interesting features about the nature of pillar cells. Pillar cells arise from the mesenchymal cells of the filament vascular bud. The embryonic cells remain completely enclosed within the basement membrane, and later become arranged in a linear fashion with the connective tissue material penetrating between the cells. After the appearance of vascular spaces between these cells, they assume the role of pillars. To begin with, the connective tissue matrix enveloping the cells is exposed to the vascular space but later it becomes incorporated in infoldings of the pillar cell plasma membrane.

Other mesenchymal cells differentiate as pericyte cells in the arteriole walls of the primary gill lamellae, and in the marginal blood channels of the secondary lamellae. Pericyte cells in such positions remain surrounded by the basement membrane. These cells have all the characteristics of smooth muscle cells and our observations support the view of Pease (1962), who described a similar situation in the case of smooth muscle cells of arteries, which have the general basement membrane as their sarcolemnal envelope. Thus the pericyte cells appear to represent undifferentiated mesenchyme cells in various stages of transformation into smooth muscle cells.

According to Fawcett (1966), "the filaments of smooth muscle cells are so thin that they are resolved with difficulty, and the cytoplasm looks surprisingly homogeneous even at moderately high magnifications. There are clusters of ribosomes, and few mitochondria ..." In the case of pillar cells, also, it is difficult to distinguish fibrils but nevertheless they can be seen.

In the gill lamellae of *Anabas* it appears that most mesenchyme cells differentiate into pillar cells and endothelial cells, but a few may develop into pericyte cells (— smooth muscle cells). There are two main possibilities regarding the development of pillar cells;

(*i*) Mesenchychymal cells of vascular bud
   (*a*) Pillar cells
   (*b*) Endothelial cells
   (*c*) Pericyte cells or smooth muscle cells
(*ii*) Mesenchymal cells of vascular bud
   (*a*) Endothelial cells
   (*b*) Pericyte cells
   (*c*) Pillar cells

Since most pillar cells possess filaments of the myo-filament type as reported by Hughes and Grimstone (1965), and Newstead (1967) also observed them in his study, it is possible that mesenchyme cells may differentiate into smooth muscle cells which later take on the shape and position of pillar cells. In a recent study of trout gills, however, the pillar cells were found to differentiate directly from filament mesenchymal cells, which assemble under the basal lamina where capillary loops between the afferent and efferent filament vessels are forming (Tovell, Morgan and Hughes, 1970). There is also evidence for contractile actomyosin-like protein in these pillar cells (Bettex-Galland and Hughes, 1972). Pillar cells have the dual function of protection against distension or collapse of the vascular spaces, and regulation of the pattern of blood flow through the secondary lamellae possibly by their differential contraction.

Many attempts have been made to explain the origin of collagen fibrils in the columns of the basement membrane (1) According to Plehn (1901), spindle-shaped cells of the lamellae form the "basal membrane". (2) Hughes and Grimstone (1965) suggested that the basement membrane and columns are produced by the pillar cells (3) Newstead (1967) thought it possible that the deposition of a filamentous basal lamina adjacent to the epithelium results from a similar interaction between some product of the epithelial cells with materials derived from a source external to the epithelium (*e.g.* fibroblasts, endothelial cells or pillar cells).

If pillar cells are modified smooth muscles cells, it is unlikely that they secrete tropocollagen and are responsible for the elaboration of collagen fibrils. It is of course possible that different layers of the so-called basement membrane have different origins. The closeness of the columns to the pillar cells suggests that at least the collagenous layer of the basement membrane develops from this source (Hughes and Grimstone, 1965). The clear and fine fibrous layers which constitute the basal lamina may be secreted by the epithelial layers.

Labyrinthine Organs and the Respiratory Membrane

A number of authors have drawn attention to similarities between the appearance of capillaries in the respiratory islets and the blood channels of the secondary lamellae. Rauther (1910) considered that typical pillar cells were present in the vascular endings beneath the epithelial surface of the labyrinth and walls of the suprabranchial chamber, Henninger (1907) thought them to be only "connective tissue rods" which penetrate between the blood channels.

Newstead (1967), however, comments that "the observations of Schultz should probably be verified with material prepared by the current, improved methods of preparation for electron microscopy". Munshi (1968) found that capillaries of the respiratory islets are simple endothelial tubes that receive further support from the pillar like cells, and he compared them with the capillaries of amphibian gills. The finest blood capillaries of these gills posses an endothelial lining in addition to the presence of pillar cells (Faussek, 1902). It was further thought by Munshi (1968) that perhaps in the respiratory islets of *Anabas*, pillar cells have the same relationship as the smooth muscle cells have with the minute arterioles. The present study has now clearly established that the so-called pillar cells of the respiratory islets are infoldings of epithelial cells and that the blood channels are endothelial tubes situated below the basement membrane. They should, therefore, be regarded as true sub-epithelial capillaries. Furthermore, in India ink injected preparations of the respiratory membrane each vessel looks like a string of black beads (Munshi, 1968). The reason for this appearance Plate 10.10(a) was not previously understood but from the electron micrographs it can be seen that endothelial cells divide each channel into systems of ring like spaces and function as valves of pillars. These valves presumably control the flow of blood through the respiratory islets. In fact, each valve may serve to deflect individual blood cells towards the respiratory surface and serve to reduce the resistance to gas transfer. The presence of mitochondria in large numbers suggests that the energy requirements of these cells may be quite high.

Examination of the fine structure of the labyrinthine organs shows that its non-vascular part resembles quite closely the structure of fish skin. The single celled chemoreceptors found in the labyrinthine organs and respiratory membrane are similar to those in the skin of fishes (Whitear, 1965, 1971). It is, therefore, well worth considering once more the views of Zograff (1888) who suggested that these structures were specialized parts of the skin. The presence of chemoreceptors in the labyrinthine organs and respiratory lining of the suprabranchial chamber is of further interest, as many attempts to detect chemoreceptors on the gill epithelium have been unsuccessful. On the gills themselves the only chemoreceptors present are the taste buds and these are mainly near the gill rakers. It would be interesting to know whether the chemoreceptors observed in the present study play any role in the control of the respiratory movements of *Anabas*, and especially whether they are responsive to changes in O_2 or CO_2 tensions of the air in the suprabranchial chambers. Recent measurements show that *Anabas* surfaces for air when the PO_2 of the gas in the suprabranchial chambers falls below 40 mm Hg (Singh and Hughes, 1973). Chemoreceptors of the type described here might be involved in such responses.

As has been pointed out, the morphological adaptation of the accessory air-breathing organs in *Anabas* are extremely detailed and electron microscopy emphasizes this still further. The air/blood pathway in the case of the respiratory islets measured from 0.12-0.3 μm. That of the labyrinthine plates was similar. These appear to be some of the thinnest diffusion barriers observed in respiratory organs, for even in the alveoli of the mammalian lung the barrier is about 0.2 to 2.5 μm, and to compensate for the relatively smaller area that their diffusing capacity greatly exceeds that of the gills (Hughes, Dube and Munshi, 1973).

Summary

The structure of the three types of respiratory organ in *Anabas testudineus* (Bloch) has been studied using electron microscopy.

The organs of gaseous exchange with water are the gills, the water/blood pathway has been shown to be composed of the same basic layers as are found in other fish. The epithelium is multilayered and contributes most of the 5-20 μm of the water/blood barrier. The epithelium overlying the marginal capillary round the free edge of the secondary lamellae is much thinner than elsewhere, being only 3.5-7 μm thick.

The air-breathing organs consist of the lining of the suprabranchial chamber and the labyrinthine plates contained in that chamber. These surfaces contain the so-called respiratory islets with complex capillary networks. It is shown that the similarities of these parallel blood channels to those of the secondary lamellae are only superficial. The presence of pillar cells separating the blood channels has not been confirmed at the higher magnifications. The separation is due to infoldings of the surface epithelium together with its attached basement membrane. The capillaries are therefore true infra-epithelial capillaries as they lie below the surface epithelium and its basement membrane.

The fine structure of the non-respiratory parts of the air-breathing organs is similar to that of the skin and includes chemoreceptor-like cells.

Evidence concerning the developmental origin of pillar cells is discussed.

We should like to thank the Smithsonian Institution, Washington, D.C., for making funds available for the travel of Professor Munshi and the supply of fish, through a co-operative programme with Banaras Hindu University. We also thank Dr. Stanley Weitzman of the Division of Fish for the helpful comments. The Nuffield Foundation also gave valuable financial support and the electron microscope was provided through a grant to G.M.H. from the Science Research Council.

Chapter 11

Cytology of Macrophages in Normal and Mercury-Treated Air-Breathing Fish, *Channa punctata* (Bloch)

Macrophages are mononuclear phagocytes present in spleen, kidney, heart and the mesentery of fish (Ferguson, 1975; Ellis *et al.*, 1976) with high pinocytic property (Ellis, 1977). Normally they are not found as a component population of circulating leucocytes. Macrophages have great importance as scavengers of the dead and foreign materials in animals (Ellis, 1977) and therefore may be used as an indicator of the pathological state of the fish. Such an assumption forms the basis of the present cytological study of macrophages.

Live specimens of *Channa punctata* (Bloch) (34-36 g) were acclimatized to laboratory conditions for 15 days, during which they were fed chopped goat liver. Groups of 10 fish were exposed separately to various concentrations of mercury in plexiglass aquaria (20 l). Ambient water was changed daily to restore the mercury level. The lethal concentration (LC_{50}) value was calculated. Twenty fish were then exposed to sublethal concentration of $HgCl_2$ (0.248 ppm) and five control and experimental fish were sacrificed on the 7th and 15th days for histochemical tests.

Acid phosphatase and Perl's histochemical tests (Pearse, 1968) were performed on smear, squash and sections of the spleen and kidney to demonstrate the density and distribution of macrophages and their intensity of reaction in the control and experimental fish.

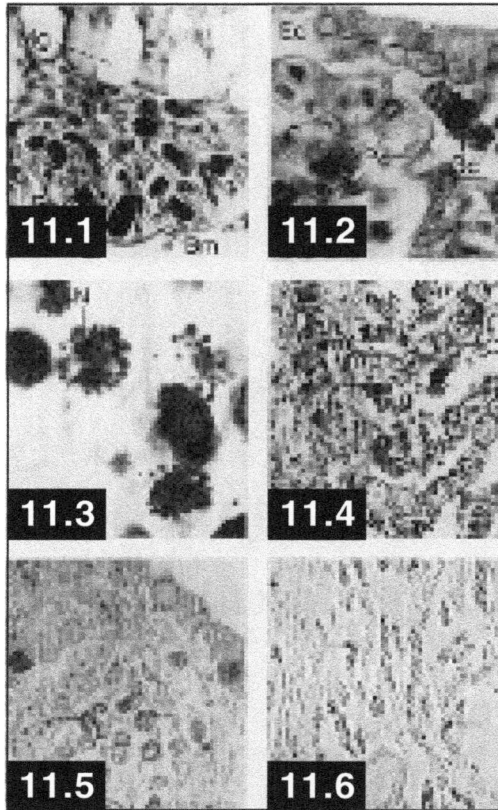

Figure 11.1: A transverse section of the primary gill lamella of *Channa striatus* showing large granular acidophil cells (Gc), the basement membrane (Bm) and the mucous glands (Mg). The granular cells are mainly present in the epithelium. Haematoxylin-eosin stain. 1500: 1.

Figure 11.2: A section of the secondary gill lamella of *Channa striatus* showing a large acidophil granular cell (Gc) in close relationship with the blood capillary (c) and the outermost layer of the epithelium (Ec); Pc pericapillary cell. Haematoxylin-eosin stain 1500:1.

Figure 11.3: An oil immersion micro-photograph of the granular cells of *Channa marulius* showing mitochondrial nature of their granules. The nucleus (N) is seen eccentric in position in one of the cells. Flemming's method for mitochondria. 1940:1.

Figure 11.4: Photomicrograph showing PAS positive granular cells in the epithelium after hot chloroform-methanol extraction of the gill tissue of *Channa marulius*; Bm basement membrane, C. capillary. 530:1.

Figure 11.5: Photomicrograph showing yellowish green stained granular cells (arrows) in the gill epithelium of *Channa marulius* by toluidine blue followed by a mixture of ammonium molybdate and potassium ferricyanide. 700:1.

Figure 11.6: Photomicrograph showing the presence of tyrosine (arrows) in the granules of the acidophil cells of labyrinthine organ of *Channa marulius*. Millon reaction for tyrosine. 700:1.

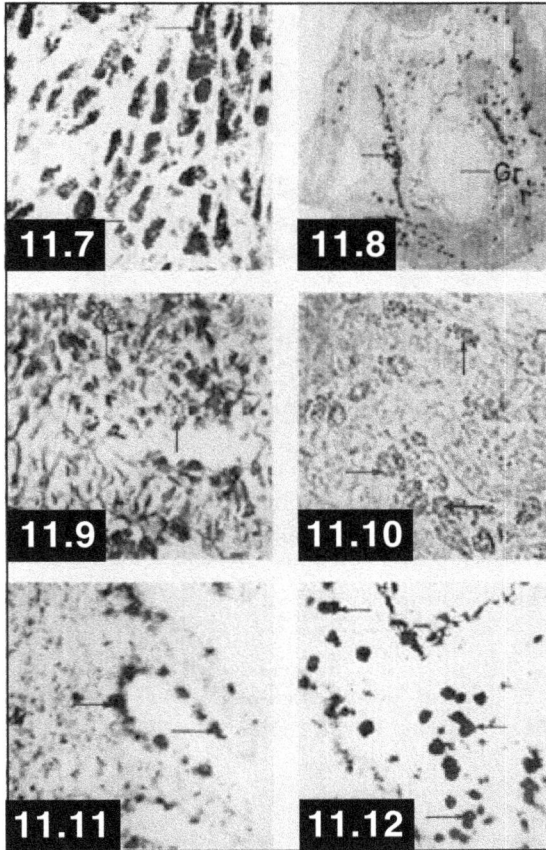

Figure 11.7: Photomicrograph showing pyronin positive granules (arrows) in the acidophil cells of the labyrinthine organ of *Channa marulius*. Methyl green pyronin test, 750:1

Figure 11.8: Photomicrograph of a transverse section of the primary gill lamella (Gr) of *Channa marulius* showing the granular cells (arrows) giving strong positive reaction with acid haematein test for phospholipids. Acid haematein method. 115:1.

Figure 11.9: Photomicrograph showing positive alkaline phosphate reaction (arrows) in the acidophil cells of *Channa marulius*. The nuclei for alkaline phosphate. 760:1.

Figure 11.10: Photomicrograph of a section of a primary gill lamella of *Channa marulius* showing after reduced black granules (arrows) of the acidophil cells. Masson-Fontana method for melanin (with hexamine silcer variant). 640:1.

Figure 11.11: Photomicrograph of a longitudinal section of a primary gill lamella of *Channa striatus* showing strong chloride reaction (arrows) in the granular acidophil cells and some of the mucous glands of the secondary gill lamellae. AgNO $_3$/HNO$_3$ test for chlorides after 2 h. of 0.65 per cent normal saline injection. 500:1.

Figure 11.12: Photomicrograph showing strong positive chloride reaction (arrows) in the granular acidophil cells of the primary gill lamella of *Channa striatus* (experimental fish). AgNO$_3$/HNO$_3$ test for chlorides after 2 h of saline injection in an experimental fish, 450:1.

Acid Phosphatase Reaction in Spleen and Kidney

In control fish comparatively larger numbers of macrophages with high acid phosphatase activity were observed in the spleen than in the kidney (Figure 11.1). When the fish were exposed to a sublethal concentration of $HgCl_2$ (0.248 ppm) for 7 days the macrophages of the spleen and kidney showed a clear increase in their acid phosphatase activity (Figures 11.2 and 11.5). This may be related to the activation of the enzyme (Hossain and Dutta, 1986). On the 15th day of exposure the acid phosphatase reaction decreased (Figures 11.3 and 11.6), suggesting that the fish were adapting to the mercury.

Pearl's Reaction in Spleen and Kidney

The Perl's reaction showed the Prussian blue granules in the splenic and renal macrophages which indicate the presence of iron in ferric (Fe^{++}) form temporarily stored in the macrophages as ferritin and haemosiderin owing to phagocytosis of affected erythrocytes. This normal phenomenon is corroborated by the mild positive reaction in the control fish (Figure 11.7). There was a marked increase in these granules in all the exposed fish but they were more densely packed in the macrophages of the fish exposed for 7 days (Figure 11.8) than in those exposed for 15 days (Figure 11.9). Increased intensity of Prussian blue reaction in the splenic and renal macrophages after exposure to mercury for 7 days suggests destruction of RBC in the blood due to mercury pollution.

It may be concluded that cytochemical studies of fish macrophages may provide a potentially useful indicator of heavy metal pollution of water bodies.

Chapter 12

Oxygen Uptake in Teleostean Fishes

1. Comparative Study of the Gill Surface Area of an Indian Shad *Hilsa ilisha* (Ham.) and A Major Carp *Labeo rohita* (Ham.)

A description of the gill structure of an anadromous clupeoid fish *Hilsa ilisha* and a riverine carp *Labeo rohita* is given. The gills of *Hilsa* were more heterogenous than those of *Labeo*. *Hilsa* had lesser number (2552) of shorter (9.18 mm) filaments, whereas, in *Labeo* of the same body weight greater number (3268) of longer (9.94mm) filaments were observed. As a result of these differences the total filament length (2343.78mm) in *Hilsa* was less than that of *Labeo* (32481.4 mm). Frequency of secondary lamellae was slightly less in *Hilsa* (59.67 mm) than that of *Labeo* (61.5 mm). The area of an average secondary lamella in *Hilsa* (0.1904 mm^2), was three times larger than that of *Labeo* (0.0594 mm^2). As a consequence of these differences, the total gill area of *Hilsa* (266275.44 mm^2) was nearly 2.25 times greater than that of *Labeo* (118673.73 mm^2) of the same body weight. Value of total gill area of *Hilsa* which is a migratory fish falls in the range of active fishes.

Introduction

The gills of fishes exhibit wide variations in their structure and respiratory surface area (Gray 1954, Oliva 1960, Hughes (1966b). A number of morphometric studies have been made, of different components of gill sieve especially of Indian airbreathing fishes (Dubale 1951, Saxena 1962, Hughes *et al.*, 1973, 1974, Munshi 1976, Hakim *et al.*, 1978). Some information is available on the structure of gills of *Hilsa ilisha* and *Labeo rohita* (Munshi, 1960) but studies on morphometrics of gill sieve of these species are meager. The present work is contemplated to make a comparative morphometric study of the gills of specimens of *Hilsa ilisha*, which is an anadromous migratory fish, and *Labeo rohita*, an Indian major carp of the same body weight (400 g) live in the ponds as well as in River Ganges.

Materials and Methods

Live specimens of *Hilsa ilisha* were collected from the river *Ganges* near Farakka Barrage and *Labeo rohita* from the same river at Bhagalpur. After taking the fresh weight of the fish, the opercula were removed and continuous flow of distilled water was passed over the gills in order to clear the mud. The heads were decapitated and fixed in aqueous Bouin's fixative and stored in the refrigerator. All the four gill arches from both the sides were dissected out separately. Each gill arch was divided right from dorsal side into sections of 16 filaments. To estimate the total filament length, the first and last filaments were also measured (Hughes and Ojha, 1986). The gill area was estimated according to the method described by Muir and Hughes (1969) and Hughes and Morgan (1973). All the measurements were made under dissecting binocular microscope and an improved variety of camera lucida (Ermascope).

Observations

Hilsa ilisha

Hilsa ilisha had four pairs of holobranchs. The gill filaments were borne by epi-cerato and hypo-branchials. The gill arch is 'Z' shaped and oriented in such a fashion that the gill lamellae of dorsal side moved away towards the auditory region and freely attached to the base of cranium by means of a membrane. The gill septum was moderately long and extended upto half the length of a primary lamella. The gill rakers were long, slender and serrated.

The filaments of both the hemibranches showed variation in their length at different regions of the gill arch. The filaments of the oral hemibranch of each gill arch were quite shorter in length than their counterparts of the aboral side. Distance between adjacent filaments was directly proportional to the length of filaments.

Secondary lamellae were arranged alternately on both the sides of the gill filament. Frequency of the secondary lamella was 59.67/mm. Higher secondary lamellar frequency was recorded on oral hemibranch. The profile of secondary lamellae varied in different regions of the gill arch.

The secondary lamellae of the base and middle parts of the filament were almost identical but at the tip region they were comparatively smaller in size.

The bilateral surface area of a secondary lamella was proportional to the length of the filaments. The average bilateral surface area of lamellae was 0.1903 mm^2 (Table 12.1).

Heterogeneity was observed in the gill area of different gill arches. Maximum area was recorded in the first gill arch (Table 12.1). The gill area decreased gradually from first to fourth gill arch. Total gill area in *Hilsa ilisha* was 266275.44 mm^2 (Table 12.1).

Oxygen Uptake in Teleostean Fishes

223

Table 12.1: Detailed measurements of the gill parameters on different gill arches from 400g specimens of *Hilsa ilisha* and *Labeo rohita*.

Gill Parameters		1st Arch Oral Hemibr.	1st Arch Aboral Hemibr.	2nd Arch Oral Hemibr.	2nd Arch Aboral Hemibr.	3rd Arch Oral Hemibr.	3rd Arch Aboral Hemibr.	4th Arch Oral Hemibr.	4th Arch Aboral Hemibr.	Total Arches Both Side	Total Arches Pseudobr. Both side
No. of filaments	Hilsa	171	171	170	170	150	154	135	155	2552	74
	Labeo	204	206	203	206	203	206	201	205	3268	
Av.fil.length (mm)	Hilsa	9.46	10.77	8.35	10.89	8.63	9.64	6.62	8.4	9.184	5.37
	Labeo	9.99	10.65	9.75	10.43	9.64	10.18	9.06	9.78	9.939	
Total fil.length (mm)	Hilsa	1619	1841.9	1419.5	1852	1295	1485	894	1312.5	23437.8	412
	Labeo	2039	2194.2	1980.5	2149.5	1956.5	2096.5	1821	2003.5	32481.4	
Sec. Lam. (mm)	Hilsa	56.42	57.19	55.85	56.37	59.11	59.14	58.81	58.63	59.67	52.4
	Labeo	59.35	58.87	61.28	60.73	63.24	62.65	63.90	62.47	61.5	
Total Sec. Lam.	Hilsa	91357.88	105356.26	79289.68	104399.6	76551.15	87821.97	52572.86	76953.78	1398606.2	21587.98
	Labeo	121021.6	129174.48	121380.04	130547.32	123728.07	131350.75	116418.27	125170.57	1997582.0	
Av.Bil. Sur.Area (mm²)	Hilsa	0.1566	0.2092	0.1861	0.2035	0.1846	0.1993	0.2064	0.2377	0.19038	0.1799
	Labeo	0.0682	0.0686	0.0609	0.0569	0.0549	0.0569	0.0565	0.0529	0.0594	
Total surface area (mm)	Hilsa	14310.62	22043.85	14755.46	21243.30	14132.59	17505.57	10854.02	18292.3	266275.44	3883.46
	Labeo	8254.67	8868.43	7320.711	7430.245	6789.474	7475.05	6575.176	6623.089	118673.73	
Area/g (mm²)	Hilsa	35.776	55.109	36.888	53.108	35.331	43.764	27.135	45.731	665.683	0.9711
	Labeo	20.636	22.171	18.30	18.575	16.974	18.687	16.438	16.438	296.684	

The pseudobranch remained free and exposed to water in the opercular cavity, attached to its inner surface. A row of 37 parallel filaments measuring 5.57 mm average length formed the pseudobranch on each side. Numerous wing like lamellae were arranged alternately on the filament. The number of secondary lamellae/mm was 52.4 on both sides of the filament. The average bilateral surface area of a secondary lamella of pseudobranch was 0.17989 mm^2 for blood to come in contact with water (Table 12.1).

Labeo rohita

Gills in *Labeo rohita* was homogenous structure with greater number of filaments than those of *Hilsa ilisha*. Two rows of filaments were borne by the epi and cerato-branchial of each gill arch. The gill septum extended half way down the length of the filaments. Gill rakers were short and flat situated on both sides of the gill arch.

Filaments on the oral hemibranch were shorter than their counterparts of the aboral hemibranch. Length of the filament was greater at the point where the hemibranch made a right angle turn to their dorsal regions. The distance between adjacent filaments was directly proportional to the length of filament.

Secondary lamellae were arranged alternately on both the sides of the gill filament. The base of secondary lamellae remained mostly embedded in the tissue of the filament. Further, they remained fused near the tip of the filaments. The number of secondary lamellae/mm was 61.49. The profile of secondary lamellae varied in different regions of the gill arch. Secondary lamellae were comparatively smaller in size and remained fused with each other. Average bilateral surface area was 0.0594 mm^2. Maximum gill surface area was recorded in the first gill arch. Gill area decreased gradually from first to fourth gill arch. The total gill surface area was 118673.73 mm^2 (Table 12.1).

Discussion

Both *Hilsa ilisha* and *Labeo rohita* have long "gill septum". In *Labeo rohita* the gill lamellae are borne only by the epi and cerato-branchial; but in *Hilsa ilisha*, in addition to epi-cerato-branchials, the pharyngo and hypobranchials also bear gill filaments. In *Hilsa ilisha*, the distal tip of the gill filaments are free from one another, but in *Labeo rohita* they unite with one another at their tips are in blocks. The gill rakers are long, slender and serrated in *Hilsa ilisha*, while in *Labeo rohita*, the rakers are short and flat leaf like structure situated on both sides of the gill arch. The extensive gill sieve apparatus in *Hilsa ilisha* has developed as a plankton filtration device.

Gills of *Hilsa ilisha* are heterogenous in nature, the length and number of filaments decreasing gradually from first to fourth gill arch. But in *Labeo rohita* the gills are homogenous and have almost the same number of filaments of same length on each gill arch. *Hilsa ilisha* has smaller number and short filaments than that of *Labeo rohita*.

In both the fishes, secondary lamellae are arranged alternately on both the sides of the gill filament. In *Hilsa* most parts of the lamellae remain free in contrast to *Labeo rohita* where the bases of the lamellae remain embedded in the tissue of the gill filaments. Further they remain fused at the tip of the filaments.

The number of secondary lamellae/mm both in *Hilsa ilisha* and *Labeo rohita* fall in the range of aquatic breathers like *Catla catla* (Kunwar 1984), *Scomber scomber* and *Coryphaena hippurus* (Hughes 1970b) of same body weight.

Gill area of *Hilsa ilisha* is more than two folds of *Labeo rohita*. Lower value of gill area in *Labeo rohita* is mainly due to the short bilateral surface areas of lamellae. The average bilateral surface area of lamellae of *Hilsa ilisha* is three folds greater than that of *Labeo rohita*.

Table 12.2: Comparison of values for gill surface area of a 400g
***Hilsa* and *Labeo* with different fish species.**

	Fish Species	Area (mm²)	References
A.	**Water breathing**		
(i)	*Active fish*		
	Katsuwonus pelamis	849679.42	Ursin 1967
	Thunnus albacares	596006.31	Muir and Hughes, 1969
	Hilsa ilisha	266275.44	Present authors
ii)	*Intermediate active fish*		
	Gray's intermediate	189377.51	Ursin 1967
	Blennius pholis	188254.95	Milton, 1971
	Scomber scombrus	166618.08	Hughes, 1970 b
	Mystus cavasius	148369.97	Singh 1979
	Stizostedian vireum Vireum	136651.37	Nimi and Morgan 1980
	Labeo rohita	118673.13	Present authors
iii)	*Sluggish Fish*		
	Salmo gairdneri	84303.19	Nimi and Morgan 1980
	Opsanus tau	637323.07	Hughes and Gray, 1972
	Mystus vittatus	57913.33	Singh 1979
	Botia lohachata	60180.17	Sharma *et al.*, 1982
B.	**Air-breathing *Lepisosteus aculatus***		
	L. asseus	32762.37	Landolt and Hill 1975
	L. platostomus		
	Clarias batrachus	24512.14	Munshi *et al.*, 1981
	Anabas testudenius	22148.37	Hughes *et al.*, 1973
	Boleophthalmus boddaerti	19680.23	Biswas *et al.*, 1981
	Macrogachua aculeautum	17553.74	Ojha and Munshi, 1974
	Channa punctata	16336.26	Hakim *et al.*, 1978
	Heteropneustes fossilis	16251.11	Hughes *et al.*, 1974A
	Channa striata	14169.48	Choudhary, 1978

Pseudobranch

In *Hilsa ilisha,* the pseudobranch is free and its lamellae project into the subopercular cavity. In *Labeo rohita* it is of glandular type where the organ is sunk below the epithelium. Muller (1839) has categorized the teleostean pseudobranchs into two types, the free type as seen in *Hilsa* and glandular as in *Labeo*. Granel (1923) and Leiner (1938) have distinguished four types of pseudobranchs that vary in their degree of isolation from the external environment.

In *Hilsa ilisha*, each filament of pseudobranch bears numerous wing-shaped lamellae. These lamellae are supplied with blood that has already been oxygenated in the hemibranch of the first branchial arch; therefore, these are non-respiratory in function (Hughes, 1984). The compound lamellar surface areas of pseudobranch is 3883.46 mm². The expired water coming out of the branchial chamber contains less amount of oxygen and the blood flowing through the pseudobranch may be having higher partial pressure. As such partial pressure difference PO_2 between blood and water is small. They may be related with ion-regulation, when they migrate from salt water to freshwater and vice-versa.

Value for gill area of *Hilsa ilisha* falls below the range of very active fishes like *Thunnus albacares*, *Katsuwonus pelamis* of the same body weight (Muir and Hughes, 1969, Ursin 1967) but it is more than two folds than that of *Labeo rohita*, which falls under the Gray's Intermediate category (Table 12.2).

6.2 Oxygen Uptake in Relation to Body Weight in *Rita rita* (Ham.) (Bagridae, Pisces) at Two Different Seasonal Temperatures

Materials and Methods

Live specimens of *R. rita* were collected from river Ganga at Kahalgaon, Bhagalpur and maintained in the glass aquaria. The fishes were fed shrimps and goat liver during the acclimatization period of 15 days.

The experimental set up was the same as adopted by Munshi and Dube (1973) in a continuously flowing glass respirometer. Oxygen concentration of inflowing and ambient water of the respirometer were determined by Winkler's Volumetric method (Welch, 1948) and were considered in calculating the rate of oxygen uptake (ml/h) and ml/kg/h).

The observations were recorded at seasonal temperatures of 21±1°C (Winter) and 31±1°C (Summer). Regression analysis using logarithmic transformation was done to show the relation between oxygen uptake and body weight.

Results

Data on oxygen uptake ($\dot{V}_{O_2}$) for eleven weight groups of fishes during winter (21±1°C) and summer (31±1°C) seasons. have been summarized in Tables 12.3 and 12.4 respectively. Relationship between body weight and $\dot{V}_{O_2}$ is given in Table 12.5.

i) Relationship between Body Weight and $\dot{V}_{O_2}$ (ml/h)

Total oxygen uptake (ml/h) through gills and skin together ranged from 2.451 to 14.143 in the weight range of 17.5 to 178.0 g during winter (21±1°C and from 3.377 to 25.208 in the weight range of 13.0 g to 162.0 g during summer (21±1°C) and winter (31±1°C).

Table 12.3: The rate of oxygen consumption ($\dot{V}o_2$) per unit time and per unit body weight (ml O_2/h and ml O_2/kg/h) against different body weight of *Rita rita* at a winter temperature 21±1ºC.

Sl.No.	Body Weight (g)	Oxygen Uptake ($\dot{V}o_2$)	
		ml O_2/h	ml O_2/kg/h
1.	17.5	2.451	140.057
2.	25.0	3.206	128.240
3.	31.0	3.765	121.452
4.	42.0	4.746	113.000
5.	63.0	6.326	100.413
6.	71.5	7.068	98.853
7.	95.0	8.679	91.358
8.	129.0	11.126	86.248
9.	146.0	12.078	82.726
10.	155.0	12.453	80.342
11.	178.0	14.143	79.455
Average	86.64	7.82	102.01

Correlation co-efficient (r=-0.9987 and -0.7697 at 21±1°C and 31±1°C respectively) showed negative correlations between the two variables.

The log/log plots of $\dot{V}o_2$ (ml/h) and body weight gave straight lines with slopes of 0.750 and 0.769 respectively during winter and summer seasons. Both the variables showed positive correlation during winter (r=0.997; P<0.001) and summer seasons (r=0.970; P<0.001).

ii) Relationship between Body Weight and $\dot{V}o_2$ (ml/kg/h)

The weight specific oxygen uptake (ml/kg/h) decreases with increasing body weight at both the temperatures.

The rate of $\dot{V}o_2$ (ml/kg/h) within the body weight range of 17.5 to 178.0 g was found to be 14057 to 79.455 at temperature 21±1°C; and 259 784 to 129.395 at temperature 31±1°C within the weight range of 13 to 162 g. The log/log plots of $\dot{V}o_2$ (ml O_2/kg/h) and body weight gave straight lines with slopes of -0.25 and -0.23 respectively during winter and summer season.

Discussion

The present study shows the effect of body size and temperatures on the oxygen uptake capacity of a freshwater catfish *Rita rita*. Different exponent values have been reported to show the relationship between $\dot{V}o_2$ and body weight. The value ranges from 0.45 to 1.0 in different animal groups but generally it is little more than 0.7 (Winberg, 1966; Parvatheswararao, 1960; Paloheimo and Dickle, 1955; Kamler, 1972; Kunwar *et al.*, 1989).

Table 12.4: The rate of oxygen consumption ($\dot{V}o_2$) per unit time and per unit body weight (ml O_2/h and ml O_2/kg/h) against different body weights of *Rita rita* at summer temp. 31±1ºC.

Sl.No.	Body Weight (g)	Oxygen Uptake ($\dot{V}o_2$)	
		ml O_2/h	*ml O_2/kg/h*
1.	13.0	3.377	259.784
2.	21.0	4.664	222.081
3.	30.0	6.337	211.233
4.	48.0	6.858	142.881
5.	58.0	7.505	129.395
6.	70.0	10.252	146.460
7.	78.9	11.948	152.204
8.	95.0	12.383	130.347
9.	112.5	14.775	131.333
10.	144.0	25.208	175.055
11.	162.0	24.042	148.407
Average	75.64	11.58	168.11

In *R. rita* the exponent value is small (0.75) during winter than in summer (0.769). The lower value could be due to cold depression effect during winter with less amount of activity. Similar observations have also been made by Parvatheswararao (1960), Ojha and Munshi (1975), Hakim *et al.* (1983), Roy and Munshi (1984) and Takeda (1990). The value of $\dot{V}o_2$ for 1 g fish was estimated to be 0.2868 and 0.4197 ml O_2/hr during winter and summer respectively indicating highest metabolic rate during summer.

The exponent value less than 1.0 indicates decrease in metabolic rate with increasing weight of the fish. Zeuthen (1953) suggested that value of regression co-efficient greater than 1.0 is exception to the rule and a value greater than 1.0 could easily result from changes in factors other than size. However, Beamish (1964) and Wares and Igram (1979) have reported greater values than 1.0. In *Rita rita* as the exponent values are less than 1.0, the weight specific $\dot{V}o_2$ (ml O_2/kg/h) decreases with increasing body weight by powers of -0.258 and -0.231 during winter and summer respectively. This indicates that with increasing body size, O_2 uptake efficiency of respiratory organ decreases and that is why the younger ones are more active.

Q_{10} is indicative of the effect of temperature on the metabolism and its value comes 2 to 3 over an interval of 10°C (Fry, 1957). Prosser and Brown (1961) stated that standard metabolism of fish increases with temperatures upto lethal levels with a rate of 2.5 times per 10°C in physiological range. Beamish (1970) found α_{10} value about 1.5 (for 100g *C. mrigala*) between 21.5 to 30.5°C, Chang and Woo (1978) studied the effect of acclimatization to temperatures in *Anguilla japonica* and suggested Q_{10} value to be 3.19 in acute temperature change, while in acclimatization to winter and summer Q_{10} value amounts to only 1.39. Here also the Q_{10} value of acclimated *Rita rita* to summer and winter came to be 1.597 which is quite nearer to the acclimated *Anguilla japonica's* Q_{10} value.

Table 12.5: Regression Co-efficient (b), intercept (a) and correlation coefficient (R) to show relationship of oxygen consumption to body weight.

Oxygen Consumption in Relation to Body Weight	Regression Coefficient		Intercept 'a'		Correlation Coefficient 'r'	
	ml O_2/h	ml O_2/h	ml O_2/h	ml O_2/h	ml O_2/h	ml O_2/h
i) At 21±1ºC	0.750	0.250 0.231	0.287	286.992	0.997 (P<0.001)	-0.999 (P<0.001)
ii) At 31±1ºC	0.769		0.420	419.363	0.970 (P<0.001	0.770 (P<0.001)

Interspecific Variations in the Circadian Rhythm of Bimodal Oxygen Uptake in Four Species of Murrels (J.S.Datta Munshi, Ajoy K. Patra, Niva Biswas and Jagdish Ojha, 1978).

Studies of bimodal oxygen uptake in four species of murrels, genus *Channa*, at different periods of the 24 h day regime showed distinct circadian rhythm in their metabolism. The metabolism of the four species remained higher in the period extending from dusk (16.00 – 18.00) to dawn (04.00 – 06.00), *C. marulius* showed the highest O_2 uptake (66.4 ± 0.5 mlO_2, kg^{-1}, h^{-1}) during midnight (24:00-02:00), *C. striatus* (78.7 ± 18.6) and *C. gachua* (95.6 ± 2.6) during early parts of the night (20:00 – 22:00) and *C. punctatus* (57.5 ± 1.1) during dusk (16:00 – 18:00). Of the total oxygen uptake *C. marulius*, *C. striatus*, *C. gachua* and *C.punctatus* extracted about 84.5 per cent, 67.7 per cent, 53.4 per cent and 86.8 per cent of oxygen through aerial routes respectively, in all the species the lowest or the second lowest rate of oxygen uptake was recorded at noon, and during the period gill breathing dominated over aerial breathing in *C. striatus* and *C. gachua*. The circadian rhythm of their oxygen uptake has been correlated with the diurnal fluctuation of metabolism of the swamp ecosystem. *A general metabolic wheel hypothesis has been postulated*.

Organismic physiology and behaviour is often rhythmic and these rhythms will persist in the laboratory in the absence of photo period and the various physico-chemical factors of the environment in which organisms live under natural conditions. Because they do persever, it is concluded that they are under the control of a so-called biological clock. Often in the artificial constancy of the laboratory, the periods of these rhythms deviate slightly from the ones displayed in nature and are referred to as circadian (Palmer, 1976).

In India there are many swampy areas infested with floating water-hyacinth, *Eichhornia crassipes*, and/or rooted *Cyperus* communities. In summer the water lodged in these areas become hypoxic and hypercarbic. The physico-chemical factors of this adverse ecological environment show rhythmic fluctuations at different hours of the day. These rhythms seem to govern the physiology and behaviour of an interesting group of air-breathing fishes which thrive well in such swampy areas. These fishes exhibit various degrees of bimodal gas exchange. The relative dependence of the fish on gill and air-breathing have been studied in a few dual-breather species (Hughes and Singh, 1970, 1971; Singh and Hughes, 1971, 1973, Singh, 1976; Lombolt and Johansen, 1976; Ojha *et al.,* 1978). However, little is known about the interspecific variations in the circadian rhythm of bimodal oxygen uptake in fishes (Patra *et al.,* 1978).

This chapter reports observations and detailed measurements of the interspecific variations in the circadian rhythm of bimodal oxygen uptake in four closely related species of murrels of the genus *Channa* and correlates them with the fluctuations in some of the physico chemical factors of their natural habitat – the swamps. Murrels belong to the family Channidae of the order Channiformes. They are widely distributed in the freshwater swamps and ponds of temperate and tropical Asia and tropical Africa. Of about 21 species, reported four, *i.e. Channa marulius* (Hamilton), *Channa striatus* (Bloch), *C. gachia* (Hamilton) and *C. punctatus* (Bloch), are commonly found in the swampy areas of northern India. All are air-breathing with a pair of suprabranchial chambers which assist the gills in gaseous exchange (Munshi, 1962).

Materials and Methods

Murrels were collected from the swamps of North Bihar, India, in July, 1977 and maintained in glass aquaria in the laboratory. The fishes were fed goat liver, small prawns and earthworms on alternate days during a minimum acclimatization period of 10 days in the laboratory. The fishes were kept fasting for 24 hours before experiments. No feeding was done during experiments.

Bimodal oxygen uptake from air and still water was measured in a closed glass respirometer containing 3 l water (initial O_2 content 4.75 mlO_2, l^{-1}; pH 7.2-7.3) and 1.1 of air. The fish had free access to air through a semicirculae hole of about 8 cm in diameter in a disc float of a thermocol material that separated the water/air interface of the respirometer. KOH in a Petri dish placed on the float absorbed CO_2. The air phase of the respirometer was connected to a manometer. Imbalance in the levels of the manometer fluid reflected uptake of oxygen when the CO_2 is absorbed by KOH. The fishes were acclimatized at least 12 hours before the reading were taken. The respirometer was plced in a constant temperature bath.

Experiments were carried out at 30 ± 1 C in summer in an air-conditioned room. A diffused light was available to the fish through the semicircular hole in the disc float. Observations showed that the fishes could locate the breathing hole more readily, because it was the only source of light.

The concentration of dissolved O_2 in the water was estimated by winkler's volumetric method (Welch, 1948). Aquatic O_2 uptake was calculated from the difference between the O_2 levels of the ambient water in the respirometer before and after the experiment and the volume of water in the respiromter. Uptake of O_2 from air was calculated from the range of imbalance of the levels of the manometric fluid in the manometer and by the use of combined gas law equation and vapour pressure (Dejours, 1975). Mean values of oygen uptake of a series of observations of adult fishes at standard temperature pressure dry (STPD) and standard deviations were calculated. pH of ambient water was measured by an electronic [pH meter (systronics). Equivalent energy utilization was calculate from the caloric equivalent of O_2

(4.8 mgO$_2$.1, Winberg, 1956). Paired t-tests were employed to test the level of significance of the differences between the sample means of the bimodal oxygen uptake during various hours of the day.

Result

$\dot{V}o_2$ (mlO$_2$, kg^{-1},h^{-1}) from air as well as water in four species of genus *channa* and their total energy cost round the clock were investigated and the data have been summarized in Table 1, and diagramatically presented in Figure 1.

1. *Channa marulius*

This species showed its maximum rate of $\dot{V}o_2$, 56.1 mlO$_2$, kg^{-1},h^{-1} from air at midnight (24:00-02:00) with a moderate high, 42.5 mlO$_2$, kg^{-1},h^{-1} in early hours of the evening (16:00-18:00). The oxygen uptake from air at early hours of the night (20:00-22:00 was significantly different (P<0.05) from that obtained during noon (12:00-14:00).

Oxygen uptake from water remained more or less constant at all periods of the day regime except at midnight (24:00-02:00) when this value was found to be the lowest.

The highest value of total $\dot{V}o_2$ (mlO$_2$, kg^{-1},h^{-1}) was observed at midnight (66.4) followed by 52.6 and 51.0 at 20:00-22:00 and 04:00-06:00 respectively. The lowest value (30.6) was recorded at noon (12:00-14:00). The differences in the total oxygen uptake between 08:00-10:00 and 16:00 – 18:00, 12:00-14:00 and 24:00 -02:00 were significant (P<0.05).

2. *Channa striatus*

The period of highest O$_2$ uptake (53.3 mlO$_2$, kg^{-1},h^{-1}) from air was in the night (20:00-22:00) with a secondary peak (41.1) in the midnight (24:00-02:00) and the lowest uptake (14.0) was recorded at noon (12:00-14:00) (Figure 1).

The highest rates of $\dot{V}o_2$ (31.3 mlO$_2$, kg^{-1},h^{-1}) from water was recorded in the morning hours (04:00-06:00), when $\dot{V}o_2$ dropped down to 13.7 in the next few hours (08:00-10:00). In the rest of the periods of the day only very slight variations were observed. The highest value for total O$_2$ consumption (78.7 mlO$_2$, kg^{-1},h^{-1}) was recorded in the night (20:00-22:00) followed by 71.7 mlO$_2$, kg^{-1},h^{-1} in the dawn (04:00-06:00). The lowest value for total $\dot{V}o_2$ (32.9) was obtained in the morning hours (08:00-10:00), and then in the noon (12:00-14:00) (Figure 12.1). There was a significant difference (P<0.05) in the oxygen uptake between noon (12:00-14:00) and midnight (24:00-02:00).

3. *Channa gachua*

The maximum $\dot{V}o_2$ (63.4 mlO$_2$, kg^{-1},h^{-1}) from air occurred in the midnight (24:00-02:00) with a second peak (62.4) in the morning (08:00-10:00). The minimum (9.9) oxygen uptake was recorded at noon (12:00 – 14:00).

Table 12.6: Interspecific variation of circadian rhythm in oxygen uptake of some air-breathing murrels at 30±1°C. Measurement of oxygen uptake was made three times for each species and period in all sections except for *C. marulius* at 24:00-02:00, for which the measurement was made two times.

Species (Mean Body Weight in g)	Hours of Day	Oxygen Uptake ($mlO_2.kg^{-1}.hr^{-1}$)						TEU*
		Aerial	%	Aquatic	%	Total	%	
Channa marulius (93.0)	04:00-06:00	39.5±0.0	(77.5)	11.5±0.0	(22.5)	51.0±0.0		0.245
	08:00-10:00	35.3±5.1	(75.5)	11.5±1.4	(24.5)	46.8±8.6		0.225
	12:00-14:00	19.0±3.7	(62.3)	11.5±0.1	(37.7)	30.8±3.8		0.147
	16:00-18:00	42.5±0.0	(78.6)	11.6±0.0	(21.4)	54.1±0.0		0.260
	20:00-22:00	41.3±0.0	(78.6)	11.3±0.0	(21.4)	52.6±0.0		0.252
	24:00-02:00	56.1±1.8	(84.5)	10.3±1.3	(15.5)	66.4±0.5		0.319
Channa striata (82.)	04:00-06:00	40.4±14.5	(56.4)	31.3±0.6	(43.6)	71.7±15.2		0.344
	08:00-10:00	19.2±1.5	(58.4)	13.7±0.3	(41.6)	32.9±1.2		0.158
	12:00-14:00	14.0±0.5	(40.7)	20.4±7.7	(59.3)	34.4±7.2		0.165
	16:00-18:00	23.4±0.9	(52.9)	20.8±9.8	(47.1)	44.3±9.2		0.213
	20:00-22:00	53.3±20.2	(67.7)	25.4±11.6	(32.3)	78.7±8.6		0.378
	24:00-02:00	41.1±1.2	(61.6)	25.5±14.0	(38.4)	66.6±13.4		0.320
Channa gachua (30.0)	04:00-06:00	39.7±5.7	(50.8)	38.4±3.8	(49.2)	78.1±9.3		0.375
	08:00-10:00	62.4±0.0	(85.6)	10.5±0.0	(14.4)	72.9±0.0		0.350
	12:00-14:00	9.9±4.1	(24.0)	31.4±7.0	(76.0)	41.3±3.7		0.198
	16:00-18:00	40.7±3.6	(72.9)	15.1±1.2	(27.1)	55.8±2.5		0.268
	20:00-22:00	51.0±0.0	(53.4)	44.6±2.6	(46.6)	96.6±2.6		0.459
	24:00-02:00	63.4±5.8	(76.2)	19.8±5.4	(23.8)	83.1±9.4		0.399
Channa punctata (93.0)	04:00-06:00	40.9±0.4	(79.2)	10.7±1.7	(20.8)	51.6±1.1		0.248
	08:00-10:00	30.7±8.9	(67.4)	15.4±2.3	(32.6)	46.0±9.7		0.221
	12:00-14:00	32.8±1.5	(75.6)	11.3±1.0	(24.4)	44.0±1.1		0.211
	16:00-18:00	49.6±2.1	(86.8)	7.9±1.8	(13.2)	57.5±1.1		0.276
	20:00-22:00	43.2±2.1	(83.4)	9.0±1.1	(16.6)	52.2±1.2		0.250
	24:00-02:00	35.4±0.6	(76.7)	10.7±0.7	(23.3)	46.1±1.0		0.221

*TEU: Total equivalent energy utilization ($Kcal.kg^{-1}.h^{-1}$).

The oxygen consumption through the gills from water was at its peak (44.6 mlO_2, kg^{-1},h^{-1}) in the night (20:00-22:00) followed by 38.4 in the early hours of the day (04:00 -06:00) and lowest (10.5) during of the day (08:00-10:00) period.

Total oxygen uptake rate was highest 95.6 mlO_2, kg^{-1},h^{-1}) in the night (20:00-22:00) and minimum (41.3) at noon (12:00-14:00). The differences in the rate of $\dot{V}_{O_2}$ between the midnight (24:00-02:00) and noon (12:00-14:00) periods was found to be statistically highly significant (P<0.01).

4. *Channa punctatus*

The highest rate of oxygen uptake through aerial route was found to be 49.6 mlO_2, kg^{-1},h^{-1} in the (dusk) evening (16:00-18:00) and the lowest (30.7) in the morning (0.8:00-10:00). In other periods of the circadian cycle there were slight variations in aerial oxygen uptake (Table 12.6). The peak value of aquatic oxygen uptake (15.4 mlO_2, kg^{-1},h^{-1}) occurred in the morning (08:00-10:00), followed by 11.3 at noon (12:00-14:00) and the lowest (7.9) at dusk (16:00-18:00). Aquaric $\dot{V}_{O_2}$ at other periods of the circadian cycle showed slight fluctuation.

The highest total O_2 uptake 57.5 mlO_2, kg^{-1},h^{-1}) was recorded in the evening (16:00-18:00) and the lowest (44.0) at noon (12:00-14:00) like all other previous species studied. The difference in the oxygen uptake values between 24:00-02:00 and 12:00-14:00 were statistically significant (P<0.05).

Discussion

The discovery of biological rhythm in oxygen consumption in air-breathing fishes is interesting. This is a physiological adaptation of air-breathing fishes in relation to the fluctuations of oxygen and carbon dioxide contents of their natural habitat.

The ability of a murrel to obtain oxygen from the water will vary with the oxygen tension of the water in swamps and the capacity of the gills to extract it from the water. In general the gills are not so well developed in the murrels (Hakim *et al.*, 1978). All the four species of *Channa* show distinct circadian rhythm in their metabolism. This rhythm seems to be associated precisely with the diel fluctuation of oxygen and carbon dioxide tensions of the water in swamps.

This study on murrels clearly indicates that there are large interspecific variations in the circadian rhythm of their total metabolic rate. Differences in the percentage of aerial oxygen uptake in the four species of murrels may be due to interspecific variations in the oxygen uptake efficiency of the bimodal gas exchange machinery (Hakim *et al.*, 1978; Ojha *et al.*, 1978).While two species (*C. striatus* and *C. marulius*) are obligate air-breathers, the other two are facultative ones. However, certain common features in their behaviour in the respirometer have been noted, viz. (i) all are more active at night as they become very restive and frequently take air-breaths: (ii) during the day the fishes breathe quietly exchanging gases mostly with gills, and behave like oxygen conformers; (iii) and interestingly enough all the species show very low metabolic rates at mid-day.

Behavioural studies of the four species indicate that they avoid bright light and hide themselves under the coverage of macrovegetation. They come out in the open waters after dusk in search of prey and are more active at night.

Two distinct microenvironments in terms of dissolved oxygen and free CO_2 were found in the open and vegetation covered water areas. Generally, dissolved oxygen was lower free CO_2 higher, pH lower and temperature lower under the water hyacinth than in the "open" water (Ultsch, 1973; Rai and Datta Munshi, 1979). A sort of diel fluctuation of O_2 and free CO_2 in the two micro-habitats have also been recorded (Rai and Munshi, 1979).

The availability of these data should enable some interesting ecological conclusions to be drawn about animals inhabiting swamps. The dissolved oxygen under macrovegetation depletes rapidly to almost zero (1.26 mlO_2, kg^{-1},h^{-1}) by 04:00h, especially in the summer, when the O_2 demand of most aquatic organisms is greatest. As such, most of the vertebrates found in swamps are either entirely dependent upon aerial breathing (snake, tortoises) or supplemental air-breathers like murrels. Further, the dependence upon aerial breathing of an organism utilizing bimodal (air and water) gas exchange can be evaluated as a function of time of day.

Thus the two metabolic systems, one of habitat (swamp) and the other of fishes, are closely interlocked with each other, one influencing the other. The general metabolism of the swamp may be contemplated as a big metabolic wheel which drives all the small metabolic wheels of different biotic communities. There is some sort of feedback mechanism also in which the metabolic activities of different biotic communities influence the whole metabolism of the ecosystem. As semblage of air-breathing fishes form an integral part of the swamp ecosystem since their origin several million years ago, the circadian rhythm has now become an inherent property of their system which they transmit even under the artificial conditions of the laboratory.

Bimodal Oxygen Uptake in Juveniles and Adults Amphibious Fish *Channa* (=Ophiocephalus) *Marulius* (J. Ojha, N. Mishra, Mahadeo Prasad Saha and Jyoti Swarup Datta Munsi, 1978)

Oxygen uptake of *Channa marulius* was studied under water with and without access to air. There was a significant increase in the oxygen uptake through gills when access to air was prevented. However, this value (0.86±0.058 mlO_2/indv./h) was quite low in comparison to the total bimodal oxygen uptake per unit time increased appreciably (4.673± 0.404 ml O_2/indv./h). In juveniles as well as in adults the air breathing dominated over aquatic breathing. This fish showed a definite circadian rhythm in the bimodal oxygen uptake in different hours of the day.

In India the ponds, creeks and also the torrential streams often dry out during summer or their water becomes muddy, highly hypoxic and hypercarbic. Under such ecologically adverse conditions there exists a very interesting group of fish which have a bimodal gas exchange mechanism where the air-breathing organ exchanges with the air, while the gill and/or skin exchanges gases with water.

Oxygen uptake through bimodal gas exchange machinery has been studied in a few Indian species of fish such as the climbing perch (Hughes and Singh, 1970; Singh and Hughes, 1973), the catfishes (Hughes and Singh, 1971; Singh and Hughes, 1971), and the mud-eel (Lombolt and Johansen, 1976). However, little is known about the importance of dual breathing in relation to body size (Johansen *et al.*, 1976).

The present work is an attempt to evaluate the relative importance of air and water breathing in the juvenile and adult air-breathing fish, *Channa marulius*. Circadian rhythm in the oxygen uptake has also been described.

Experimental Animals

Channa (=Ophiocephalus) marulius (Ham, 1982) is an amphibious freshwater fish locally known as Bhaura. It belongs to the order Channiformes and the family Channidae. This species is distributed in freshwaters of India, Pakistan, Ceylon and China. The juveniles have a brilliant orange band passing from tip of snout over the eyes to the tip of the caudal fin, while in the adult forms there are 4 to 5 round black patches below the lateral line. In both cases there is a large black prominent ocellus at the upper part of the base of the caudal fin. It has special air-breathing organs in the form of a pair of suprabranchial chambers situated dorsal to the gill arches and lined internally by a vascular respiratory membrane. This fish extracts O_2 from water through its gills and from time to time it comes to the surface to gulp air. Gaseous exchange takes place between the air contained in the suprabranchial chambers and the blood that circulates in the respiratory epithelium lining these chambers.

Live specimen of *C.marulius* were collected from local ponds of Bhagalpur, transported to and maintained in glass aquaria (40 l) in the animal house of the department.

The fish were fed on chopped goat liver on alternate days during a minimum acclimatization period of 15 days in the laboratory.

Methods

The rate of oxygen consumption through the gills from flowing water was measured under two experimental conditions: (i) when access to air was allowed and (ii) when it was prevented. In the first series of experiments a rectangular respirometer made up of acroplex was used. This respirometer was approximately of two litres capacity with a small air chamber at the top, similar to that used by Hughes and Singh (1970) for *Anabas*. For the second series of experiments, a cylindrical glass respirometer of (2 l) capacity was used (Munshi and Dube, 1973). In this respirometer the air-breathing was prevented and therefore, the fish used only the gills for gaseous exchange. In the third series of experiments oxygen consumption from air and still water was measured in a closed glass respirometer containing 3 l of water (initial O_2 content, 6.0 to 6.8 mg O_2/h; pH 7.2) and 1 l of air. The fish had free access to air through a small semi-circular hole (2" diameter) in a disc float of thermcol. Carbosorb (B.D.H.) or KOH in a petridish placed on the float absorbed CO_2, the air phase of the respirometer was connected to a differential aqua-manometer. Imbalance of manometer follows uptake of oxygen when the CO_2 is absorbed.

The concentration of dissolved O_2 in the water was estimated by Winkler's volumetric method (Welch, 1948). In the first two series of experiments the oxygen uptake through the gills was calculated from the difference in O_2 content of the inflowing and outflowing water and the rate of water flow through the respirometer. The experiments were conducted at $30 \pm 1°C$. In the third series of experiments the oxygen uptake through the gills was calculated from the difference between the O_2

levels of the ambient water in the respirometer before and after the experiment and the volume of water in the respirometer.

Oxygen uptake from air was measured and calculated from the range of imbalance of the levels of Kerosine oil in the manometer and by the use of the combined gas law equations and vapour pressure (Dejeours, 1975). Mean values of oxygen uptake $mlO_2/$ indiv./h) at STPD and standard errors were calculated.

Bimodal oxygen uptake of an adult specimen (93 g) was measured at different hours of the day to ascertain the circadian rhythm and the relative importance of air and water breathing.

Equivalent energy utilization was also estimated by applying an oxy-caloric equivalent of 4.8 Kcal per litre of oxygen (Winberg, 1956). The total oxygen uptake was considered to represent the entire metabolic demand.

The paired 't' test was employed to test the significance of the differences between the two sample means of the uptake of oxygen through gills under surfacing allowed and surfacing prevented conditions.

'Drowning experiments were also performed on juvenile and adult specimens to ascertain whether the fish is an obligate or facultative air breather.

Results

Measurements of the bimodal oxygen uptake ($\dot{V}_{O_2}$) under different experimental conditions in juvenile and adult *Channa marulius* have been summarized in Tables 12.1–12.4.

A. Oxygen Uptake from Continuously Flowing Water with Free Access to Air (Juveniles)

In this experiment the fish extract oxygen from the water through their gills and also come to the surface to gulp air, the mean rates of oxygen uptake through the gills per unit time and body weight for an average 15.2 g fish are 0.39 ± 0.038 $mlO_2/$indiv./ h and 25.77 ± 1.986 $mlO_2/$Kg/h respectively. Results summarized in Table 12.1 show that the average ventilatory frequency in normoxic water is 44 counts/min.

B. $\dot{V}_{O_2}$ from Continuously Flowing Mormoxic Water when Access to Air was Prevented (Juveniles)

In this experimental condition the fish show restlessness to begin with but after about two min they settle down to rest. The mean rates of oxygen uptake through gills per unit time and body weight for an average 15.2 g fish are 0.863 ± 0.058 $mlO_2/$ indiv./h and 57.93 ± 3.977 $mlO_2/$Kg/h respectively (Table 12.1). The opercular frequency is also higher (60/min) in this condition. The oxygen uptake through the gills and the ventilatory frequencies are significantly higher ($P<0.05$) than the values obtained in the previous experimental condition.

C. Bimodal Oxygen Uptake from Still Water and Air

In this experiment the water and air-sac is in contact over a small surface areas (2" in diameter) and the fish extracts oxygen from the still water (3 l) contained in the respirometer and is also allowed to gulp in air from the air phase of the closed respirometer.

In juvenile fish, out of the total oxygen uptake of 136.1 ± 8.92 mlO$_2$/kg/h, about 82 per cent (113.27 ± 9.36 mlO$_2$/kg/h) is contributed by the air-breathing organs and only 18 per cent (222.83 ± 2.27 mlO$_2$/kg/h) by the gills. The quotient between aquatic and aerial breathing also suggests predominance of aerial over aquatic breathing. The average equivalent energy utilization from the total oxygen comes to 0.0104 (k.Cal/h).

In adult fish the oxygen uptake per unit time is more (4.673 ± 0.494 mlO$_2$/indiv./h) in comparison to the juveniles (2.04 ± 0.14 mlO$_2$/indiv./h). However, the oxygen uptake per unit body weight is less (56.99 ± 4.93 mlO$_2$/kg/h) in adults than in juveniles (136.1 ± 8.92 mlO$_2$/kg/h). This indicates a decreasing trend of the metabolism with increasing body weight. Because of higher total oxygen uptake per unit time, adult fish have a more equivalent energy utilization (0.0224 k.cal/h) than juveniles (0.0104 k.Cal/h). In adult specimen too the aerial dominates over aquatic breathing.

D. Circadian Rhythum in the Bimodal Oxygen Uptake in Adult Specimens

A definite circadian rhythm is noticed in the oxygen uptake in this fish. Maximum oxygen uptake (56.10 mlO$_2$/kg/h) through the air breathing organs is recorded during mid-night (00-02 hours) and the minimum (19.05 mlO$_2$/kg/h) during the noon (12-14) hours. The rest of the hours show little variation in the oxygen uptake through air. There is a little variation in the aquatic oxygen uptake in different hours of the day. However, minimum oxygen uptake through water is recorded during mid-night when the aerial breathing is at its peak. Accordingly the oxygen uptake through the bimodal gas exchange machinery is maximum (66.40 mlO$_2$/kg/h) during mid-night and a minimum is recorded at noon (30.57 mlO$_2$/kg/h). Dusk and other hours of the day show no significant variation. Because of greater oxygen uptake the equivalent energy utilization is also maximum during mid-night (0.0296 k.Cal/h) and minimum during mid-day (0.0136 k.Cal/h). Other hours show little variations. In every hour the aerial breathing dominates over aquatic breathing as indicated by present aerial breathing and the quotient between aquatic and aerial oxygen uptake.

'Drowning' experiments reveal that juveniles (11-20 g) survive total submersion in normoxic water (6.8 mgO$_2$/lit; pH-7.2) for a longer time whereas the larger weight group of fish succumb to total submersion under the same experimental condition.

Discussion

Channa marulius has a bimodal gas exchange mechanism. It extracts oxygen from water through its gills and uses a pair of suprabranchial chambers in exchanging gases with air. The oxygen uptake efficiency of the bimodal gas exchange machinery varies with the age and with various experimental conditions. Under surfacing allowed conditions juvenile *C. marulius* extract little oxygen (0.391 mlO$_2$/indv./h) from running water and rely on air-breathing to compensate the oxygen required for total metabolic activities. Whereas in surfacing prevented conditions the fish has to extract adequate oxygen from water alone for total metabolism. Under this experimental condition there is a significant increase (P<0.05) in the uptake of oxygen from water alone. The oxygen uptake increases about 2 folds (0.863 mlO$_2$/indv./h) to provide at least minimum oxygen required for the total metabolism of the fish. In this condition juveniles remain comfortable and at least for the 4 hours of the experimentation they show no sign of asphyxiation. However, there is an apparent increase in the ventilatory frequency. It is presumed that the significant increase in the oxygen uptake in surfacing-prevented condition may be associated with an increased ventilatory frequency. This increased ventilatory frequency may in turn ventilate more water thus enabling the gills to extract more oxygen per unit time.

Air-breathing forms exhibit various degrees of bimodal gas exchange mechanism. Juvenile *C. marulius* extract about 82 per cent of the total oxygen uptake through air-breathing organs. Oxygen uptake from still water is low (0.34 mlO$_2$/indv./h) in comparison to that obtained for running tap-water under surfacing allowed (0.391 mlO$_2$/indv/h) and surfacing-prevented (0.863 mlO$_2$)/indv./h) conditions. Under surfacing prevented condition the oxygen provided by the gills is more than twice compared with their contribution underarial condition. However, this amount is quite low in comparison to the total oxygen uptake through bimodal gas exchange machinery (2.04 mlO$_2$/indv./h). It is therefore, suggested that the air-breathing organs in this fish are more efficient than those of other air-breathing fish. The adult fish too rely more on air-breathing (76 per cent) than on water breathing. There is an apparent increase in oxygen uptake per unit time through bimodal gas exchange machinery (4.673 mlO$_2$/indv./h) in adult fish. This results in a more equivalent energy utilization in adult in comparison to juvenile. The decrease in the oxygen uptake through bimodal gas exchange per unit body weight indicates a decrease in metabolism as the fish grow in size.

Some amphibious fish rely more on aquatic breathing and only supplement gas exchange from air and are known as facultative air-breathers, others are obligatory air-breather and get drowned if not allowed to breathe atmospheric air. The 'drowning'experiments conducted on juveniles and adult *C. marulius* reveal that the former seems to be facultative air-breathers whereas the latter are obligate ones. It is presumed that the gills and skin of juveniles are efficient enough to meet the minimum oxygen requirement for total metabolism of the fish and therefore the fish do not asphyxiate even in submerged condition with continuous flow of normoxic water. Whereas, in adults the oxygen uptake efficiency of gills and skin decrease with increasing body weight and therefore does not cope with the increasing oxygen demand of the fish and that therefore in the higher weight group of fish succmb in submerged condition even in well oxygenated water. From these findings it is concluded that the juveniles of *C. marulius* are facultative air-breather where as the adults are obligate ones. Similar intraspecific variation in asphysiation time has been reported in *Anabas testudineus* (Munshi and Dube, 1973). However, *Clarias batrachus* (Munshi et al., 1976); *Saccobranchus fossilis* (Munshi et al., 1978) and *Colisa fasciatus* (Ojha et al., 1977) do not asphyxiate under surfacing prevented conditions.

Circadian Rhythm

Most freshwater as well as marine fish show a cyclic pattern in daily activity. Most species are more active at certain times of the day than at others (Schwassmann, 1971). Like other fish species, *C. marulius* also shows a clear circadian rhythm in their bimodal oxygen uptake. Measurements on the oxygen uptake at different hours of the day reflect maximum metabolism during midnight (00-02 hours) and minimum during mid-day (12-14 hours). The increase in the metabolism during mid-night is due to the increase in the aerial breathing (84-49 per cent). This fish reduces its metabolism during mid-day. This decrease is apparently due to decrease in the air-breathing (62.31 per cent). The variations of their metabolism during different hours of the day may be due to the obligate air breathing habit, as the adult depends more on air-breathing and their gills play a secondary role in the O_2 uptake. It is interesting to correlate the metabolic rhythm in the laboratory with that in the natural ecosystem where these fish live. It has been seen that in senescent swamps dissolved oxygen becomes practically nil and under such adverse ecological circumstances air-breathing fish come out of their niches to rest on the moist banks and rely on aerial breathing. Whereas during day time they remain hidden below the water hyacinth (*E. crassipes*) communities. In nature, the biological rhythm is under the influence of food capture, play, rest, sleep and also the fluctuations of the physico-chemical factors of the ambient water. Whereas the respirometer presents a stable environment and here also the fish show circadian rhythm in their metabolism. It is most probable that the metabolic rhythm has become the habit of the fish and will be repeated in any environment whether natural or artificial.

As fish show a definite circadian rhythm in the bimodal oxygen uptake, it is suggested that in addition to various factors, *viz.* size and temperature, and at the different hours of the day should also be considered in any physiological measurement.

Summary

Oxygen uptake through bimodal gas exchange machinery under various experimental conditions was measured in an air-breathing fish, *Channa* (*=Ophiocephalus*) *marulius*. In surfacing-prevented condition the oxygen uptake through the gills in juveniles (15.2 ± 0.746 g) was about twice (0.863 ± 0.058 mlO$_2$/ indv./h) than in surfacing allowed condition (0.391 ± 0.038 mlO$_2$/indv./h). The significant increase in the oxygen uptake in the former experimental condition was perhaps associated with increased ventilatory frequency. However, this value was quite lower than the value obtained for total bimodal oxygen uptake in juvenile (2.04 ± 0.14 mlO$_2$/indiv./h) and (4.673 ± 0.404 mlO$_2$/indv./h) in adult specimens. Inspite of comparatively lower oxygen uptake through the gills under surfacing prevented conditions the juveniles did not asphysiate, whereas adult ones succumb to total submersion under continuously flowing normoxic water. From these findings it was concluded that juveniles and adults were facultative and obligate air-breathers respectively. In juveniles as well as adults the air-breathing dominated over water breathing. Because of greater oxygen uptake, adults had a more equivalent energy utilization (0.0224 k.Cal/h) than juveniles (0.0104 k.Cal/h). This amphibious fish showed a definite circadian rhythm in its bimodal oxygen uptake. In adults the peak hours of the total bimodal oxygen uptake (6.175 mlO_2/indv./h) was during mid-night (00-02 hours). A significant lower rate (2.843 mlO_2/indv./h) was noticed during mid-day (12-14 hours). However, other hours of the day did not show much variation. In all hours of the day air-breathing dominated over water breathing with a maximum (84.5 per cent) during mid-night and a minimum (62.3 per cent) during mid-day.

Oxygen Uptake Capacity of Gills and Skin in Relation to body weight of the Air-breathing Silluroid Fish, *Clarias batrachus* (Linn.)

Oxygen consumption through gills and skin in relation to body weight was estimated in the air-breathing catfish, *Clarias batrachus*, under two experimental conditions, *viz.* (i) when access to air was allowed and (ii) when air-breathing was prevented. There was a positive correlartion between $\dot{V}_{O_2}$ (ml/hr) and body weight in both experimental condition. Oxygen consumption by a power of 0.869 when access to air was allowed whereas the power was slightly less (b=0.841) when air-breathing was prevented. As the values for exponent (b) were less then 1.0, the weight specific V_{0_2} (ml/kg/hr) decreased with increasing body weight. The decrease was more marked (b= -0.180) in fishes which were not allowed air than in those where access to air was allowed (b=0.148) under normal condition of water and air-breathing the rate of $\dot{V}_{O_2}$ (ml/kg/hr) via gills and skin from water ranged from 39.7 ± 3.21 to 76.7 ± 9.01 and this increased to 42.17 ± 6.2 to 105.9 ± 8.33 when air-breathing was prevented. The increased in the rate of $\dot{V}_{O_2}$ was perhaps associated with the increase in the volume of water irrigating the gills per unit time.

Fishes are primarily water breathers where gills and skin take part in the metabolic gas exchange. Many species of teleostean fishes inhabiting waters of low O_2 and high CO_2 content have, however, accessory respiratory organ which enable them to breathe atmospheric air. About 140 species of air-breathing fishes exhibit various degrees of bimodal gas exchange (Rahn and Howell, 1974). Of these fishes, the three living dipnoan genera (*Protopterus, Lepidosiren, Neoceratodus*) are of special interest because they are closest to the line of evolution from which the tetrapods are believed to have originated. Comprehensive studies have been made on the respiratory physiology of these genera (Johansen and Lenfant, 1967; Lenfant *et. al*, 1966; Lenfant and Johansen, 1968). Currently more attention is being devoted to the adaptation of teleosts that breathe air (Hughes and Singh, 1970, 1971), but few studies have been made of the relation to oxygen consumption and growth in air-breathing fishes (Munshi and Dube, 1973).

The air-breathing catfish, *Clarias batrachus*, lives in freshwater pools of East India which have low O_2 and high CO_2 contents. This catfish has special air-breathing organs which allow direct gas exchange with atmospheric air. The accessory respiratory organs comprise (i) suprabranchial chambers situated dorsally to the gill cavities and lined by respiratory membrane; (ii) fans borne on each gill arch and (iii) the respiratory tree or dendritic organs borne by the second and fourth gill arches (Munshi, 1961). In water the gills and skin perform the gaseous exchange. Time to time the fish comes to the surface to obtain air, and passes it into the suprabranchial cavity that contains the characteristic dendritic organs. Gaseous exchange takes place in this cavity between the inhaled air and the blood that circulates through the dendritic organs.

The purpose of the present study was to investigate the oxygen uptake capacity of gills and skin in relation to body weight of the air-breathing catfish, *Clarias batrachus*, when it was allowed to breathe air and when air-breathing was prevented.

Methods

Live specimens of *Clarias batrachus* were collected from local ponds and maintained in large glass aquaria with continuous flow of water. The fishes were fed on goat liver pieces on alternate days during a minimum acclimation period of 15 days in the laboratory.

The rate of oxygen consumption through gills and skin was measured in fishes of different body weights under two experimental conditions, viz (i) when access to air was allowed and (ii) when it was prevented. Two types of self-designed spirometer were used. In the first series of experiments a rectangular plexiglass spirometer was used. It was approximately of three litres capacity with a small air chamber at the top as designed by Hughes and Singh (1970). In this spirometer, the fish had free access to air in addition to gill and skin-breathing. A small air chamber at the top of the spirometer was constructed to minimize gaseous exchange at the air-water interface. For the second series of experiments, the large cylindrical glass spirometer designed by Munshi and Dube (1973) was used. In this spirometer, airbreathing was prevented and the fishes could use only gills and skin for gaseous exchange.

The fishes were introduced in their respective spirometer which were connected to a large constant level water tank to maintain the flow of water under constant hydrostatic pressure. Water flow through the spirometer was adjusted according to the size of the fish to avoid possible suffocation and stress. The fishes were acclimated to the spirometer at least 12 hours before the readings were taken. Water samples were collected from the bottles connected at the end of the spirometers. Concentration of the dissolved oxygen in the samples were determined by Winkler's volumetric method (Welch, 1948). Oxygen consumption per unit time and body weight through the gills and skin were calculated by the differences in the oxygen levels between ambient water and that supplied to the spirometers along with the rate of water flow and the weight of the fish. The experiments were conducted at $26 \pm 1°C$. Drowning experiments were also peformed in different weight groups to observe the effect of size on asphysiation time.

Characteristics of the regression lines relating the logarithm of oxygen consumption to log body weight were calculated by the method of least squares.

Results

Data for oxygen consumption ($\dot{V}_{O_2}$) for 9 weight groups of *Clarias batrachus* and the regession analysis showing the relationship between $\dot{V}_{O_2}$ and body weight.

1. Oxygen Consumption from Water with Free Access to Air

In the experimental conditions, the fish comes to the surface at 3 to 15 minutes intervals to breath air. Some fishes have been observed to remain submerged for 30 minutes. Under such experimental conditions $\dot{V}_{O_2}$ from water ranged from 0.85 ± 0.22 to 4.65 ± 0.25 ml/hr within the body weight range of $17.3 - 95.7$ g, when the ambient O_2 content of water was in the range of 4.4 to 5.7 ml/lt. Oxygen consumption (ml/hr) increased with increasing body weight, not linearly but in a parabolic function. The relationship between the two variable could be expressed in the allometric forms

$$\dot{V}_{O_2} = a.W^b$$

Or

$$\text{Log } \dot{V}_{O_2} = \log a + b. \log W$$

Where $\dot{V}_{O_2}$ = oxygen consumption; a = antilog of the intercept (log a) indicating the rate of oxygen consumption for a 1 g fish; W = body weight; b = regression coefficient.

(A) $\dot{V}_{O_2}$ (ml/hr) vs. Body Weight

The log/log plot of $\dot{V}_{O_2}$ (ml/hr) and body weight gave a straight line with a slope of 0.869 and the intercept -1.017, and the value of $\dot{V}_{O_2}$ for 1 g fish was estimated at 0.096 ml/hr. Therefore, the relationship between $\dot{V}_{O_2}$ (ml/hr) and body weight with 95 per cent confidence limits may be represented by the equations;

$$\text{Log } \dot{V}_{O_2} = -1.017 + 0.869 \log W$$

$$\dot{V}_{O_2} = 0.096 \ W^{0.869}$$

$$0.549 < b < 1.189$$

The two variable were highly correlated (r=0.925; p< 0.01)

(B) $\dot{V}_{O_2}$ (ml/kg/hr) vs. Body Weight

When the data for oxygen consumption (ml/kg/hr) were plotted against body weight in log/log coordinates, a straight line was obtained with a slope of -0.148 and intercept 2.014. The value of $\dot{V}_{O_2}$ for 1 g fish was estimated at 103.28 ml/kg/hr. Therefore, the relationship between the two variables with 95 per cent confidence intervals may be represented by the equations:

$$\text{Log } \dot{V}_{O_2} = 2.014 - 0.148 + \log W$$

$$\dot{V}_{O_2} = 103.28 \ W^{-0.148}$$

$$-0.459 < b < 0.182$$

$\dot{V}_{O_2}$ (ml/kg/hr) showed a negative correlation with body weight (r= -0.783; p< 0.05)

2. Oxygen Consumption from Water when Access to Air was Prevented

At the start of this experiment the fishes were restless due to the presence of air in the suprabranchial chambers. After eliminating the air the fish settled and were breathing without any sign of struggling. Mean oxygen consumption under this experimental condition ranged from 0.97 ± 0.07 to 5.90 ± 0.40 ml/hr (Table 12.1).

(A) $\dot{V}_{O_2}$ (ml/hr) vs Body Weight

The log/log plots of $\dot{V}_{O_2}$ (ml/hr) against body weight displayed a slope of 0.841 and the intercept was estimated at –0.874. $\dot{V}_{O_2}$ for 1 g fish was estimated at 0.134 ml/hr. Therefore, the relationship between the two variables with 95 per cent confidence limits may be represented by the equations :

$$\text{Log } \dot{V}_{O_2} = -0.874 + 0.841 + \log W$$

Or

$$\dot{V}_{O_2} = 0.134 \, W^{0.841}$$
$$0.446 < b < 1.235$$

There was a high correlation between $\dot{V}_{O_2}$ (ml/hr) and body weight (r=0.885; p < 0.01).

(B) $\dot{V}_{O_2}$ (ml/kg/hr) vs. Body Weight

The log/log plots of $\dot{V}_{O_2}$ (ml/kg/hr) against body weight under this experimental condition gave a slope of -0.180 and the calculated intercept was 2.163, The values for 1g fish was estimated at 145.6 (ml/kg/hr). Therefore, the relationship between the two variables with 95 per cent confidence limits may be expressed by the equations:

$$\text{Log } \dot{V}_{O_2} = 2.163 - 0.180 + \log W$$

Or

$$\dot{V}_{O_2} = 145.6 \, W^{-0.180}$$
$$-0.560 < b < 0.201$$

There was, however, no close correlation between the weight of the fish and the oxygen consumption per unit body weight (r= -0.389; p > 0.05).

Drowning experiments revealed that *Clarias batrachus* can survive in air-saturated water for 5 days and that size and weight of the fish have no pronounced effect on asphyxiation time.

Clarias when exchanging gases with tap water and air, the opercular frequencies ranges from 28-36/min but when access to air was prevented, the opercular frequencies increased to 40-50/min.

Discussion

The slope of the regression line relating oxygen consumption per unit time to body weight showed differences among different species of teleostean fishes. Job (1955) found 0.85 for *Salvelinus fontinalis*; Winberg (1956) obtained 0.81 as an average for a number of species; and Paloheimo and Dickle (1965) suggested 0.80 as characteristic for most teleost species. Zeuthen (1953) observed that the exponent "b" was always lower than 1.0. Exceptions to this general rule occurred only for narrow size ranges. Therefore, the metabolic rate can only decrease with the evolution of body size (Pricocine and Wiame 1946). The above mentioned values for the slope of the regression line are based on observations in pure water breathing teleosts. Munshi and Dube (1973) have investigated the oxygen uptake capacity of the gills in relation to body weight in the air-breathing climbing perch, *Anabas testudineus,* and reported 0.67 as the exponent for $\dot{V}_{O_2}$ through gills when the fish had free access to air. Lowering of the exponent was correlated with the air-breathing habit of the fish. Under similar experimental conditions Munshi and Hakim found b = 0.79 for $\dot{V}_{O_2}$ through gills against body weight in the tropical air-breathing snake head, *Channa (Ophiocephalus) punctatus.* In another air-breathing siluroid fish *Saccobranchus (= Heteropneustes) fossilis,* Munshi *et al.* (1975) observed a still higher exponent (b=1.0) for $\dot{V}_{O_2}$ from water against body weight under similar experimental conditions. In comparison, the obtained value of the slope (b=0.869) for $\dot{V}_{O_2}$ from water to body weight was higher than in *Anabas* and *Channa* but lower than in *Saccobranchus fossilis.* Thus, with increasing body weight the fish can extract oxygen from water more efficiently than *Anabas* and *Channa* do. *Saccobranchus* seems to be still more efficient in this respect. Hughes and Singh (1971) observed that large *Saccobranchus* kept in continuous flow of water may live without air breathing for periods of 6-12 hr. As in *Clarias,* the exponent for $\dot{V}_{O_2}$ (ml/hr) to body weight was less than 1.0 oxygen consumption through gills and skin (ml/kg/hr) will decrease with increasing body weight. It was estimated statistically that $\dot{V}_{O_2}$ per unit body weight decreases by a power of – 0.148 with increasing body weight. In *Anabas* and *Channa* very similar conditions exist because in both fishes the value for the exponent (b) is less than 1. However, in these two air-breathing fishes, $\dot{V}_{O_2}$ (ml/kg/hr) decreases at a faster rate than in *Clarias batrachus.* In *Saccobranchus fossilis,* oxygen consumption per unit body weight does not show a decreasing tendency with increasing body weight because the slope of the regression line relating $\dot{V}_{O_2}$ (ml/hr) and body weight is near 1.0.

Further lowering of the exponent was observed by Munshi and Dube (1973) in *Anabas testudineus* (b-0.53), Hakim *et al.* (1976) in *Channa punctata* (b=0.62) and Munshi *et al.* (1976) in *Saccobranchus fossilis* (b – 0.76), when the fishes were not allowed to atmospheric air and were forced to water-breathing. Under similar experimental conditions *Clarias batrachus* also displayed a slope of b = 0.841 of the regression line relating $\dot{V}_{O_2}$ ml/hr to body weight. These findings indicate that in general the gills and skin work best when the fishes have free access to air. When the fishes are not allowed air, their gills and skin are in stressed condition even in well-oxygenated water and are not so efficient in extracting oxygen from water. As the slope of the regression line is under 1 in these air breathing fishes, oxygen consumption per unit body weight (ml/kg/hr) will decrease with increasing body weight. In *Clarias batrachus* the decrease of $\dot{V}_{O_2}$ (ml/kg/hr) has been estimated at a power of -0.180.

Oxygen Consumption from Air and Water in *Heteropneustes (=Saccobranchus) fossilis* (Bloch) in Relation to Body Weight at three Different Seasons

Oxygen consumption through the gills and the accessory respiratory organs in relation to body size and seasonal temperature (winter 20.0 ± 1°C; rainy, 25.0 ± 1°C;

and summer, 30.5 ± 1°C) has been studied in the air-breathing fish, *Heteropneustes* (=*Saccobranchus*) *fossilis*, using longarithmic transformation.

The exponent is higher (b=0.83) in the winter than in the rainy season (b=0.76) and minimal in summer (b=0.62) when total $\dot{V}_{O_2}$ is measured. The smaller fish appear to be more sensitive to temperature changes (as indicated by the exponent values) than the larger ones but the latter show marked heat depression at 30.5 °C.

In this fish, the gills are well developed for extracting oxygen from water. Aquatic breathing comprises about 60-63 per cent of total oxygen-uptake, while only 37-40 per cent of oxygen is derived from air through the accessory respiratory organs.

The higher rate of oxygen consumption per unit body weight in small fish has been related to their higher metabolic rate at all temperatures.

Oxygen consumtion ($\dot{V}_{O_2}$) via air-sacs, gills and skin is highly correlated with the surface area of the respective respiratory organs as indicated by the high correlation cosufficient (r) at all temperatures studied.

Rübner (1883) seems to have been the first investigator to work on the size/metabolism relationship in animals when he found the metabolic rate of dogs to be proportional to W^{0-67}. Subsequently, Morgulis (1914), Smith (1935a,b), Leiner (1937), Zeuthen (1947) also studied this relationship. In recent years, several investigations have been concerned with the rate of oxygen consumption ($\dot{V}_{O_2}$) as a function of body weight in fish (Parvatheswararao 1959, 1960, Hemingsen 1960, Saunders 1963, Brett 1965a, Paloheimo and Dickie 1965, Kamler 1972, Khamleva 1973, Munshi and Dube 1973, Lombolt and Johansen 1976, Hughes 1977, 1978, Munshi *et al.*, 1978). Several workers have also studied the influence of body size and environmental temperature on $\dot{V}_{O_2}$ (Schaeperclaus 1933, Sumner and Lauham 1942, Fry 1947 and Hart 1948, Philips *et al.*, 1960. Rozin and Mayer 1961, Brett 1965, Sage 1971, Graham 1972, Harret and Jordana 1973, Ojha and Munshi 1975, Hakim *et al.*, 1983).

In the present study an analysis has been made to determine the influence of body weight and environmental temperature on the metabolism of *Heteropneustes* (=*Saccobranchus*) *fossilis* (Bloch).

Materials and Methods

Animals

Live specimens of healthy *Heteropneustes* (=*Saccobranchus*) *fossilis* of various body size, collected from swamps of North Bihar in different seasons were maintained in large glass aquaria with running water. They were fed goat liver on alternate days for 15 days, and were fasted for atleast 12 hrs before experiments. The fish were acclimatized to the experimental conditions in the respirometer at least overnight.

Experiments

A fish was placed in a glass respirometer containing 3L still fresh tap water. The ambient O_2 content of the water varied at different temperature (at 30.5 °C, O_2 content 5.0-6.3 ppm and pH 6.9-7.3; at 25 °C, O_2 content 6.0-7.1 ppm, pH 7.1-7.5; at 20 °C, O_2 6.5-8.0 ppm, pH 7.0-7.7). The fish had free access to air through a small semi-circular hole in a floating disc of thermocol covering the water/air interface of the respirometer. KOH in a petridish placed on the float absorbed CO_2. Thus the fish could exchange gases through the gills and skin with water, and also come to the surface to gulp air into the suprabranchial chambers, and thence into the air sacs situated between the vertebrae and body myotomes. Observations showed that the fish located the breathing hole of the partition wall of the respirometer more readily when it was the only source of light. Experiments were conducted at various seasonal temperature $30.5 \pm 1^\circ C$ (summer); $25.0 \pm 1^\circ C$ (rainy) and $20.0 \pm 1^\circ C$ (winter). The fish did not show any specific movements in the respirometer, rather remained more or less stationary, and as such no record of its 'random' activity was maintained.

Oxygen consumption was estimated, taking into account the room temperature, water-bath temperature, water pH, barometric pressure and vapour pressure. Mean values of $\dot{V}_{O_2}$ in ml kg^{-1}h^{-1} at standard temperature Pressure Dry (STPD) and standard errors were calculated.

$\dot{V}_{O_2}$ through the gills and skin was calculated from the difference between O_2 content of the ambient water in the respirometer before and after the experiment and the volume of water in the respirometer. The concentration of dissolved oxygen in the water was estimated by Winkler's volumetric method (Welch 1948). O_2 consumption from air was measured from the rise in the level of K. Oil in the vertical graduated tube and by use of the combined gas law equations and vapour pressure (Dejours 1975).

Areas of the respiratory surface (gills skin and air-sacs) for different weight group of fish were estimated from the equations derived for this species by Hughes *et al.* (1974).

Results

The surfaces at which gas exchange is known to occur are the gills, the air-sacs and the skin (Hughes and Singh, 1971). The two air-sacs are formed as backward extensions of the suprabranchial chambers. Electron microscopy suggests that they are modified gill structures (Hughes and Munshi, 1973, 1978).

Mean values for $\dot{V}_{O_2}$ (ml/hr/fish) from air and still water simultaneously for various body weights at different temperatures are shown in Tables 12.1–12.3 and plotted on log/log coordinates in Figures 12.1–12.3.

In general, the weight-specific metabolic rate in animals decrease with increasing body size (Hamingen 1960). The relationship is expressed by the equation.

$$\dot{V}_{O_2} = aW^b$$

Or in the logarithmic form: $\log_{10} \dot{V}_{O_2} = \log_{10} + b \log_{10} W$ (1)

Where $\dot{V}_{O_2}$ is the rate of O_2 uptake and W is whole-body weight; a is the Y-intercept of a log/log plot of $\dot{V}_{O_2}$, vs W and gives some indication of the relative magnitude of $\dot{V}_{O_2}$, b is the slope of such a plot (Figures 12.1 and 12.2).

A. Oxygen Consumption as a Function of Body Size and Temperature

(*a*) Oxygen consumption from air; when *Heteropneustes fossilis* was kept in the respirometer with free access to air, from air at 30.5±1 C ranged from 0.5996

Table 12.7: Respiratory areas* and oxygen consumption from air and water for different weight groups of Heteropneustes (= Saccobranchus) fossilis at 30.5±1°C in summer season.

Mean Body Weight of Fish (g)	Respiratory Surface Areas			Mean Oxygen Rate $\dot{V}O_2$ ml/h from Still Water with Access to Air			VO_2 from Air (Per cent)	VO_2 from Water (Per cent)
	Air-Sac Area (mm²)	Gill+Skin Areas (mm²)	Air-Sac+ Gill+Skin Area (mm²)	$\dot{V}O_2$ through Air-sac (ml/hr)	$\dot{V}O_2$ through Gill+Skin (ml/hr)	$\dot{V}O_2$ through Air-sac+Gill+ Skin (ml/hr)		
13.00	796.93	6216.60	7013.30	0.866	1.204	2.070	41.844	58.156
23.66	1184.60	9443.50	10629.00	1.387	1.536	2.923	47.450	52.550
35.00	1533.10	12413.00	13951.00	2.251	1.675	3.926	57.335	42.665
41.50	1718.30	13981.00	15703.00	2.459	3.062	5.522	44.540	55.460
51.00	1969.50	16145.00	18119.00	2.443	4.083	6.526	37.432	62.568
67.00	2359.30	19533.00	21899.00	2.570	3.664	6.234	41.226	58.774
71.66	2466.70	20472.00	22946.00	2.748	4.886	7.634	35.997	64.003
85.00	2761.80	23063.00	25834.00	3.362	5.235	8.597	39.109	60.891
93.00	2931.20	24558.00	27499.00	3.495	5.444	8.939	39.097	60.903
Av. 53.42				2.398	3.421	5.819	42.67	57.33

* Based on allometric relationships given in Hughes, Singh, Guha, Dube and Munshi (1974).

Table 12.8: Respiratory areas* and oxygen consumption from air and water for different weight groups of Heteropneustes (= Saccobranchus) fossilis at 25.0±1°C in rainy season

Mean Body Weight of Fish (g)	Respiratory Surface Areas			Mean Oxygen Rate $\dot{V}O_2$ ml/h from Still Water with Access to Air			VO_2 from Air (Per cent)	VO_2 from Water (Per cent)
	Air-Sac Area (mm²)	Gill+Skin Areas (mm²)	Air-Sac+ Gill+Skin Area (mm²)	$\dot{V}O_2$ through Air-sac (ml/hr)	$\dot{V}O_2$ through Gill+Skin (ml/hr)	$\dot{V}O_2$ through Air-sac+Gill+ Skin (ml/hr)		
13.00	796.93	6216.60	7013.30	0.866	1.204	2.070	41.844	58.156
23.66	1184.60	9443.50	10629.00	1.387	1.536	2.923	47.450	52.550
35.00	1533.10	12413.00	13951.00	2.251	1.675	3.926	57.335	42.665
41.50	1718.30	13981.00	15703.00	2.459	3.062	5.522	44.540	55.460
51.00	1969.50	16145.00	18119.00	2.443	4.083	6.526	37.432	62.568
67.00	2359.30	19533.00	21899.00	2.570	3.664	6.234	41.226	58.774
71.66	2466.70	20472.00	22946.00	2.748	4.886	7.634	35.997	64.003
85.00	2761.80	23063.00	25834.00	3.362	5.235	8.597	39.109	60.891
93.00	2931.20	24558.00	27499.00	3.495	5.444	8.939	39.097	60.903
Av. 53.42				2.398	3.421	5.819	42.67	57.33

* Based on allometric relationships given in Hughes, Singh, Guha, Dube and Munshi (1974).

Table 12.9: Respiratory areas* and oxygen consumption from air and water for different weight groups of *Heteropneustes* (= *Saccobranchus*) *fossilis* at 20.0±1°C in Winter season.

Mean Body Weight of Fish (g)	Respiratory Surface Areas			Mean Oxygen Rate $\dot{V}O_2$ ml/h from Still Water with Access to Air			VO_2 from Air (Per cent)	VO_2 from Water (Per cent)
	Air-Sac Area (mm²)	Gill+Skin Areas (mm²)	Air-Sac+ Gill+Skin Area (mm²)	$\dot{V}O_2$ through Air-sac (ml/hr)	$\dot{V}O_2$ through Gill+Skin (ml/hr)	$\dot{V}O_2$ through Air-sac+Gill+ Skin (ml/hr)		
16.83	945.46	7444.80	8390.60	0.314	0.963	1.277	24.571	75.429
26.75	1284.80	10289.00	11575.00	0.480	1.073	1.553	30.894	69.106
34.00	1502.40	12164.00	13673.00	0.561	1.304	1.865	30.025	69.975
46.83	1861.40	15211.00	17077.00	0.718	1.480	2.198	32.668	67.332
57.00	2119.90	17449.00	19574.00	0.737	2.172	2.909	25.342	74.658
64.50	2300.60	19021.00	21377.00	0.910	3.106	4.016	22.666	77.334
72.50	2485.80	20639.00	23132.00	1.138	3.193	4.331	26.276	73.724
84.33	2747.40	22937.00	25692.00	1.064	3.336	4.400	24.173	75.827
96.50	3003.70	25199.00	28214.00	1.345	3.466	4.811	27.959	72.041
107.16	3219.30	27111.00	30342.00	1.536	3.507	5.043	30.457	69.543
Av. 60.64				0.880	2.360	3.241	27.50	72.50

* Based on allometric relationships given in Hughes, Singh, Guha, Dube and Munshi (1974).

Table 12.10: Intercept (A), regression (b) and correlation coefficient to show the relationship of oxygen consumption ($\dot{V}o_2$) from water via skin and gills and from air via air-sac to their respective body weights. The standard deviation (SD) of the interest and regression coefficients are given.

Experimental Conditions	Temperature of Ambient Water	Parameter	Intercept (A)		Regression Coefficient (b)		Correlation Coefficient (r)
			Estimated Value	S.D.	Estimated Value	S.D.	
In still water with access to air	30.3.±1ºC	Air-sac (A)	−0.188	0.02324	0.783	0.081	0.9123 P<0.001
–do–	–do–	Gill+skin (G+S)	0.460	0.01975	0.520	0.069	0.8643 P<0.100
–do–	–do–	Air-sac+gill+skin (A+G+S)	0.497	0.01975	0.624	0.069	0.8895 P<0.001
–do–	25.0±1ºC	Air-sac (A)	−0.754	0.01703	0.664	0.064	0.9388 P<0.001
–do–	–do–	Gill+skin (G+S)	−0.930	0.02720	0.849	0.103	0.9229 P<0.001
–do–	–do–	Air-sac+gill+skin (A+G+S)	− 0.552	0.01378	0.763	0.044	0.9770 P<0.001
–do–	20.0±1ºC	Air-sac (A)	−0.509	0.01049	0.819	0.043	0.978 P<0.001
–do–	–do–	Gill+skin (G+S)	−0.122	0.02168	0.840	0.088	0.9184 P<0.001
–do–	–do–	Air-sac+gill skin (A+G+S)	0.029	0.01581	0.01581	0.064	0.9542 P<0.001

Table 12.11: Intercept (A), regression (b) and correlation (r) coefficient to show the relationship of oxygen consumption ($\dot{V}o_2$) from water via air-sac, gill+skin and air-sac+gill+skin to their respective respiratory area. The standard deviation (SD) of the intercept and regression coefficients are given.

Experimental Conditions	Temperature of Ambient Water	Parameter	Intercept (A)		Regression Coefficient (b)		Correlation Coefficient (r)
			Estimated Value	S.D.	Estimated Value	S.D.	
In still water with access to air	30.5±1°C	Air-sac (A)	−3.7583	0.02302	1.18602	0.12153	0.9556 P<0.001
–do–		Gill+skin/$\dot{V}o_2$	−2.7848	0.01949	0.74460	0.09783	0.9299 P<0.100
–do–		Air-sac+gill+skin/$\dot{V}o_2$	−3.2660	0.02000	0.89928	0.09990	0.9385 P<0.001
–do–	25.0±1°C	Air-sac (A/$\dot{V}o_2$)	−2.9265	0.01703	1.00362	0.09690	0.9687 P<0.001
–do–		Gill+skin/$\dot{V}o_2$	−4.6007	0.02739	1.21642	0.14778	0.9523 P<0.001
–do–		Air-sac+gill+skin/$\dot{V}o_2$	−3.9538	0.01378	1.10667	0.07497	0.9841 P<0.001
–do–	20.0±1°C	Air-sac (A/$\dot{V}o_2$)	−3.9912	0.01095	1.23796	0.06588	0.9711 P<0.001
–do–		Gill+skin/$\dot{V}o_2$	−4.7467	0.02168	1.20218	0.12602	0.9584 P<0.001
–do–		Air-sac+gill+skin (A+G+S)	−4.6582	0.01549	1.20019	0.08972	0.9768 P<0.001
In flowing water with access to air	25.0±1°C	Gill+skin/$\dot{V}o_2$	−7.3639	0.01500	1.89810	0.0696	0.9914 P<0.001

to 3.9095ml O_2/hr for fish of body weights from 18.5g. for highly weights in the range 128.0 to 136.5g, gradually declined from 3.9095 to 2.2206ml O_2/hr (Table 12.1). At 25.0±1 C, the gradually increased from 0.866 to 3.495 ml O_2/hr in the weight range 13.0 to 93 0g.

In the winter season (20.0±1°C) the results were quite different. From air gradually increased from 0.313 to 1.536 ml O_2/hr for body weights from 16.83 to 107.16g.

The regression coefficients of the log/log W plot (Table 12.4) gave different values for the three seasons. In winter at 20.0±1 C the slope was greatest (0.819), and a minimal value (0.664) was observed in the rainy season (25.0±1 C). In summer at temperature of 30.5±1 C), it was 0.783.

Good correlation was obtained between body weight and $\dot{V}_{O_2}$ at different temperatures in the three seasons studied (0.912, P<0.001 at 30°C ± 1°C, 0.940, P<0.001 at 25°C ± 1°C and 0.980, P<0.001 at 20°C ± 1°C)

The oxygen consumed by the fish from air was less than 50 per cent of the total $\dot{V}_{O_2}$ from air and water combined at all temperatures studies.

(b) $\dot{V}_{O_2}$ from water when the fish had access to air; when the fish were kept in air saturated still water with the free access to air at 30°C ± 1°C, the $\dot{V}_{O_2}$ from water varied from 1.256 to 4.083 ml O_2/hr with the gradual increase in body weights. Above 118.5 g the $\dot{V}_{O_2}$ decreased from 4.083 to 2.931 ml O_2/hr for body weights from 118.5 to 136.5g (Table 12.1).

$\dot{V}_{O_2}$ from water at 25°C ± 1°C showed a gradual increase from 1.204 to 5.44 ml O_2/hr for the range, 13.0 to 93 gram.

Again, when $\dot{V}_{O_2}$ from water at 20°C ± 1°C was measured, a gradual increase in rate of oxygen consumption from 0.9632 to 3.5075 ml O_2/hr was obtained for body weights of 16.83 to 107.16 g.

Different response pattens by fish of different body sizes in these seasonal temperatures were reflected in the slopes (b) of the regression lines obtained for the log/log plots (Table 12.4). The 'b' values were almost identical (0.840 and 0.839) at 25°C ± 1°C and 20°C ± 1°C respectively, but a very low value (0.520) was found at 30.5°C ± 1°C.

Good correlations between body weights and $\dot{V}_{O_2}$ were present at all three seasons.

The $\dot{V}_{O_2}$ through gills and skin was quite high (more than 50 per cent of total $\dot{V}_{O_2}$) in comparison to the aerial breathing through air-sacs at the winter season (20°C ± 1°C) $\dot{V}_{O_2}$ through the gills was quite significant.

(c) Total $\dot{V}_{O_2}$ from both air and water; when a fish was kept in a respirometer with air-saturated water having free access to air, it took oxygen from both media.

(d) The oxygen consumption rate ranged at different seasonal temperature from 1.8561 to 5.1528 ml O_2/hr fish at 30.5°C ± 1°C, 2.07 to 8.94 at 25°C ± 1°C and 1.277 to 5.0435 mlO_2/hr at 20°C ± 1°C within the body weight of 18.5 to 136.5g, 13.0 to 93.9g and 16.83 to 107.16g respectively (Tables 12.1–12.3).

The average $\dot{V}_{O_2}$ from air, water and the total from both media were calculated at 30°C ± 1°C, 20°C ± 1°C and 20°C ± 1°C which gave values as given below:

1.879, 2.630, 4.973; 2.398, 3.421, 5.819, 0.880, 2.360 and 3.241 mlO_2/hr for average body weights of 75.136, 53.424 and 60.640 g respectively.

The log/log plots of $\dot{V}_{O_2}$ vs body weight gave straight lines with slopes of 0.624, 0.763 and 0.833 at 30.5°C ± 1°C, 25.0°C ± 1°C and 20°C ± 1°C temperatures respectively. This equation (1) after substitution became :

$$\text{Log } \dot{V}_{O_2} = \log 0.1069 + 0.833 \log W \text{ ...at } 20°C \pm 1°C \tag{2}$$

$$\text{Log } \dot{V}_{O_2} = \log 0.2802 + 0.763 \log W \text{ ...at } 25°C \pm 1°C \tag{3}$$

$$\text{Log } \dot{V}_{O_2} = \log 0.3140 + 0.624 \log W \text{ ...at } 30°C \pm 1°C \tag{4}$$

Equations (2), (3) and (4) indicated that with unit increase in the body weight of *Heteropneustes*, the $\dot{V}_{O_2}$ also increased by the fractional powers of 0.624, 0.763 and 0.833 at 30.5°C ± 1°C, 25.0°C ± 1°C and 20.0°C ± 1°C respectively and thus oxygen consumption/g body weight remained higher in smaller fish than in larger ones.

High correlation coefficients with statistically significant values observed at 30.5°C ± 1°C, 25.0°C ± 1°C and 20.0°C ± 1°C were 0.90 P<0.001, 0.98 P<0.001 and 0.95 P<0.001 respectively. The correlation coefficients again indicated that the rates of oxygen consumption were closely related to the weight of the fish.

B. Relationship Between Respiratory Surface Area and Oxygen Consumption ($\dot{V}_{O_2}$) from Still Water when the Fish had Free Access to Air

(a) At ambient temperature of 30.5°C ± 1°C in summer. The log/log plots of respiratory surface area of air-sac, gill +skin and airsac+gill+skin and the corresponding oxygen consumption gave straight lines with slopes of 1.186, 0.7446 and 0.8993 respectively and the estimated intercepts were found to be -3.7583, -2.7848 and – 3.266. Thus the relationship between the two variables can be represented in logarithmic forms as:

$$\text{Log } \dot{V}_{O_2} = -3.7583 + 1.186 \log A \text{ (air-sac). air} \tag{5}$$

$$\text{Log } \dot{V}_{O_2} = -2.7848 + 0.7446 \log A \text{ (gills +skin) ...water} \tag{6}$$

$$\text{Log } \dot{V}_{O_2} = -3.266 + 0.8993 \log A \text{ (airsac +gills+skin) ..air +water} \tag{7}$$

The correlation coefficients (r=0.9556, P<0.001; r=0.9299 P<0.001; r=0.9486 P<0.001 for (5), (6) and (7) respectively) were observed to be high at all parameters.

It seems that larger fish depend more on aerial breathing than the smaller ones at high temperature of 30.5°C ± 1°C as indicated by the high exponential values (b=1.186) obtained for air-sac areas versus oxygen consumption.

(b) *At ambient temperature of 25.0°C ± 1°C during the rainy season.* When the data on $\dot{V}_{O_2}$ (ml/hr) through air-sac, gill+skin and air-sac+gill+skin were plotted against the respiratory surface areas of these organs, it gave straight lines with slopes of 1.0036, 1.2164 and 1.1067 and the estimated values of the intercepts on Y-axis were found to be -2.9265, -4.6007 and -3.9538. The relationships are summarized in the following equations:

$$\text{Log } \dot{V}_{O_2} = -2.9265 + 1.0036 \log A \dots \text{airsac} \tag{8}$$

$$\text{Log } \dot{V}_{O_2} = -4.6007 + 1.2164 \log A \dots \text{gill+skin} \tag{9}$$

$$\text{Log } \dot{V}_{O_2} = -3.9538 + 1.1067 \log A \dots \text{airsac + gill + skin} \tag{10}$$

The O_2 consumption and the respective respiratory surface areas are highly correlated (r=0.9687 P<0.001, r= 0.9523 P<0.001, r=0.984 P<0.001).

(c) At ambient temperature of 20.0°C ± 1°C in winter: The log/log plots of respiratory areas (A+G+S) and the oxygen consumptions gave straight lines with slopes of 1.238, 1.202 and 1.20 and the estimated intercepts were – 3.991, -4.7467 and 4.658. The relationships are represented in the logarithmic form as follows :

$$\text{Log } \dot{V}_{O_2} = -3.991 +1.238 \log A \text{ (A)} \tag{11}$$

$$\text{Log } \dot{V}_{O_2} = -4.7467 +1.202 \log A \text{ (G+S)} \tag{12}$$

$$\text{Log } V_2 = -4.658+1.20 \log A \text{ (A+G+S)} \tag{13}$$

High correlation coefficients (r = 0.971 P<0.001, r=0.9584 P<0.001, r =0.9769 P<0.001 were observed in all cases.

Discussion

The data obtained in the present study of routine oxygen consumption in *Heteropneustes (=Saccobranchus) fossilis* is of interest for environmental biologists. Data showing the effect of temperature and body size on the relative importance of aquatic and aerial breathing is summarized in Tables 12.1–1.3.

Metabolism is correlated inversely to body weight and the slope of the regression line varies from 0.55 to 1.0 depending on many factors (Prosser and Brown 1962). In fish of both arctic and tropical regions the value of the exponent (b) obtained by fitting a straight line to a logarithmic transformation of the data is nearly 0.8 (Smith 1935b, Zeuthen 1947, Scholander *et al.,* 1953, Zeuthen 1953, Job 1955, Winberg 1956, Hemingsen 1960, Paloheimo and Dickle 1966).

In *Heteropneustes* the exponent for $\dot{V}_{O_2}$ in relation to body weight varies during different seasons. In winter (temp. 20.0°C ± 1°C), aquatic breathing (via gill+skin) increases a little more rapidly (b=0.84) than aerial breathing (via air-sac) (b=0.82). In winter, the dissolved oxygen in ambient water is sufficiently high to maintain the daily requirement of the fish. More strikingly in the rainy season (temp. 25.0°C ± 1°C), aquatic breathing increases more rapidly (b=0.84) than aerial breathing (b=0.66). Therefore, it is evident that at 25.0 ± 1°C as the fish grow in size they depend increasingly on water than air for their oxygen uptake. In nature, the ambient water also becomes more aerated due to frequent rains. However, quite different results were obtained in summer (temp. 30.5 ± 1°C), when the ambient temperature has adverse effect on aquatic breathing. In this season, with increase in size the fish rely more on aerial breathing (b=0.78) than aquatic (b±0.52). In summer the oxygen dissolved in water becomes depleted in the swamps which may compel the fish to depend mostly on aerial breathing.

The rate of increase of total $\dot{V}_{O_2}$ in relation to body weight is higher in winter (b=0.83) than in the rainy season (b=0.76) and is minimal in summer (b=0.62). The high temperature during summer produces a reduction in slope of the total $\dot{V}_{O_2}$ /body weight regression line. It seems that the bigger fish (128.0 g and 135.5 g) are heat depressed, whereas the younger fish are better able to tolerate temperature changes and also higher temperatures.

Parvatheswararao (1960) reported similar findings for *Puntius sophore* in which he described the heat depression at high temperature (30.0° and 35.0°C) and cold depression at 15.0°C as the body size increases.

But, just the reverse results were obtained in this species at 20.0 ± 1°C, where the fish seemed to be better cold adapted as the slope of the regression line is higher (b= 0.83). A similar result was obtained by Hakim, *et al.* (1983) where the term "Cold adapted" was applied to *Channa punctata* at 20.0°C because of the higher exponent (b=0.79) in winter than in summer (temp. 31.0°C) when the slope was 0.60.

In *Heteropneustes* $\dot{V}_{O_2}$ from water and air show variable proportions. Percentage uptakes of oxygen from water and air at three different ambient temperatures are summarized in Tables 12.1–12.3. It is evident from these tables that this species obtains about 57-70 per cent of this total oxygen requirement through gills and skin from water, and oxygen uptake from air through the respiratory sacs is only about 30-43 per cent. The dependence on gill breathing is quite evident and corroborates the earlier findings of Hughes and Singh (1971).

It is interesting to see that the fish consumed more oxygen at 25°C than at 30°C. This requires explanation. The possibility is that in nature the fish might have been acclimatized at some level higher than 25°C, which caused a higher rate of oxygen consumption in these fish at 25°C under laboratory conditions.

Again, when $\dot{V}_{O_2}$ /g body weight is studied at different temperatures, the intercept (a) increases with increase of ambient temperature, being highest (a=0.3) in summer and lowest (a=0.11) in winter, *i.e.* the intercept values (a) are also temperature dependent.

Munshi *et al.* (1978) reported in their studies on *Heteropneustes fossilis* high exponential values for both juvenile (b= 1.084) and adult fish (b=0.986) when they were allowed free access to air in a constant flow respirometer at 26 ± 1°C. Although these slopes may not be significantly different, they speculated that juveniles depend more on air-breathing for their total $\dot{V}_{O_2}$ and that with growth and maturation they prefer to live more on aquatic breathing. In still water with free access to air, in the present investigations the $\dot{V}_{O_2}$ /weight curves gave slope values of b=0.840 and 0.664 for aquatic and aerial breathing respectively at 25.0°C ± 1°C.

From this study it may be concluded that the gills are the most important sites for O_2 uptake in the normal habitat. The accessory respiratory organs which are used for oxygen uptake from air do not appear to be as important as the gills. It is evident, however, that the air-sacs serve a vital role which enable these fish to meet their total oxygen demand under adverse conditions of water oxygen supply.

Study of the morphometrics of the gas exchange surfaces by Hughes *et al.* (1974) has revealed that with unit increase in body weight, the air-sac, gill + skin and air-sac+ gill+skin are increased by a power of 0.6619, 0.6982 and 0.6944 respectively. Therefore, from the regression coefficient values it can be stated that of these respiratory surfaces, the gill+skin area increases slightly more rapidly than those of air-sacs and air-sacs+gill+skin. So, it is likely that the gills of *Heteropneustes fossilis* provide sufficient gas exchange surface to supply enough O_2 for the metabolic activites of the fish.

The present findings on the relationship between the respiratory surface area and the O_2 consumption from water supports the above-mentioned ideas only at 25.0°C ± 1°C during the rainy season when the water levels of ponds and swamps increase and the O_2 content of the water is high and temperature is low. At this temperature with unit increase in the respiratory area of the gills, the O_2 consumption from water increases by a power of 1.216 which is slightly higher than aerial breathing through air-sacs. The oxygen uptake is almost directly propertioral to the gill area.

In contrast, the equations at 20.0°C ± 1°C and 30.5°C ± 1°C suggest lower slopes for the regression lines relating $\dot{V}_{O_2}$ to gill+skin area (b=1.2021 and 0.7446). Such lower values may indicate a reduction in diffusing capacity of gill+skin in larger specimens at low and high temperatures (20.0°C and 30.5°C).

Oxygen Uptake Capacity of Gills in Relation to Body Size of the Air-breathing Fish *Anabas testudineus* (Bloch), J.S.D. Munshi and S.C. Dube, 1973)

Oxygen consumption through the gills in relation to body weight was estimated in the air-breathing fish, *Anabas testudineus* under two experimental conditions, viz (i) when air-breathing was allowed and (ii) when it was prevented. The lower value of the slope (b=0.67) of the line for oxygen consumption to body weight (when air-breathing was allowed) in comparison to water-breathing teleosts (b=0.85) was related to the air-breathing habit of the fish and the increase in the relative proportion of air-breathing surface with the increase in body weight of the fish. A further lowering of the slope (b=0.5259) on preventing air-breathing has been related to the decrease in the efficiency of the respiratory membrane of gills under stressed conditions. The observed capacity of smaller fishes to survive for longer periods depending upon gill breathing slope than those of large fishes has been related to the decrease in the oxygen uptake capacity of gills as fishes grow in size. The higher rate of oxygen consumption per unit body weight in small fishes than in larger ones has been related to their higher metabolic rate and the larger surface area of gills per unit body weight available for gaseous exchange.

In recent years, several investigations have dealt with the function of teleostean gills and the respiratory surface has been studied in relation to body weight in several species of teleosts (Price, 1931; Gray 1954; Hughes, 1966, 1970, 1972; Muir and Hughes, 1969) to understand the nature of growth of the gill-sieve. Similarly, the rate of oxygen consumption as a function of body weight has been studied in many teleost species (Smith, 1935; Scholander *et al.,* 1953; Job 1955; Winberg 1956; Parvatheswararao 1960; Saunders 1963; Paloheimo and Dickle 1965; Brett, 1965). These results indicated that the relationship of oxygen consumption to weight bears the same relationship as gill area to body weight. Thus, it has been suggested that some relationship might exist between the size of the respiratory surface and the rate of oxygen consumption among fishes (Winberg, 1956). As most of these growth studies were restricted to water-breathing fishes (having only gills for gaseous exchange), nothing is known about this relationship in air-breathing fishes, where gaseous exchange takes place both from air and water through the accessory respiratory organs and gills. Recently, Hughes *et al.* (1973, 1974) studied the size of the respiratory surface of gills and air-breathing organs in relation to body weight in *Anabas testudineus* and *Saccobranchus* (= *Heteropneustes) fossilis.* In the present study an analysis has been made of the oxygen uptake capacity of the gills in relation to body weight of the air-breathing climbing perch, *Anabas testudineus,* when it is free to breathe both air and water and when air-breathing has been prevented. These results will be discussed as related to body weight, the extent of gill area and the asphyxiation of *Anabas* when not allowed to breathe air.

Methods

Living specimens of different sizes of *Anabas testudineus* (Bloch) were collected from Purnea and nearby places in Bihar, India and maintained in large glass aquaria, with continuous flow of water. The fishes were fed on goat liver pieces on alternate days during a minimum acclimation period of one month in the laboratory.

The rate of oxygen consumption through the gills was measured in *Anabas* of different body weights under two experimental conditions, *viz.* (1) when access to air was allowed and (2) when it was prevented. Two types of self-designed spirometers were used. In the first series of experiments, a rectangular aeroplex box of approximately three litre capacity, having a small air chamber at the top was used; here the fish was free to breath air in addition to gill-breathing. The air chamber at the top of the spirometer was constructed to minimize gaseous exchange at the air-water interface. The fish becomes adapted to this spirometer in a few hours and then goes freely upto the air-chamber to inhale air. For the second series of experiments, a big cylindrical glass spirometer was constructed with arrangements for continuous water flow through it and the removal of enclosed air.

The fish was introduced in the spirometer which was connected to a large constant level water tank to maintain the flow of water under constant hydrostatic pressure. The water entered the spirometer at one side and the rate of its flow per unit time was measured as it left the other side. The flow of water through the spirometer was adjusted according to the size of the fish, so that its metabolism did not reduce the oxygen content of the ambient water to below 80 per cent of the air saturation level. The fish was acclimated to the spirometer at least overnight before the observations has begun. Two bottles were connected at the two ends of the spirometer to collect the water supplied and the ambient water.

Concentration of the dissolved oxygen in the samples collected was estimated by Winkler's volumetric method (Welch 1948). The difference in the oxygen levels between the ambient water and that supplied to the spirometer, together with the rate of water flow and the weight of the fish were used to calculate the rate of oxygen consumption per unit body weight of the fish. Subsequent readings were taken at one hour intervals. Opercular frequencies were counted for a period of one minute at intervals of one hour and the amplitude of their movements was estimated by visual observation. The observations were recorded at $27.5 \pm 1°C$. Regression analysis using logarithmic transformations was made in order to find out the relationship of body weight with oxygen consumption through the gills of *Anabas*.

Results

Anabas testudineus is a habitual air-breathing fish and has well developed organs for respiration (Munshi, 1968). When kept in water it comes to the surface at irregular intervals to inhale air in addition to normal gill-breathing.

The interval between the air-breaths vary from two minutes to fifteen minutes depending on the concentration of oxygen dissolved in the water. In water containing 4.2 to 5.6 ml/lit. dissolved oxygen and with free access to air, *Anabas* shows 30-45 opercular movements per minute. When access to air is prevented, the fish becomes restless in search of air, but the restlessness slowly decrease and the fish settle down at the bottom of the spirometer. However, the amplitude and frequency of opercular movements increase to 60-65 per minute.

Mean oxygen consumption per unit time of different weight groups of fishes under the two experimental conditions are summarized in Table 12.12. Results indicated a marked effect of body size on the rate of oxygen consumption. When oxygen consumption has been plotted against body weight on log/log co-ordinates and the scores were fitted by the least square regression method, it gave a straight line. Thus the relationship between the rate of oxygen consumption and the body weight obeyed the equation:

$$X = aW^b \qquad\qquad (1)$$

Or $\log X = \log a + b \log W,$

where

   X: Rate of oxygen consumption (ml/hr)

   ∞: rate of oxygen consumption for 1 g fish (ml/hr)

   b: Regression coefficient (slope)

   W Body weight (g)

1. Oxygen Consumption from Water when the Fish has Free Access to Air

When *Anabas* was kept in the spirometer with free access to air, its oxygen consumption from water ranged from 0.71 to 2.83 ml/hr with an increase in the body weight from 8.0 to 68.0 g. The log/log plot of the rate of oxygen consumption to weight gave a straight line with a slope of 0.6661 and the estimated value of oxygen consumption for one gram fish was found to be 0.1764 ml/hr (Table 12.12). Thus, equation (1) after substitution becomes.

$$\text{Log } X = \log 0.1764 + 0.6661. \log W \tag{2}$$

Equation (2) indicates that with an unit increase in the body weight of *Anabas*, the rate of oxygen consumption will increase by a fractional power of 0.6661 and thus oxygen consumption per g body weight remains higher in smaller fishes (88.99 ml/kg/hr) than in large ones (41.74 ml/kg/hr) (Table 12.12). There was a good correlation between body weight and oxygen consumption rate, the correlation coefficient being 0.9947 (Table 12.13).

Table 12.12: Rate of oxygen consumption via gills of *Anabas testudineus* different in size under two experimental conditions.

Mean Oxygen Consumption Rate with Access to Air			Mean Oxygen Consumption Rate with No Access to Air		
Body Weight (g)	$\dot{V}O_2$ (ml/hr)	$\dot{V}O_2$ (ml/kg/hr)	Body Weight (g)	$\dot{V}O_2$ (ml/hr)	$\dot{V}O_2$ (ml/kg/hr)
8.0	0.71	88.99	8.0	1.48	185.34
17.0	1.08	63.47	17.0	2.89	169.94
24.0	1.51	62.82	29.0	3.52	119.25
34.0	2.01	59.12	33.0	3.02	91.38
44.0	2.20	40.96	42.0	3.82	90.93
55.0	2.53	45.92	54.0	4.02	74.45
68.0	2.83	41.74	65.0	5.26	80.93

2. Oxygen Consumption from Water with Surfacing Prevented

Under such conditions, the fish could obtain oxygen only from water. The marked increase in the ventilatory movements of the fish resulted in an increased oxygen consumption. Mean oxygen consumption under this particular experimental condition ranged from 1.48 to 5.26 ml/hr for the different weight groups of *Anabas* (Table 12.1). This was much more than the oxygen consumed by *Anabas* of practically the same body weights under the first experimental conditions. The log/log plot of the rate of oxygen consumption against body weight gave a straight line on fitting the scores by the least square regression method. The slope of the line was 0.5259, much less than the value obtained when air-breathing was allowed (Table 12.2). The estimated value of a, *i.e.* of the rate of oxygen consumption for one g fish was 0.5453 ml/hr. Thus, after substitution, the equation is

$$\text{Log } X = \log 0.5453 + 0.5259. \log W \tag{3}$$

Indicating that the rate of oxygen consumption of *Anabas* under such conditions increases by a fractional power of 0.5259 with every unit increase in body weight. Thus, the rate of increase in oxygen consumption from water alone with the increase in the size of *Anabas* was comparatively less when air-breathing was not allowed than when the fish was breathing both air and water. However, the rate of oxygen consumption per unit body weight remained high (80.93 to 185.34 ml/kg/hr) when air-breathing was prevented than when it was allowed (Table 12.1). Oxygen consumption per unit body weight was higher in small fishes than in large ones (Figure 12.5).

The correlation coefficient (r= 0.9569) again indicated that the rate of oxygen consumption was closely related to the weight of the fish.

Discussion

The rate of oxygen consumption through the gills of *Anabas* gives a straight line when the logarithm of the rate is plotted against the logarithm of body weight. Thus, the relationship of the metabolic rate to body weight in this air-breathing fish was similar to that found in other water-breathing teleosts. The slope of the logarithmic line for standard metabolism against body weight varied slightly among different species of teleosts, Job (1955) found 0.85 for *Salvelinus fontinalis*: Winberg (1956) obtained 0.81 as an average value for a number of species and Paloheimo and Dickie (1965) suggested 0.80 to be the value characteristic for most species of teleosts. In comparison to these values, the obtained value of the slope (0.67) for oxygen consumption through the gills to body weight (log/log) was much less when the fish was allowed to breath both air and water than under normal conditions. This was obviously due to the air-breathing habit of the fish which supplements the total oxygen consumption of the species. Measurements of the surface of gills and of the accessory respiratory organs of *Anabas* (Hughes *et al.*, 1973) indicated that the rate of increase in the area of accessory respiratory organs (b=0.713) was higher than that for the gills (b=0.615) during their growth. Therefore, it is expected that in large fishes the proportion of oxygen uptake through the air-breathing organs will dominate over that of the gills. Secondly, the thickness of the blood-water barrier in the secondary gill lamellae increase in large fishes (b=0.2272; Dube and Munshi, 1973a). Such an increase in the thickness of the barrier in secondary gill lamellae of tench has been suggested to decrease their diffusing capacity as the body size increases (Hughes, 1972). It is, therefore, suggested that in larger fishes there will be an increasing dependence on the air-breathing organs for extracting oxygen from the environment which results in a lowering of the slope of oxygen consumption through the gills to the body weight line.

When *Anabas* is not allowed to breathe air, its ventilatory movements increase in both frequency and amplitude which seems to be a homeostatic attempt (Hughes, 1964) to compensate the functioning of the accessory respiratory organs in meeting the total oxygen demand of the fish. Under normal conditions of water and air-breathing, the rate of oxygen consumption via the gills of *Anabas* ranges from 41.74 to 88.99 ml/kg/hr, which increases to 80.93 to 185.34 ml/kg/hr when air-breathing is prevented (Table 12.12). This increase is up to a certain extent due to the increased activity of the fish. Such an increase in the rate of oxygen consumption associated with an increase in the ventilatory movements was observed also by Saunders (1962) in carps, white suckers and brown bullheads and by Hughes and Singh (1970) in *Anabas*, which they related to the increase in the effectiveness of gaseous exchange through the gills due to the increased volume of water irrigating the gills per unit time.

The slope of log oxygen consumption to log body weight line was 0.5259 when air-breathing was not allowed. This was less than the value of 0.6661 found when air-breathing was allowed. This reduction in the slope of the logarithmic line indicates that under stressed conditions the capacity of the gills to take up oxygen decreases as the fish grow in size. As a result, the rate of oxygen consumption did not increase satisfactorily with the increase in body weight and the logarithmic line was depressed towards the upper end of body size.

The respiratory surface area of the gills (Hughes *et al.*, 1974), when expressed as the area across which each ml of oxygen is transferred per minute (surfacing prevented), the ratio of the surface area to oxygen consumed per unit time was less in small fishes, and increased with the increase in body weight (Table 12.3). This means that more gill surface is required for the uptake of each ml of oxygen per unit time in large fishes, and perhaps the respiratory membrane of the gills of small fishes is more efficient in oxygen uptake than that of large ones.

The drowning experiments conducted on *Anabas* by Boake (1865-66), Dobson (1874), Das (1927) and Ghosh (1934) indicated that fishes belonging to this species become asphyxiated within 12 minutes to 3 hours and 30 minutes, when not allowed to breathe air. Hora (1935), on the other hand, observed that while some of the specimens of *Anabas* were asphyxiated within 15 minutes, others survived indefinitely under similar conditions. Thus, his observations were in agreement with the results of Boake (1865-66) in that the period of drowning varies with different individuals. The cause of this intraspecific variation in asphyxiation time during drowning experiments was not clear in the above mentioned studies. In the present study it has been observed that the asphyxiation time for *Anabas*, when not allowed to breath air, varies with the size of the fish, and as the body size increases the asphyxiation time decreases. This again supports the previously discussed view that the capacity of gills to take up oxygen decreases as the fish grows in size and, as a result, gill-breathing by itself can sustain a larger fish for a shorter period than a small one. This was further supported by the morphometric determination of the diffusing capcity of gills of *Anabas* (Dube and Munshi, 1973b).

Table 12.13: Respiratory area for unit volume of oxygen consumed (ml O$_2$/min) Anabas of different sizes (computed by extrapolation of regression lines for gill area and oxygen consumption rate against body weight).

Body Weight (g)	Gill Area/ml Oxygen Consumed/min (mm²)
10.0	382.0
20.0	398.6
30.0	416.8
40.0	426.3
50.0	433.9
60.0	437.6
70.0	445.6
80.0	451.8
90.0	456.7
100.0	460.7

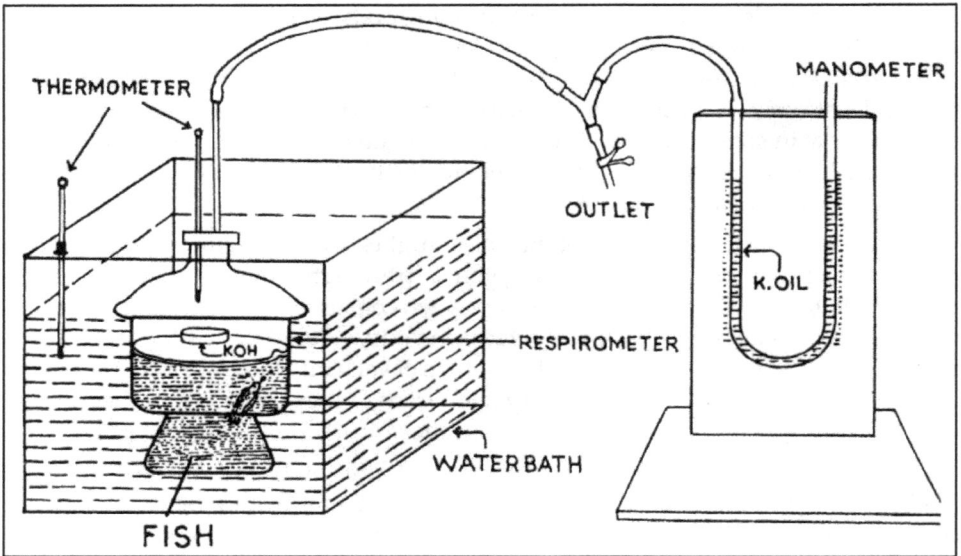

Figure 12.1: Experimental apparatus for measuring oxygen uptake through the biomodal gas exchange system of *Channa gachua*.

Hughes and Singh (1970) found oxygen consumption (from water) of *Anabas* at 25 ± 1°C to vary from 45.44 to 61.88 ml/kg/hr, when the fish was allowed to breath both air and water, and from 70.7 to 85.0 ml/kg/hr when air-breathing was prevented. When these values have been compared with the present observations (Table 12.1), the latter seem to be slightly higher, which was related to the higher temperature (27.5 ± 1°C) of the ambient water during the present observations.

Oxygen consumption per unit body weight was higher in smaller fishes but as their weight increases, the rate of oxygen consumption decreases sharply in the beginning up to about fifty g body weight and thereafter very slightly. In water-breathing fishes this phenomenon has been related to the higher metabolic rate of smaller fishes (Fry, 1957). It may also be related to the larger surface area of gills per unit body weight of *Anabas* (Hughes *et al.*, 1973) available for gaseous exchange in smaller fishes than the large ones.

Hughes and Singh (1970) reported that specimens of *Anabas* weighing about 29 to 51 g consume about 53.6 per cent of their total oxygen requirement through the air-breathing organs. Dube (1972) found that the air-breathing surface in fishes of this size comprises about 42 to 43 per cent of the total respiratory area. Thus it is suggested that larger fishes having a larger respiratory surface will naturally depend much more on air than on gill breathing because of the higher diffusing capacity of air-breathing organs than of gills (Hughes *et al.*, 1973). It will be interesting to investigate the relative rates of oxygen consumption through gills and air-breathing organs in different weight groups of *Anabas*.

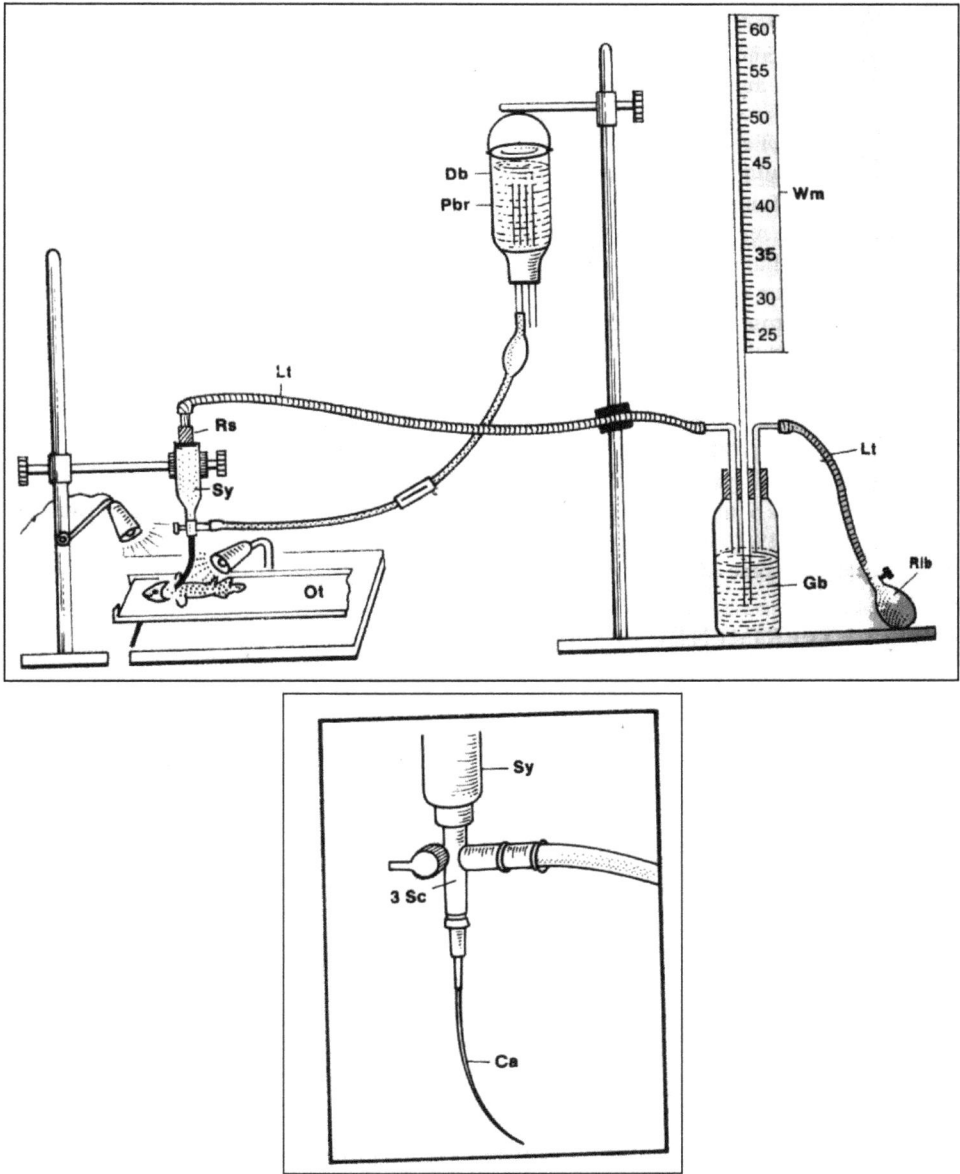

Figure 12.2: Experimental set up for the vascular corrosion cast preparation. (Inset details for canula attachment).

Abbreviations: Ca: Canula; Db: Drip bottle; Gb: Glass bottle; Lt: Latex tubing; Ot: Operation tray; Pbr: Phosphate buffered ringer; Rib: Rubber inflation bulb; Rs: Rubber stopper; 3Sc: 3 way stop cock; Sy: Syringe; Wm: Water manometer.

(Ghosh and Dattamunshi, 2003).

Chapter 13
Phylogeny

This chapter presents an attempt to assess the main phyletic trends in teleostean fishes, based primarily on study of the living forms. Results indicate the necessity of a major re-grouping of teleostean orders, and this also is attempted.

The work stemmed from talks and correspondence between Greenwood, Rosen, and Myers at the time subsequent to the XVth International Congress of Zoology in Washington in 1963, especially during discussions of the then unpublished results of Greenwood's study of the osteoglossiform fishes and of Rosen's work on the atherini a group of fishes (Greenwood, 1963; Rosen, 1964). Active cooperative work was not initiated until early in 1964. When Weitzman joined the group.

The main burden of the investigation has been carried by Greenwood and Rosen closely followed by Weitzman, who is almost wholly responsible for the strictly ostariophysan lineages. Myers has acted principally as adviser and editor, and has contributed in large part to the compilation of the list of family names.

Part of the earlier findings of Rosen and Greenwood were communicated orally by them to A.S.Romer during a conference in London early in 1964, for use by the latter in the new edition of his "Vertebrate Paleontology".

Traditionally, studies such as ours have been based on morphology, especially the skeleton, which is the only complete organ system available for detailed comparison with fossils. However, with the varity of both primitive and advanced teleosts living today, we are most emphatically of the opinion that approaches other than morphological ones would be exceedingly fruitful in the investigation of teleostean interrelationships.

It must be not be imagined, however, that the full informational content of teleostean morphology has been extracted. Only the barest beginnings have so far been accomplished, even within the realm of osteology. We doubt if more than the external anatomy of 95 percent of the species of living teleosts has been examined, and for many families there has so far been little or no deeper study. Researchs on the nervous, digestive, muscular, and vascular systems of teleosts are scattered and mostly uncoordinated, and relatively few of them have been done with any specific systematic objectives in view, despite the fact that pioneer work, such as that of Freihofer (1963) on a single nerve complex has uncovered a wealth of information bearing upon phyletic relationships.

History

The latest widely accepted general classification of teleostean fishes is that of Berg (1940). The second edition, edited by Sevetovidov (Berg.1955), and its German translation (Berg.1958), is the arrangement of the teleosts materially altered. Berg's teleostean groupings like those of Jordan (1923), closely reflect the conclusions reached by Regan in a long series of brilliant papers culminating in his brief general exposition (Regan, 1929). Regan's teleostean papers, in turn, were built upon the much earlier foundations laid by Boulenger (1904) and Woodward (1901) and in the nineteenth century, by Gunther, Cope (1871), Gill (1872,1893). In fact, except for relatively minor revisions, shifts and splitting, most of the major groups of living teleosts recognized by Berg do not depart in any revolutionary way from those recognized 70 to 90 years ago by Gill.

More than 20 years, ago Woodward (1942) published a prophetic little paper on the beginnings of teleostean fishes, in which he advanced the view that the Teleostei, long reorganized as a natural, monophyletic group, in reality had evoled as a number of distinct, lineages (1960).

The problems faced in attempting a "natural" taxonomic classification of a large and varied group of organisms have been widely discussed (see Rensch, 1960, and especially Simpson, 1961). The problems are greatest when the meaningful part of the fossil record is relatively scanty, as it is in the teleosts, and distinct lineages from diverse holostean ancestors in the Mesozoic. Indeed, during the past 35 or 40 years, it has become generally recognized by paleo-ichthyologists that the holosteans themselves represent merely a stage or level of organization into or through which numerous actinopterygian lines passed during their evolution from separate stocks of Late Palaeozoic or Early Mesozoic palaeoniscoid derivatives. Woodward's paper was not widely noted by students of living teleosts, but Bertelsen and Marshall(1956). in discussing the mirapinnids, explicity supported the view that different teleostean lineages have attained certain comparable grades of organization. The idea of teleostean polyphytelism was expressed by Bertin and Arambaurg (1958) in their extensive account of the group. However, their treatment of the living forms appeals to us as chiefly another reshuffling of long-recognized entities, improved here and there by Arambourg's extensive knowledge of fossil teleosts, but marred by an unfortunate lack of familiarity with many recent groups as well as by such egregious errors as acceptance of Y.Le Danois' (1961) imprecise and unacceptable work on the tetraodontiforms.

Thus we are left at present day with no general classification of teleostean fishes that has utilized those modern concepts of phyletic classification that have become common in the study of mammalian evolution (e.g., Simpson, 1945) Yet the feeling has been growing among our group and others that many of the most generally recognized teleostean orders are no more than catch ally for separate lineages which have attained a comparable state of specialization or complexity (see particularly Gosline, 1960). Relationships must be inferred largely on the basis of the morphology of living stocks. The question of "horizontal" versus "vertical" classifications in such groups becomes essentially that or typological versus phyletic taxonomy, which Simpson (1960, pp.46-66) has adequately discussed.

Teleostean classification, up to and including not only Berg's work, but also a very large part of that of Bertin and Arambourg (1958), has been arrived at primarily by methods that are essentially typological in nature an attempt first to define orders and other higher taxa and then to speculate upon their origin, albeit in the light of the known fossils. In the mammals, the prepondeant weight of the fossil over other evidence long ago forced mammalogists to the phyletic type of classification. No such revolution in teleostean classification has occurred up to the present day.

Of the present authors, Myers has thought for a number of years that the varied "order" of "isopondylous" or "clupeiform" fishes is apolyphyletic assemblage; Greenwood (1963) has already begun the demonstration that such is true; while Rosen (1964 and other works) has begun the dismemberment of Regans "percomorphs" and the preliminary demonstration of the affinities of certain perchlike groups with the relatives of the salmonoids. Moreover, all of us (see especially Weitzman, 1964 and out discussion below on the gonorychoid fishes) have more recently begun to think that the ostariophysan fishes may be far older than was previously believed and contain separate lineages running back to a generalized Mesozoic teleostean. Our prime purpose, then has been to separate and point out what we believe to have been the main and subsidiary phyletic lineages of teleosts and the often parallel or converging trends that characterize the evolution of these lines. We have not been especially interested in the definition of living (or fossil) "groups" as such. Definitions of higher taxa, even those based on deep and extensive study of living assemblages, are rarely very full or precise and are seldom used except by those who wish to amend them. However, unless one wishes to abandon the principle that taxonomic classify – living species will approach or surpass 30,000. The most numerous additions to the total may be expected in the deep seas and in the excessively rich freshwater fauna of tropical America.

Fishes of teleosteans type (Leptolepididae) first appear in the known fossil record in the Middle Triassic. Some of these are so advanced in the details of their structure that we can speculate that the shift from the holostean to the teleostean level began much earlier in some forms. However, the death of Early Mesozoic fossils of teleostean type except in marine Triassic and Jurassic beds in the area of the Tethys Sea, may be related to a freshwater origin of many teleostean lines in regions where fresh – water fish bearing deposits are undiscovered. The many teleosts known from the deposits of epeiric seas laid down during the Late Cretaceous indicate that several lineages had by then attained an organization similar to that of living forms. This statement is especially true of the elopoid and berycoid lines. However, the absence in Cretaceous deposits of several important lines of teleostean development (notably the salmonids and ostariophysans, which give considerable evidence of an age comparable to that of the elopoieds) again leads to the suspicion that much teleostean evolution was going on in Mesozoic freshwaters- evolution of which we as yet have no trace.

By the Eocene, or possibly even the Paleocene, teleostean marine shorefaunas bore a striking resemblance to modern assemblages, a fact that again is wholly unlike the situation in mammals. Since that time, a number of teleostean families appear to have undergone comparatively little change.

We have not excluded fossils from consideration, although we do not place them in our formal classification. A number of important fossils are discussed in the expository comments preceding our classification. Paleo-ichthyologists who deal extensively with teleostean fossils are quite aware that the classification of living teleosts must be understood before the fossil record can be properly interpreted. Caution should reflect what can be determined of phylogeny (as some people do), taxa that are obviously polyphyletic must be broken up and a new classification must be adopted.

The classification that we now propose is based on an analysis of what we consider to be the predominant evolutionary trends in the teleosts. By basing the definitions of groups solely on these trends, we have tried to free teleosts from classification as much as possible from the confining influence of typology.

The families recognized, and their placement, follow Berg's (1940) arrangement but with amendations based on works published subsequently and on unpublished information supplied by our colleagues.

Teleostean Diversity and Age

Although the teleosteans are far from being the well circumscribed group that Johanees Muller (1846) and his successors belived them to be, they are at the present time well separated from the living holosteans and chondrosteans. The term"teleostean" (or "teleost") has meaning, even if it represents merely the final grand stage which so many diverse lines of actinopterygian (or teleostome) fishes have attained, and within which the actinopterygians have flowered into the largest (hence, by some definitions, the most successful) of all major vertebrate groups.

Their diversity is astounding. Estimates of the number of living species vary from somewhat under 20,000 up to 40,000. The facts that discovery of new species and genera is still commonplace, and that new forms of considerable, evolutionary importance (*e.g.*, *Denticeps* and many recently described deep sea forms) are still being discovered at a surprising rate, demonstrate that we are further from a reasonably complete knowledge of living teleosts than we are of any other large, non-piscine, vertebrate group. Bailey (1960) estimated the present total to be somewhat fewer than 17,000 species and Myers (1958) estimated that the eventual total number is very high.

Nature of Major Groupings Adopted

We propose,that separate evolutionary routes toward the acanthopterygian grade had been traversed by relatively unrelated lines, and we also propose that the malacopterygian level probably was attained polyphyletically from holostean of pholidophoroid type.

The principal innovations in this classification are the separation of the teleostean fishes into three divisions and the realignment of taxa among eight superorders. Various smaller groups (suborders and families) are redistributed among orders.

In our conception, each of the three divisions represents a distinct phyletic lineage derived from the holostean level of organization. It is presumed that in Division I, a primitive elopiform ancestor has produced principally the eels and eel-like fishes and perhaps also the herring –like fishes. However, we know of no evidence to rule out the possibility that the herring and their allies had an independent origin from among the pholidophoroids. For the present, and because elopomorphs and clupeomorphs appear more closely related to one another than to other groups, we have followed a conventional alignment of these fishes.

In Division II there have evoled only two series of unusually specialized and predominantly freshwater radiations, the Osteoglossiformes and Mormyriformes. Neither of these orders could possibly have been involved in the ancestry of other teleostean.

By contrast with the other divisions, Division III contains the bulk of the living teleostean fishes. There have evolved within this division several radiations leading to more than one organizational level and to the dominant groups of extant freshwater and marine species, namely, the Ostariophysi and Acanthopterygii.

In the discussions below of the major structural and development of divisional trends, where we outline the reasons for erecting the various new orders and super orders, a separate analysis is given of suborders-Stomiatoidei and Alepocephaloidei of Division III because of their previous placement near groups here included in Division I. The ordinal and subordinal composition of the super order Ostariophysi is also given in detail. The one superorder not discussed at length is the Atherinomorpha. It does not fall readily within the concepts of either of two adjoining groups, the Paracanthopterygii and Acanthopterygii, although it includes forms at more or less the same organizational level as the fishes of those superorders. The main structural and developmental characteristics of the Atheriniformes, the only contained order of the Atherinomorpha, were described in some detail by Rosen (1964).

Among the various living primitive groups of teleostean, only elopids (Division I) and salmonids (Division III) are sufficiently generalized to be suitable morphologically as basal types for the major teleostean radiations. Other primitive groups, for example, the clupeiforms and osteoglossiforms, are too specialized for this role. We realize that elopids, of all teleosts, possess the greatest assemblage of holostean characters. At the same time, we recognize that their larval and other specializations, and the absence of certain snout and jaw structures, put at least the living elopids off the maincourse of teleostean evolution. Salmonids, on the other hand, have none of these limitations and thus seem better to fulfill the requirement of a morphotype that may have given rise to the major radiations of Division III. A question that naturally arises from this conclusion concerns the possibilities that elopids and salmonids might have arisen polyphyletically from the holostean level or monophyletically from a single holostean or early teleostean entity. Elopids are still so close to the holostean grade, however (indeed, some workers consider them to be holosteans), that a common ancestor of those two modern groups, if it existed, were itself likely to have been an advanced type of holostean fish. If this view is supported by paleontological evidence, as we think it ultimately will be, the salmonids and elopids would represent separate attainments of the teleostean grade and thus would be, examples of *polyphyletism*, at least at the teleostean level.

Major Trends within the Divisions and Superorders

Division - I

Fishes of ancestry at or near the holostean level of organization in which each contained group except for the eel-like fishes has one or more very primitive members.

Characteristic and often primitive trends include:

1. The development, particularly in the compressed, silvery, marine fishes, of a short broad and arcuate maxilla equipped with large movable supramaxillae in association with a high coronoid process on the dentary and articular.

2. The development of maxillary teeth that are seldom excluded from the gape, - even partially.

3. The development of parasphenoid and pterygoid teeth.

4. The development in the basicranium of numerous separate intraosseous passages for the parts of the fifth and seventh cranial nerves and certain major blood vessels.

5. The development of a full complement of intermuscular bones.

6. Caudal fin, when present, with hypural supports on one to four centra.

7. The development of a functional ductus pneumaticus.

8. The development of an otophisic connection not involving the intercalation of bony elements.

9. The development of an ethmoidal commissure of the cephalic lateral–line system.

10. The development of a confluence between the preopercular and infraorbital cephalic lateral line canals, hence the formation of a recessus lateralis.

11. The development of a leptocephalous larva.

ELOPOMORPHA

1. Principally marine fishes of diverse form, most of the modern species eel-like.
2. Gular plate in non-eel like representatives.
3. Branchiostegals usually very numerous.
4. Mesocoracoid arch present only in the non-eel like forms.
5. Hypurals, when present, on three or more centra.
6. Ethnoidal commissure present or in modified state in many groups
7. Opercular series often reduced or even absent.
8. Larva, when known, as leptocephalus.

CLUPEOMORPHA

1. Silvery compressed fishes, usually marine, with caduceus scales.
2. Branchostegals numbering as high as 15, but usually fewer.
3. Intracranial diverticula of swim bladder forming bullae within the ear capsule.
4. Mesocoracoid arch invariably present.
5. Hypurals on one to three centra.
6. Cephalic lateral-like canals extending over operculum; usually no lateral-line pores on trunk.
7. Recessus lateralis present.

Division - II

Fishes of ancestry at or near the holostean level of organization in which all contained members have retained numerous primitive characteristics of the jaw suspension and shoulder girdle, and have developed complexly ornamented scales. Characteristically primitive trends, and some of the divisional specialization, include:

1. The fusion of the premaxillae into a single bone.
2. The development of a simple, well – toothed maxilla (edentulous in the Mormyriformes), generally contributing to the gape but partially excluded in a few genera.
3. The development of parasphenoid, glossohyal and pterygoid teeth.
4. The fusion of various elements in the palatopterygoid arch, the palatine and vomer fused in Mormyriformes.
5. The development of paired tendon bones (uncalcified in hiodon) on the second hypobranchial and basibranchial, in all genera.
6. The loss in many species, of multiple intraosseous passages in the prootic bone for the fifth and seventh cranial nerves and certain major blood vessels.

7. A reduction in two phyletic lines, of the caudal fin and its confluence with the dorsal fins.

8. The development, in those genera with a distinct caudal fin, of a supporting skeleton of which the elements are not readily homologized with similar elements in fishes of other divisions. There is a reduction in the number of hypural elements which are apparently supported by two and a half centra in all except one genus (*Hiodon*), in which three centra are involved and there is a full complement of hypurals.

9. A reduction in size, or loss, of the suboperculum.

10. The development of upper intermuscular bones only.

11. The development of a functional ductus pneumaticus.

12. The development of an otophysic connection not involving the intercalation of bony elements in the young or adults of all groups except the Osteoglossoidei in which no connection exists.

13. The development of distinctly separated preopercular and infraorbital canals and the occurrence in a single genus (Pantodon) of a suprapreopercular bone.

14. The development of somatic electric organs in one order.

OSTEOGLOSSOMORPHA.

1. Fresh-water and predominantly tropical fishes of extraordinarily diverse body form and size, and including one form in which the pectorals are greatly expanded so as to give the impression of a flying fish. Most species are insectivorous or piscivorous.

2. Premaxillae ankylosed to form a single median bone in one order Mormyriformes and in one genus of the Osteoglossiformes (Pandon); the premaxillae firmly bound to the ethmovomerine region in all genera.

3. Primary bite of mouth between parasphenoid and glossohyal and basihyals.

4. Head of palatine without maxillary process.

5. Branchiostegals three to five, in two cases 11 and 13.

6. Subtemporal fossae present in only a few genera.

7. Expansive suprascapulars in all except the Osteoglossoidei.

8. A lateral cranial foramen in most species (Osteoglossoidei excepted).

9. Hypurals, in fishes with distinct caudal fins, reduced in number by fusion of the upper elements and, apparently supported on at least two an and a half centra in all.

Division – III

Fishes mostly of distinctively teleostean level ancestry, only a single basal group having obviously holostean affinities.

Characteristic trends include:

1. The lowering of the center of gravity and the approximation of the center of buoyancy with the center of mass.
2. The development of a large, frequently mobile premaxilla that completely or partially excludes the maxilla from the gape.
3. The loss of maxillary teeth and functional supramaxillae.
4. The loss of parasphenoid and pterygoid teeth.
5. The development of an oesopharyngeus superior and the development of retractors branchialia attached to the third to sixth vertebrate.
6. The development in the basicranium of a common passage (trigeminofacialis chamber) for the fifth and seventh nerves and orbital artery and head vein.
7. Loss of the supraorbital bone.
8. The reduction in size of the infraorbital bones
9. The reduction in the number of scale bones in the dorsicranium.
10. The loss or reduction of certain temporal fossae but the enlargement of the postemporal fossa and the loss of its roof.
11. The covering of the posterior part of the dorsocranium by epiaxial body muscles.
12. The elevation of the pectoral fin base on the side.
13. The forward migration of the pelvic girdle and its linkage with the pectoral girdle.
14. The reduction in the number of pectoral radials
15. The reduction in the numer of vertebrae and of pelvic
16. Reduction of intermuscular bones
17. Reduction of the hypural bones to a single unit on a terminal half – centrum.
18. The varied specialization of caudal finshape
19. The development of an adipose fin in several primitive lines.
20. The development of fin spines and ctenoid scales.
21. The development of an otophysic connection involving the intercalation of bony elements.
22. The disappearance of the ductus pneumaticus.
23. The development of distinctly separated preopercular and infraorbital canals, hence the frequest occurrence of a suprapreopercular bone.
24. The development along the trunk of a ramus lateralis accessories of the seventh nerve.

PROTACANTHOPTERYGII

1. Predominantly slender, predatory fishes; many generalized and some specialized forms in freshwater.

2. Photophores in oceanic representatives.

3. Widespread trend toward exclusion by premaxillae of the maxillae from the gape.

4. Widespread trend toward the development of premaxillary processes

5. Palatopremaxillary and ethmomaxillary ligaments present in numerous representatives.

6. Upper jaw slightly protrusile in a few cases

7. Glossohyal teeth usually prominent.

8. Branchiostegals very numerous in many instances, reduced to two or three in some cases.

9. Hyoid and branchiostegal skeleton approaching paracanthopterygian and acanthopterygian form.

10. Paired proethmoids present in many cases often simulating ascending premaxillary processes.

11. Few species with opercular spines or serrations.

12. Mesocoracoid pesent in generalized lines only.

13. Baudelot's ligament to first vertebra.

14. Occasional trends for the pelvic fins to advance; pelvics commonly of more than six rays.

15. Occassional trends toward elevation of the pectoral fin base on flank.

16. Vertebrae commonly more than 24, precaudal elements commonly 15 or more.

17. Hypurals on one to three centra, but a basis acanthopterygian caudal skeleton development in some representatives and a paracanthopterygian type in others; caudal fin commonly with more than 15 branched rays.

18. Adipose fin present in most species.

19. Suprapreopercular (canal –bearing ossicle above uppermost part of preopercular canal) present in generalized representatives.

OSTARIOPHYSI

1. Predominantly freshwater fishes of worldwide distribution on the continents and adjoining archipelagoes, of extraordinarily diverse form and habits encompassing numerous well – toothed predatory and vegetarian types and toothless detritus and microphagus types, many of both categories with well- development of barbells.

2. Upper jaw protrusile in numerous species.

3. Major trends toward reduction in number (or absence) of jaw teeth.

4. Lower pharyngeal bones usually well development.

5. Branchiostegals generally few in number but as many as 15 in some species.

6. Pelvic fins abdominal.

7. Hypurals on one centrum

8. Fin spines present in numerous instances.

9. Scales present or absent, when present cycloid in most instances, ctenoid in a few, and in certain forms replaced by dense, bony plates.

10. Adipose fin in many groups

11. Otophysic connections involving the intercalation of bony elements in all.

12. Swim bladder primitively subdivided, reduced in many species.

13. Supra-opercular (ossicle above upper most part of preopercular canal) in numerous speices.

PARACANTHOPTERYGII

1. Mostly marine, stout, soft-bodied fishes inhabiting deep water or when in shallow water being nocturnal or occurring in cryptic habitats.

2. Virtual loss of photophores.

3. Feeding mechanism adapted for carnivorous diet in all species.

4. Ascending process of premaxilla often joined to premaxilla by flexible cartilage, or absent; premaxilla with an articular process in all cases, and with a lateral (maxillary) process inmost cases.

5. Ethmomaxillary and palatopremaxillary ligaments well developed.

6. Upper jaw not protractile.

7. Mm.levator maxillae superioris well developed or modified and consolidated with part of muscle adductor mandibulae.

8. Superficial division of muscle adductor mandibulae reduced or absent.

9. *Muscle adductor arcus palatine* covering floor of orbit.

10. Ceratohyal and epihyal ankylosed.

11. Branchiostegals not exceeding six in number, the blade like elements with an anteroproximal in most species; the four blade like elements on the inner side of the depressed anterior section of ceratohyal.

12. Upper and lower pharyngeal bones well developed and toothed.

13. No subocular shelf on infraorbital bone.

14. Extrascapular bones present, often forming solid roof for post emporal fossae.

15. Parietals meeting in midline or closely approaching one another in most species and frequently housing a posttemporal commissure of the cephalic lateral-line system.

16. Intercalar very extensive in a numerous species.

17. Mucous canals prominent on head of most species.

18. Baudelot's ligament to first vertebra, or to basicranium where first vertebra is fused to basioccipital.

19. Modified epipleural ribs ("endocleithra)" from exoccipitals to cleithrum in several species.

20. Mesocoracoid absent; pectoral radials two to 13, often hour glass-shaped very long, and extending well beyond the scapulocoravoid margin.

21. Pelvic fins thoracic, jugular, or mental in all but one species, with occasionally as many as17 rays.

22. Pleural ribs often reduced, frequently absent.

23. Caudal skeleton, when present, with two large hypurals on separate vertebrae in most, or the two fused together into a single unicentral unit.

24. Fin spines developed or not

25. Ctenoid scales developed in some species.

26. Swim bladder frequently subdivided and connected by diverticulae to parapophyses of precaudal vertebrae, in some instance an otophysic connection involving the intercalation of bony elements.

27. Numerous species viviparous.

ATHERINOMORPHA

1. Generally small surface-feeding fishes, principally in fresh and brackish water, some marine, most freshwater species with pronounced secondary sexual dimorphism in size, color, and in fin shape and function many species with bony external male genitalia developed from anal, pelvic, or pectoral fin, or some combination of these.

2. Upper jaw protractile in many species, without true ascending processes, and supported by a foundation of loose connective tissue and a complex maxillary process, without palato premaxillary or ethmo-maxillary ligaments.

3. Mm.levator maxillae superioris absent.

4. Superficial division of adductor mandibulae and well developed with a tendon to the lower maxillary shaft.

5. Upper and lower pharyngeal bones well developed, dentigerous, the upper bones consisting of a large plate made up of pharyngobranchials 3 and 4 and smaller modified pharyngobranchial 2; pharyngobranchial 1 present but obsolescent in only a few instances.

6. Ceratohyal and epihyal joined together by dorsal lamella.

7. Branchiostegals 4 to15 in number.

8. Mesethmoid usually bilaminar, invariably discoidal or scalelike.

9. Infraorbital series reduced to two, rarely three, elements.

10. Opercular bones unarmed.

11. Pectoral radials four in number, cuboidal, recessed within excavation in scapulocoracoid margin.

12. Supracleithrum, when present, discoidal, confined within dorsal tip of cleithrum.

13. Baudelot's ligament to basicranium.

14. Pelvic girdle abdominal, subabdominal, or thoracic.

15. Vertrbral number high in mosts species, precaudal number 20.

16. Caudal skeleton with two large hypural plates on Terminal half centrum, with no instance more than four hypurals, of which two are invariably broad and fan shaped.

17. Fin spines present or not.

18. Ctenoid scales in relatively few species.

19. Numerous viviparous species, some with unique encapsuled or unencapsuled spermatophores.

20. In oviparous species, egg large, demersal with adhesive filaments, and without oil globule.

21. Embryo with heart displaced forward anterior to head.

ACANTHOPTERYGII

1. Fishes of extremely variable form and habits, principally in salt water and principally benthic and littoral.

2. Photophores very uncommon.

3. Feeding mechanism extremely varied, permitting the utilization of numerous food sources.

4. Upper jaw protractile in many species, with a premaxilla having ascending, articular, and lateral (maxillary) processes.

5. Palatopremaxillary and ethmomaxillary ligaments present, but in some cases modified.

6. Mandibular levator maxillae superioris muscle absent in all but one genus (Polymixia).

7. Superficial division of the *M. adductor mandibulae* well developed.

8. *M. adductor arcus palatini* usually confined to posterior wall of orbit.

9. Upper and lower pharyngeals well developed and toothed.

10. Hyoid bar with ankylosed ceratohyal and epihyal; distal depressed section of ceratohyal and epihyal with four bladelike branchiostegals, the hairlike anterior branchiostegals, when present, on inner surface of depressed distal section of ceratohyal.

11. A subocular shelf present on the infraorbital series in numerous species.

12. Infraorbital bones frequently in contact with preoperculum

13. Bones of head commonly with numerous pungent spines.

14. Opercular apparatus armed in many species.

15. Baudelot's ligament usually attached to basicranium rarely (Polymixidae and some scorpaenidae) to first vertebra.

16. Supracleithrum extending above cleithral tip in most members of the group.

17. Mesocoracoid absent; pectoral radials not exceeding four in number, often hourglass-shaped

18. Pelvic fins, if present thoracic or jugular in position pectorals inserted high on the sides.

19. Pelvic fin is typically consisting of a spine and five articulated rays except in berycoids and a few other forms.

20. Pleural ribs usually well developed.

21. Vertebrae commonly numbering 24, with usually equal numbers of caudal and precaudal elements, except in some elongate and in most fresh- water forms.

22. Hypural bone virtually always emanating from a single centrum; when on two centra, the hypurals no fewer than six in number, in no case formed as two hypural plates as in the Paracanthopterygii.

23. Caudal branched rays in most species 15,17 in more primitive members of the group.

24. Fin spine present in most species.

25. Ctenoid scales common.

26. Presumably uniformaly physoclistic.

27. Otophysic connections rare, in no case involving the intercalation of bony elements.

28. Viviparity uncommon.

29. Egg shape and bucyancy highly variable.

The Phylogeny of the Anabantoidei

The anabantids represent the most primitive family of the Anabantoidei. The presence of large extrascapulars, the toothed prevomer, palatine and parasphenoid, and the articulation of the metapterygoid with the symplectic process of the hyomandibular are indicative of the initial stage in anabantoid phylogeny, *Anabas testudineus* is known from Pliocene and Pleistocene deposits of Java (Sanders, 1934). A well-preserved specimen of *Osphronemus goramy* has been found in Sumatran marl Shales dating from the early Tertiary (Sanders, 1934). According to Romer (1945), a wide variety of teleosts evolved in the Cretaceous, and all major teleost groups were established by the Eocene. The possible occurrence of the highly specialized *Osphronemus goramy* in the early Tertiary suggests that the four anabantoid families were already differentiated by that time. The ancestral anabantoids seem to have originated from a percoid stock during the Upper Cretaceous or Paleocene, in the tropical portion of the Oriental region. The great abundance and diversity of the recent Anabantoidei in Asia supports the hypothesis that the geographical origin of this group has been most likely tropical Asia (Darlington, 1957). It is also generally accepted that the representation of *Anabas* (erroneously taken to include Ctenopoma) in Africa is evidence of a late post-Tethys immigration of the anabantids into Africa (Kossswig, 1954 and Steinitz, 1954). But it has been shown that *Anabas* and *Ctenopoma* are widely separated genera. *Anabas* is not represented in Africa. *Anabas* is certainly not the direct ancestor of *Ctenopoma*, since it possesses several specialized features not found to *Ctenopoma*. The hypothesis that *Anabas* and *Ctenopoma* are derivatives of a common ancestral anabantoid stock is supported with strong osteological evidence. It is assumed that this ancestral form evolved during the Upper Cretaceous or Paleocene in the tropical part of the Oriental region, from where it radiated into Africa. The invasion into Africa probably occurred in the early Tertiary (Darlington, 1957), Darlington also pointed out that from the Mesozoic into the tertiary the Tethys Sea stretched across the whole of southern Europe and Asia to the East Indies. But the Tethys was not a stable sea, and the region was constantly changing with changing barriers and bridges, and an immense changing frontage between salt and freshwater. There is considerable evidence from mammalion fossils in the Upper Eocene and Lower Oligocene of EI Faiyum (Northern Egypt) that Africa and Eurasia were united in the late Eocene. Picard (1943) also pointed out that land seems to have existed in the higher parts of Transjordan and Cisjordan since the Upper Eocene. The ancestral anabantoid form seems to have invaded Africa during the Upper Eocene. It differentiated into the genus *Ctenopoma* in Africa, whereas in Asia it gave rise to (1) *Anabas*, known since the Pliocene and Pleistocene (2) the osphronemids, known from the early Tertiary (Sanders, 1934), (3) the helostomatids and (4) the belontiids. No fossil records exist of the latter two families. Steinitz (1954) pointed out that the majority of the fishes found in both Asia and Africa supports the hypothesis of a connection between the two continents during the Miocene. Kosswig (1954) favors a Pliocene land connection, Steinitz remarked that a continuous land mass between the Asiatic and African continents became established only during the Lower Miocene. It should be emphasized, however, that the Tethys Sea was unstable, and that the bridges and barriers were constantly changing. The ancestral anabantoid form might have been able to cross barriers which were in accessible to other freshwater fishes. *Anabas* and *Ctenopoma* are extraordinarily hardy fish, surviving in spite of very poorly oxygenated water because of their remarkable air-breathing capacity. They are able to live for prolonged periods outside the water. *Anabas testudineus* has more than usual salt tolerance (Weber and de Beaufort, 1922, Giltay, 1933, Saunders, 1934, and Delsman, 1951). Numerous specimens are collected daily from the Bay of Djakarta (Java), *Anabas* behaves as a secondary division freshwater fish and is, very likely, able to cross narrow sea barriers as reported for *Aplocheilus, Rasbora* and *Puntius* (Myers, 1951 and Darlington, 1957). The assumption that *Anabas* has been carried by man across Wallace's line to the eastern part of the Indo-Australian Archipelago (Myers, 1937) is difficult to support. *Osphronemus goramy*, which is as hardy a fish as *Anabas* but is far superior as a food fish, has not been carried across Wallace's line by man *Osphronemus* would probably be preferred above *Anabas* for transportation by the natives. It seems therefore less questionable to ascribe the wide distribution of *Anabas* to its exceptional physiological characteristics. Similar physiological traits would enable the ancestral anabantoid form to invade Africa during the Eocene whether there was a continuous land bridge between Africa and Asia as proposed by Darlington (1957) or not, as hypothesized by Steinitz (1954). The presence of a broad connection, Lemuria, between Africa, Madagascar, and India in the Tertiary has been conclusively disproved (de Beaufort, 1951).

Following the paleogeographic suggestions by Steinitz (1954) we may assume that the ancestral anabantid stock migrated into Africa during the Miocene. After the complete breakdown of Africa's land connection with Asia by the Erythrean Rift Valley in the Pliocene the ancestral stock differentiated into the genus *Ctenopoma* in Africa and *Anabas* in Asia. The presence of *Osphronemus goramy* in the early Tertiary does not support this hypothesis, unless we assume that the ancestral anabantid had already split into the three more specialized families before or during the period of radiation.

Darlington (1957) favors an early Tertiary migration of the ancestral anabantoid stock into Africa. The westbound migration of the ancestral form would have taken place simultaneously with the differentiation of the osphronemids, helostomatids and belontids in Asia. Parallel evolutionary patterns are found in the Ostariophysi (Darlington, 1957). The evolution of the Anabantoidei corresponds with the general eolutionary patterns of freshwater fishes on the Asiatic and African continents. Africa has been relatively stable, favourig the survival of archaic forms, whereas Asia has somehow favoured the evolution of new groups of fishes. There is decisive evidence against the direct derivation of the osphronemids, helostomatids, and belontiids from *Anabas*. *Anabas* probably evolved from the ancestral stock after Africa's land connection with Asia was broken down during the Lower Pliocene (Steinitz, 1954). *Osphronemus goramy* was already present during the Lower tertiary; these facts indicate that *Osphronemus* evolved before *Anabas*, from a common anabantid stock. The osteological and zoogeographical evidence seems to indicate that the archaic recent Anabantidae, the specialized Osphronemidae, Helostomatidae and Belontiidae arose during the Lower Tertiary from a common ancestral anabantoid stock. No attempt has been made to reconstruct the hypothetical common ancestor. The set of characters found in both *Ctenopoma* and *Anabas* may be regarded as those which the group owed to its common ancestor. The common ancestor constructed by abstraction from known forms represents an archetype as understood by the *Naturphilosophen* and Owen, and may differ considerably from the actual common ancestor.

The genus *Sandelia* isolated on the southern tip of South Africa (Barnard, 1943), has been derived from *Ctenopoma*. *Sandelia* exhibits several specializations, as for example the loss of one branchiostegal ray, the reduction in size of the suprabranchial cavity, the simple first epibranchial, and the separated nasals. But the toothed palatine and prevomer, the spineless serrate opercular, and the arrangement of the pharyngeal processes on the basioceipital indicate the close relationship with *Ctenopoma*. There is also strong zoogeographical evidence for the direct derivation of *Sandelia* from *Ctenopoma*. The present distributional pattern of the African anabantoids indicate that *Ctenopoma* has spread along a principal route which leads from an extensive, favourable area of the African tropics into a smaller, less favourable area of temperate South Africa. The expanding *Ctenopoma* population apparently crossed the climate barrier with difficulty and the group across the barrier became isolated and differentiated independently from its ancestors. As Darlington (1957) remarked, the South African fauna as a whole has been formed by southward extension of parts of the tropical African fauna. It appears that *Anabas* and *Ctenopoma* evolved by schistic (Smith, 1956 and 1960) evolution or splitting from a common ancestor. Geographic isolation seems to have brought about differences between the two separated populations. *Sandelia,* on the other hand, evolved by phyletic evolution from *Ctenopoma.* A progressive change in some features can be followed in the series *Ctenopoma muriei* from central tropical Africa. *Ctenopoma multispinis* from the Zambezi River system and *Sandelia* from temperate South Africa: the lachrymal is weakly serrate in *muriei*, not serrate in *multispinis* and *Sandelia*; suborbitals extend to nearly half the distance between orbit and preopercular in *muriei*, suborbitals narrow in multispints, suborbitals very narrow in *Sandelia*; a gradual but progressive reduction in size of the suprabranchial cavity and an increase in length of the ascending process of the premaxillary in the sequence *muriet, multispinis,* and *Sandelia.*

During the Lower Tertiary three divergent lines radiated more or less simultaneously from the ancestral anabantid stock. In the first line, the *osphronemids,* the transverse processes of the parasphenoid are lost, the ectopterygoids retained, the suborbital shelf becomes greatly reduced, the basipterygia and first pelvic fin ray greatly elongate, and the metapterygoid is divested of its articulation with the symplectic process of the hyomandibular. The second line is represented by the highly specialized helostomatids, in which transverse process and pharyngeal processes of the parasphenoid are retained, the ectopterygoid is not lost, the dentition is completely lost, the dentary and angular acquire a highly movable hinge joint, and the metapterygoid does not articulate with the symplectic process of the hyomandibular. The third line represents the largest group, the belontids, in which the ectopterygoids are lost and the transverse process of the parasphenoid are not retained. The three groups seem to have originated by *schistic evolution* from the same ancestral adaptive type, entering into diverging adaptive zones. *Osphronemus goramy* made a rather sudden appearance in the Lower Tertiary, most likely the Upper Eocene, and has not changed since. The fossil specimen resembles the recent forms in great detail (Sanders, 1934). If we assume that the fossil is actually *Osphronemus goramy* then the species did not change during a period of approximately 36 million years. The rather sudden appearance of the osphronemids from the ancestral anabantids and the complete absence of intermediate stages seem to indicate that the change from the carnivorous ancestral adaptive type to the large-sized omnivorous forms might have taken place quickly. This evolutionary change at an exceptionally high rate might be considered as tachytelic. Tachytely often occurs when populations shift from one major adaptive zone to another (Simpson, 1953). In this particular case the shift is from purely predaceous feeding habits to omnivorous adaptive types. The very elongate first pelvic fin ray, proven to be gustatory and tactile, probably evolved as a concomitant of the feeding behaviour. The possible tachytelic line leading toward the differentiation of the osphronemids became bradytelic. *Osphronemus* has not changed morphologically since the Upper Eocene. *Osphronemus* is rather broadly adapted; it can live on a variety of foods (Weber and de Beaufort, 1922), is able to survive in poorly oxygenated water, and has a high salt tolerance. Environmental factors may therefore fluctuate greatly without requiring any adaptive change for survival. These factors may have contributed to the arrested evolution of *Osphronemus* since the Upper Eocene.

Helostoma temmincki represents the most specialized member of the anabantoids. The morphological gap between *Helostoma* and the ancestral stock is exceptionally large. The edentulous condition, the unique movable hinge joint between the dentary and angular, and the elaborate filtering apparatures in the hyobranchial region are, undoubtedly, advanced features adaptive to a highly specialized feeding behaviour. Transitional forms between *Helostoma* and the predaceous ancestral stock are lacking. This sudden evolutionary change from the predaceous adaptive zone to the planktonic adapted forms seems to indicate that an all-or-none reaction is involved whereby no intermediate stages persisted. Once the direction of evolution of the feeding mechanism was established, natural selection continuously acted in that direction until completion of the trend. We are possibly confronted with a beautiful example of *quantum evolution*. The all-or-none element in *quantum evolution* arises in general from discontinuity between adaptive zones. It seems that the direction of the relatively rapid change in the evolutionary history of *Helostoma* was rigidly adaptive as a result of strong selection pressure. In this case quantum evolution has, possibly, led to a group of fish whose adaptive zone and correlated morphology are such that a family rank should be established. After penetration into the specific adaptive zone during the Lower Tertiary, the evolution of the helostomatids seems to have been arrested. *Helostoma* seems to be perfectly adapted to its narrow zone. It has been pointed out by Simpson (1953) for animals and Stebbins (1949) for plants that the lines with arrested evolution were specialized when their evolution was arrested. The evolutionary pattern of *Helostoma* seems to fit the generalization that the most slowly evolving groups seem to be very highly and specifically adapted to a particular zone.

The third evolutionary line originating from the ancestral anabantoid stock give rise to the Belontiidae. The evolutionary pattern of the belontiids is mainly a combination of phyletic and schistic evolution. The Belontiinae have several features in common with the ancestral anabantoid stock, among which the closely cojoined *basipterygia* and the articulation of the metapterygoid with the symplectic process of the hyomandibular are the most important. *Belontia* seems to resemble an intermediate stage between the ancestral anabantoid stock and the more advanced belontiids. Some specialized features not found in any other belontiid, as for example the posterior extension of the suprabranchial cavity supported by the first six epipleurals and the high supra-occipital crest, indicate strongly that *Belontia* forms an early, sterile, side branch of the main evolutionary line. The main line, originating from the same ancestral stock as the Belontiinae, leads to a progressively changing series, which consists of *Betta, Trichopsis* and *Macropodus*. The latter genera represent the subfamily Macropodinae, which differs from the Belontiinae in lacking a posterior extension of the suprabranchial cavity and a distinct supraoccipital crest, and the loss of the articulation between the metapterygoid and the symplectic process of the hyomandibular. These differences suggest that the Belontiinae and Macropodinae represent two divergent lines. In the belontiine line there is a tendency toward deepening of the skull by the differentiation of a high supraoccipital crest and a posterior extension of the suprabranchial cavity, whereas in the macropodine line the deepening of the skull is the result of the development of pharyngeal processes on the parasphenoid and basioccipital. We can trace the following gradual and progressive changes: reduction in dentition of the parasphenoid, deepening of the body, increase to protrusility of the jaws, and increase in size of the pharyngeal process of the parasphenoid and basioceipital. These progressive changes are evidently nonrandom and cumulative and apparently adaptive. The morphological changes are correlated with the changing feeding habits. The *Macropodinae* are omnivorous and there is a trend toward specialization for eating small-sized food. It should be emphasized that the macropodine evolutionary pattern is not purely phyletic, but the longest line leading to *Malpulutta* takes a branch from *Trichopsis*, which was produced by splitting as indicated by the aberrant elongate and tubular nasals.

The Trichogasterinae forms a homogenous group in which the trends present in the Macropodine reach their peaks. The group probably originated from a Macropodus-like stock, by an intensification of the trends already present in the Macropodinae. The occupation of contiguous and successively higher and narrower zones by a sequence of populations splitting off from the next lower zones to have taken place during the evolution of the Macropodinae and Trichogasterinae. The progressive specialization culminated in the penetration of the narrow adaptive zone by the herbivorous Trichogasterine, which feed on a diet of small plant parts, epiphytes, plankton, filamentous algae and detritus. It seems that all zones remain in existence and are occupied. In the line leading from *Betta* to the Trichogasterinus, we can trace a sequential occupation of zones decreasing in width. *Betta* is omnivorous, whereas the Trichogasterine are herbivorous and nearly planktonic. The morphological differences between the Macropodinae and the Trichogasterinae are of such magnitude that the rank of subfamilies should be recognized.

The position of the anabantoids among the Perciformes is still questionable. The ancestors of the anabantoids are still unknown, but are perhaps to be found among the Percoidei, which occupies a central place in the order Perciformes. The anabantoids arose during a major episode of proliferation of teleostean suborders, families, and genera. During this "explosive phase" of adaptive radiation (Upper Cretaceous and Lower Tertiary) of teleosts there was, probably, a considerable phyletic divergence among the evolving lineages, penetrating into a great variety of adaptive zones by *successive quantum shifts*. Among the crowding lines parallelism and convergence were at work. The phylogeny of teleosts and phylogeny in general are very poorly understood. In this discussion I have attempted to analyze some of the interwoven modes and factors of the evolution of the anabantoids. But as Simpson states, relative to adaptive radiation. "The total process cannot be made simple, but it can be analysed in part. It is not understood in all its appalling intricacy, but some understanding is in our grasp, and we may trust our own powers to obtain more" (Karel F. Liem, 1963).

Epilogue

Respiratory Properties of the Blood and Control of Breathing in Certain Indian Air-Breathing Fishes

The air-breathing fishes are one of the most important links in studying the transformations and adaptations that took place during evolution of the aerial mode of respiration of higher vertebrates (amphibians, reptiles, birds and mammals) from a simple mode of aquatic respiration. Water and air are very different respiratory media. Waterbreathing animals have to spend a large amount of energy in extracting oxygen from water but the elimination of CO_2 is easily achieved through aquatic routes (Hughes, 1966; Rahn, 1966; Singh, 1976).

Hughes and Singh (1970 a,b; 1971) and Singh and Hughes (1971, 1973) have clearly shown that among some of the Indian air-breathing fishes, the air-breathing organs are capable of absorbing O_2 very efficiently from air but the task of eliminating a major portion of metabolic CO_2 is still performed efficiently through aquatic routes such as gills (*Anabas testudineus*) and/or skin (*Clarias batrachus*, *Heteropneustes fossilis* and *Monopterus* (=*Amphipnous cuchia*). However, in certain other Indian air-breathing fishes, *e.g.*, *Notopterus* sp. Air-sacs function as lungs and take care of O_2 absorption while CO_2 elimination through air-sacs is similar to the primitive air-breathing fish, *Amia calva* (Johannnnsen, *et al.*, 1970), advanced lung fishes *e.g.*, Protopterus aethippicus (Johansen and Lenfant, 1968; McMahon, 1970) and Lepidosiren paradoxa (Johansen and Lenfant, 1967). Further, Hughes and Singh (1970a, 1971), Singh and Hughes (1971), Lenfant and Johansen (1972) Munshi *et al.* (1976), Singh (1976) observed that these air-breathing fishes are capable of adjusting (lowering or increasing) their metabolic rates according to the characteristics of respiratory media and also in response to different temperatures. This involves considerable changes in ventilation and blood perfusion in relation to active and resting physiological responses by the fish. The evolution of such respiratory and metabolic characteristics enable these fishes to adapt to adverse environmental conditions (*e.g.*, hypoxic or hypercarbic water and drought conditions) including the ability of some fishes to migrate from a drying pool/swamp to a wet pool/swamp.

In view of the above characteristics of these fishes, the respiratory properties of their blood have been examined in relation to the evolution of blood sensitivity to high CO_2 and low pH levels in the water and increased buffering capacity in the blood etc., which help in adaptations to aerial modes of respiration (Lenfant, *et al.*, 1967; Srivastava, 1968 a,b; Lenfant and Johansen, 1968; Dube and Munshi, 1973; Singh, 1976; Banerjee, 1981, 1986 Singh and Hughes (1995 a,b). While studying O_2 dissociation curves in magur (*C. batrachus*) and (*H. fossilis*), Singh and Hughes (1995 a,b) observed characteristic respiratory adaptations of the blood of these fishes to O_2 deficient and CO_2 rich environments.

Willmer (1934), Lenfant and Johansen (1968), Johansen, *et al.* (1968) Hughes and Singh (1970a,b), Johansen (1971), Singh and Hughes (1973) and Singh (1976) studied in detail the pattern of breathing and/or perfusion in several groups of air-breathing fishes. Singh and Hughes (1973) and Singh (1976) proposed a hypothesis for the control of breathing in the climbing perch on the basis of respiratory and cardiac responses recorded in this very interesting air-breathing fish. This chapter deals with certain interesting aspects of respiratory properties of blood in some of these Indian air-breathing fishes.

Respiratory Properties of the Blood

The blood of the air-breathing fishes in general has high haemoglobin content and O_2 capacity, besides several other characteristics suitable for their bimodal and/or aerial mode of respiration. Table 1 summarised the respiratory characterisitics of the blood of several Indian air-breathing fishes. They generally show a high content of haemoglobin and consequently high O_2 capacity of the blood, although there are a few exceptions, but surprisingly the Hb content and O_2 capacity of the blood of lung fishes are lower and are comparable to values for waterbreathing fishes. However, the O_2 affinity of the blood of air-breathing fishes (both teleosts and dipnoi) is high compared to water-breathers such as trout and carp.

The P_{50} value of the blood of *C. batrachus* is 16.2 mmHg (at pH = 7.58) (Figure 1). The blood of *H. fossilis* has a comparatively high affinity for O_2 as the P_{50} is only 6.7 mmHg even at lower blood pH values (*e.g.* at pH = 7.105) (Figure 2). However, 100 per cent oxygen saturation of *Clarias* blood occurs at much lower Po_2 values (53 mmHg in presence of Pco_2 = 0.3 mmHg) whereas 100 per cent saturation of *H. fossilis* blood occurs at a Po_2 of 70 mmHg at a similar Pco_2 value.

The oxygen dissociation curves of *C. batrachus* at different Pco_2 values are plotted in Figure 1. They clearly show that the curves were not shifted and affected very

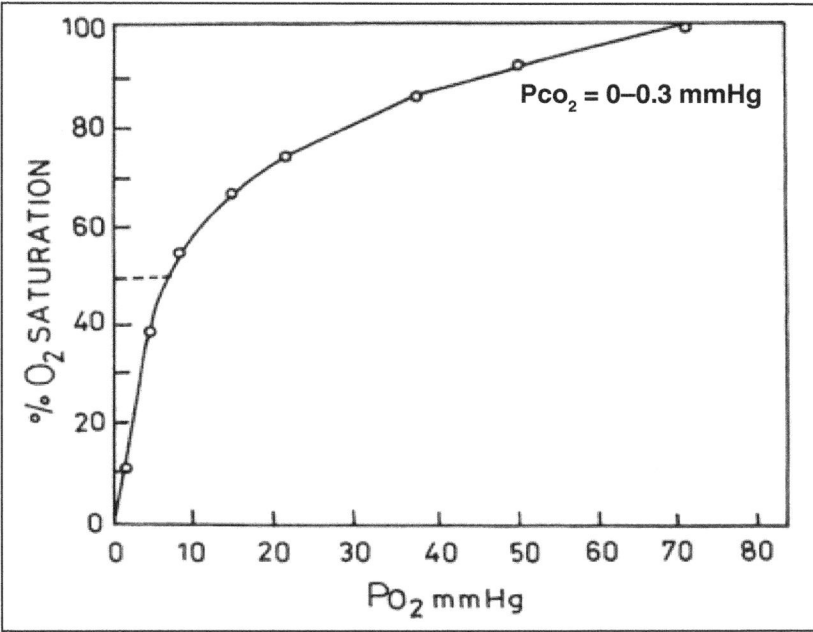

Figure 1

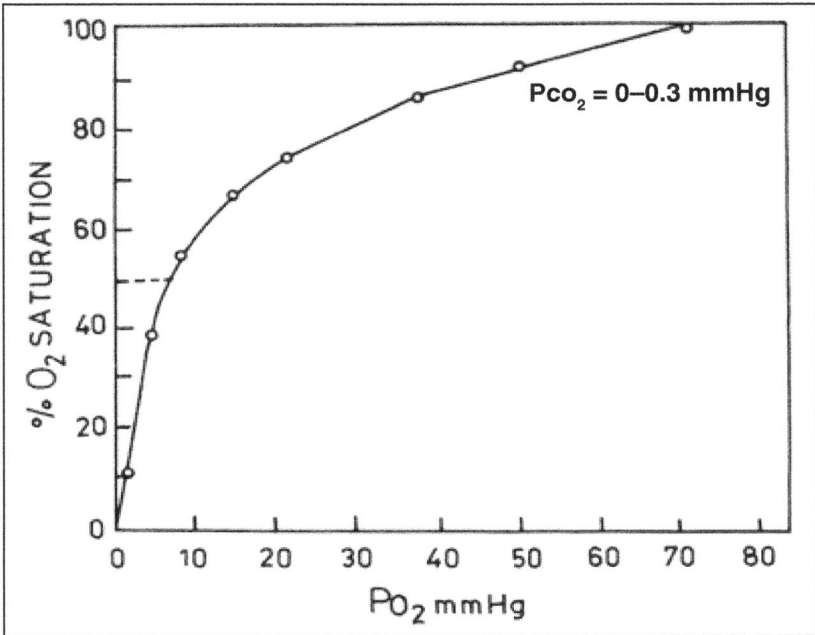

Figure 2

Table 1: Certain characteristics of the blood of Indian air–breathing fishes. The values of lung–fishes and water–breathing fishes are given for comparison.

Species	T (°C)	Hb (g per cent)	RCV (per cent)	MCHC (per cent)	O₂ Cap (Vol per cent)	P₅₀ (mmHg)	Bohr Effect	Root Effect	HCO⁻ (mM/L)	Buffering Capacity	References
C. batrachus	25	12.59	36.2	34.78	16.87	16.2 (at pH 7.58)	-0.30 (at pH 7.58)	Some	12.973 (at pH 6.96)	2.55	Singh and Hughes, 1995b
H. fossilis	25	11.6	43.8	26.48	15.52	6.7 (at pH 7.105)	–	–	–	–	Singh and Hughes, 1995a
A. testudineus (wt. 9–21.3 g)		10.98–15.74	41.51	32.38	14.7–21.1	–	–	–	–	–	Sinha and Kumar, 1991
M. cuchia		21.6–25.9	40.3–50.3	47.2–71.3	28.9–34.7	–	–	–	–	–	Mishra et al., 1984
C. punctatus		14.8–15.2	30.2–32.4	46.97–48.60	19.83–20.0	–	–	–	–	–	Banerjee, 1981
C. striatus		13.5–13.8	26.4–28.2	48.9–51.3	18.1–18.5	–	–	–	–	–	Banerjee, 1984
P. aethiopicus		6.2	25.0	–	6.80	11.5 (Pco₂ = 6.2 mmHg) (at pH= 7.5–8.0)	-0.47	None	20.0	1.52	Lenfant and Johansen, 1968
Lepidosiron	28	6.64	28.0	–	8.25	–	-0.295	None	27.2	2.43	Lenfant et al., 1970
S. gairdneri	20	8.0	37.2±1.1	–	10.72	40.0 (at pH= 7.62–7.23)	-0.59	Large	–	–	Eddy, 1971; Houston and Dewilde, 1969

T: Temperature; Hb: Haemoglobin; RCV: Red cell volume; MCHC: Mean compuscular haemoglobin content; $O_{2cap.}$: Oxygen capacity; P_{50}: Oxygen partial pressure for 50 per cent saturation of blood.

much upto high Pco_2, values of 22.6 mmHg but were significantly shifted to the right in the presence of 38.0 mmHg Pco_2. This indicates a low (-0.3) Bohr-effect and also slight. Root-effect since 100 per cent O_2 saturation was not achieved at the higher Pco_2 value even in the presence of 80 mmHg or higher Po_2 values.

The blood of *C. batrachus* shows very good buffering capacity with high bicarbonate concentrations (15-17 mM/litre) at pH values 7.0-7.1. The slope of the curve was 0.8 when the blood was equilibrated with various concentrations of oxygen in the presence of CO_2 with a partial pressure of 22.6 mmHg. However, interestingly, two slopes were observed with values of -0.75 and -4.17 when the blood was equilibrated with gas mixture having higher Pco_2 values at 38 and 50.3 mm Hg. The curve showed a steep rise indicating high bicarbonate levels (10-17 mM/litre blood) at pH 7.0-6.92 giving a relatively high buffering capacity to the blood.

The respiratory properties of the blood of *C. batrachus* closely resemble those of air-breathing fishes such as the lung fishes (Table 1) but compared to the electric eel, *E. electricus* (Johansen *et al.*, 1965) have slightly less buffering capacity, a lower Bohr-effect, although its oxygen affinity is greater in the presence of higher Pco_2 tensions.

The high affinity for O_2 of *H. fossilis* blood confirms that it should have greater limits of tolerance to hypoxic environemts than *C. batrachus* or Anabas and indeed this was the case as observed by Hughes and Singh (1970 a,b) and Singh and Hughes, 1971). Moreover, Hughes and Singh (1971) reported that the RQ in air-exposed *Heteropneustes* was lowered, indicating that the retention of some CO_2 occurs during such conditions. Further, it was observed that the blood pH value of the air-exposed fish was lowered considerably, confirming that CO_2 was not efficiently eliminated in such conditions. Retention of CO_2 during air exposure in a variety of air-breathing teleosts has also been reported *e.g.* Clarias (Singh and Hughes, 1971), *Anabas* (Hughes and Singh, 1970a,b), Electrophorus electricus (Farber and Rahn, 1970), Protopterus aethiopicus (Lenfant and Johansen, 1968).

The O_2 dissociation curves of *Clarias* at higher Pco_2 values show a low Bohr-effect and slight Root-effect, as well as high buffering capacity (Table 4) and clearly establishes the suitable respiratory properties of its blood for inhabiting and thriving in swampy hypercarbic waters (Willmer, 1934, 1976). Such O_2 dissociation curves clearly show that more O_2 can be loaded and unloaded by the blood of fish inhabiting hypercarbic waters. A similar situation was noted for the O_2 dissociation curves of the South American air-breathing fish, yarrow (*Erythrinus erythrinus*) and as well as other species (Johansen, 1970) in relation to swampy environments.

Therefore, it is concluded that the blood of *C. batrachus* has evolved excellent respiratory properties suitable for its normal habitat which is often swampy, hypercarbic and hypoxic freshwater of tropical Asia (Munshi and Hughes, 1992).

Singh and Sahoo (1995) observed that the plasma protein concentration in *C. batrachus* ranged from 3.65-3.71 g/100 ml (Table 2) which is higher than average values of 1.68-3.0 for many water-breathing fishes but falls in the wider range of 1.68-6.19 g per cent reported by Satchell (1971). The plasma proteins contribute to the osmolarity of plasma, they maintain colloidal osmotic pressure and also regulate the movement of water across the capillaries. They supply proteins required for the cells and also contribute in the buffering power of the blood. Plasma proteins are also concerned with the defence of the body as well as serving to transfer vitamins, hormones and certain inorganic ions such as iron. Such medium to relatively high values of plasma protein concentration in *Clarias* indicate the suitability of its blood to bimodal and or aerial modes of respiration.

Table 2: Blood parameters of *C. batrachus* fed with isonitrogenous control feed (49 per cent of fishmeal, CP 37.0 per cent) and balanced experimental feed (24.5 per cent fishmeal fortified diet, CP = 37.3 per cent).

Control Feed				Experimental Feed			
Hb	Hematocrit (per cent)	Total RBC Count (x 10⁶/ mm²)	Plasma Protein Conc. (g/100ml)	Hb	Hematocrit (per cent)	Total RBC Count (x 10⁶/ mm²)	Plasma Protein Conc. (g/100ml)
11.61** ± 0.56	31.88* ± 1.18	1.97* ± 0.14	3.71* ± 0.37	10.92** ± 0.39	31.27* ± 0.49	1.99* ± 0.11	3.65* ± 0.19

* No significant difference in the two groups of fish (P>0.05).

** Significantly different in the two groups of fish (P>0.01).

Source: Singh and Sahoo, 1995.

Table 3: Respiratory properties of the blood of *C. batrachus* (unanaesthetized) and living in water with access to air and when exposed to air (anaerobic venous blood samples were used for pH and Po$_2$ determinations Values for six fish).

	Fish in Water				Fish Exposed to Air for 3 Hours		
WPo$_2$	pH of Water	pH (Venous Blood)	Po$_2$ (Venous Blood mmHg)	RCV (per cent)	pH (Venous Blood)	Po$_2$ (Venous Blood mmHg)	RCV (per cent)
68.0	6.5	7.300	6.0	35.0	7.138	4.0	36.0
104.0	6.5	7.297	2.0	33.0	7.275	2.0	3.0
100.0	6.5	7.357	4.0	–	7.223	12.0	–
98.0	7.770	7.150	9.0	–	7.247	13.0	–
52.0	7.380	7.093	10.0	–	–	–	–
52.0	7.380	7.165	11.0	–	–	–	–

WPo$_2$: Partial pressure of oxygen in water; Po$_2$: Partial pressure of oxygen.

Source: Singh and Hughes, 1995b.

One of the striking observations made on the venous blood of *C. batrachus* was that in six fish the venous blood Po$_2$ ranged from 2.0-11.0 mmHg when fish lived in water with access to air and 2.0-13.0 mmHg when measurements were made from fish exposed for three hours to air (Table 3). This clearly indicates substantial unloading of blood oxygen to tissues (about 90 per cent) since 10 mmHg Po$_2$ reprsents only in 20 per cent oxygen saturation of the blood equilibrated with air (Pco_2 =22.6 mmHg and pH 7.1-7.28 (Figure 1). If we examine the blood supply to and from the gills and the accessory air-breathing organs, gill fans, dendritic organs and the respiratory membrane of the supra-branchial chambers we find that all these are situated between the gill circulation and cephalic circulation. The dorsal aorta carries the oxygenated blood supply to various organs hence there is no mixing of oxygenated blood with venous blood; and the venous blood is returned to the heart by the pulmonary vein and the ductus cuvieri from anterior and posterior cardinal veins. The blood supply to the air-sac of *H. fossilis* is from the fourth afferent branchial artery only, and the oxygenated blood from the air-sac is returned to the dorsal aorta through the fourth efferent branchial artery (Munshi and Hughes, 1992). Thus, the organs and tissues are supplied with oxygenated blood by the branches of dorsal aorta and again no mixing of oxygenated and deoxygenated blood occurs in *H. fossilis*. This explains our observations that the venous blood is almost deoxygenated (Po_2 = 2.0-13.0 mmHg) and also has lower pH value (7.01-7.357) than the fully oxygenated blood pH (7.105-7.58).

Table 4: Values for P_{50}, pH and calculated concentrations (H^+) and (HCO_3^-) in blood of *C. batrachus* when equilibrated with gas mixture having different levels of Pco_2.

Pco_2 in Air (mmHg)	P_{50} (mmHg)	pH	(H^+) (mM/Litre)	(HCO_3^-) (mM/Litre)
0.3	16.2	7.584	26.06	0.358
22.6	18.7	7.116	75.93	8.600
38.0	24.3	6.983	104.0	10.454
50.3	27.3	6.960	109.6	12.973
74.5	–	6.918	120.2	17.325

However, in *A. testudineus* the blood supply to the cephalic and other parts of the body is different since the oxygenated blood from the accessory respiratory organs is returned to the heart (sinus venosus) where it is mixed with the venous blood before being supplied to the tissues (Munshi and Hughes, 1992). In *Anabas*, the labyrinthine organs are supplied by blood from the first and second afferent branchial arteries and the oxygenated blood is collected by the jugular vein into the anterior cardinal vein which in turn returns the blood to the heart where it is mixed with the venous blood brought by the ductus cuvieri. The dorsal aorta is supplied by the oxygenated blood brought by the third and fourth afferent branchial arteries only. The third gill is very much reduced and fourth gill arch in *Anabas* bears no secondary lamellae (Munshi, 1976) hence, no oxygenation of blood occurs here. This clearly shows that in *Anabas* tissues are supplied with various degress of mixed blood under two conditions (i) when living in water with access to air, and (ii) when exposed to air. This means the dorsal aortic blood supply will have higher levels of CO_2 due to mixing of oxygenated and venous blood supplies (Singh, 1976). This indicates that Anabas should have respiratory circulatory properties where oxygenated blood should accommodate comparatively high Pco_2 values and low Po_2 lvels. The high haemoglobin content of the blood (11.0-19.8 g per cent) and its high oxygen-carrying capacity are adaptations by this fish to adverse conditions such as hypoxia and hypercarbia. Its tolerance to very high CO_2 content in water (20-33 vol per cent) and water at acidic pH (5.8-6.0) clearly indicates not only its low sensitivity but also high tolerance to severe hypercapnia (Singh, 1976). Singh and Hughes (1973) observed that in *Anabas*, the Pco_2 value of gas in the supra-branchial changes rose to the levels of 20-33 mmHg, indicating retention of CO_2 during aerial gas exchange, which was confirmed by Hughes and Singh (1970a). The oxygen dissociation curve of *Anabas* deserves further study to explore its relative insensitivity to increasing CO_2. the measurements of the magnitude of Bohr-effect, Root-effect and buffering capacity would be most rewarding in understanding the circulatory adaptability of this fish to swampy conditions, drought, and the effects of long emersions to its survival capabilities. In conclusion, it may be said that the respiratory properties of the blood of *C. batrachus*, *H. fossilis* and *A. testudineus* are well suited to their normal habitat which is often swampy, hypercarbic, and hyposic freshwater of tropical Asia.

References

Banerjee, V. (1981) : Erythrocyte number and haemoglobin content in *Channa punctata* (Bloch) in relation to heart weight and bodyweight Biol.Bull, India 3(3), 187-189.

Banerjee, V. (1984) : Thrombocytes in *Clarias batrachus* (L) with special reference to body weight, Sex and Season. Biol. Bull, India 6(3): 290-295.

Banerjee, V. (1986) : Hematology of a freshwater eel *Amphipnous cuchia* (Ham) Erythrocyte dimensions with special reference to body length. Sex and Season Comp. Physiol. Eel, 11(3) 68-73.

Dube, S.C. and Munshi, J.S.D. (1973) : A quantitative study of the erythrocyte and haemoglobin in the blood of an air-breathing fish *Anabas testudineus* (Bloch) in relation to its body size, Folia Haematol 100 (4):436-446.

Eddy, B. (1971) : Blood gas relationships in the rainbow trout *Salmo gairaneri* J. Exp. Biol.55, 695-711.

Farber, J. and Rahn, H. (1970): Gas exchange between air and water and the ventilation pattern in the electric eel. Respir. Physiol. 9:151-161.

Houston, A.H. and De Wilde, M.Anne (1969) : Environmental temperature and the body fluid system of the freshwater teleost III. Hematology and blood volume of thermally acclimated brook trout, Salvelinus fontinalis Comp. Biochem Physiol, 38 : 877-885.

Hughes, G.M. (1966) : Evolution between air and water, Ciba Foundation Symposium on Development of the Lung. Pp. 64-80 (eds. De Reuck and Porter). Churchill Press, London.

Hughes, G.M. and Singh, B.N. (1970a) : Respiration in an air-breathing fish, the climbing perch *Anabas testudineus* Bloch. I. Oxygen uptake and carbon dioxide release into air and water, J. Exp. Biol. 53:265-280.

Hughes, G.M. and Singh, B.N. (1970b) : Respiration in an air-breathing fish, the climbing perch *Anabas testudineus* Bloch. II. Respiration patterns and control of breathing J. Exp. Biol. 53 : 265-280.

Hughes, G.M. and Singh B.N. (1971) : Gas exchange with air and water in an air-breathing catfish *Saccobranchus fossilis* J. Exp. Biol. 55, 667-682.

Johansen, K. (1970) : Air-breathing fishes. In : Fish Physiology W.S. Hoar and D.J. Randall (eds.). Academic Press, New York and London, Vol. IV pp. 361-411.

Johansen, K. (1971) : Comparative Physiology; Gas exchange and circulation in fishes Am. Rev. Physiol. 33 : 569-611.

Johansen, K and Lentant, C. (1967): Respiratory function in the South American lung fish. *Lepidosiren paradoxa* (Fitz.). J. Exp. Biol. 46: 205-218.

Johansen, K. Hanson, D. and Lenfant, C. (1970): Respiration in a primative air breather, *Amia calva. Respir. Physiol.* 9:162-174.

Lenfant, C. and Johansen, K. (1968): Respiration in the African lungfish, *Propterus aethiopicus.* J. Exp. Biol. 49: 437-452.

Lenfant, C., Johansen, K. and Hansen, D. (1970) : Bimodal gas exchange and ventilation perfusion relationships in lower vertebrates. Fed. Proc. 29: 1124-1129.

Lenfant, C. and Johansen, K (1972): Gas exchange in gill, skin, and lung breathing. Respir. Physiol. 14: 211-218.

Munshi, J.S.D. (1976): Gross and fine structure of the respiratory organs of air-breathing fishes. In respiration of Amphibious Vertebrates, pp 73-104. (Ed. G.M. Hughes). Academic Press, London-New York.

Munshi, J.S.D. and Hughes, G.M. (1992) : *Air-breathing fishes of India: their structure, function and Life-history.* Oxford and IBH Publishing Co. Pvt. Ltd., New Delhi.

Munshi, J.S.D., Sinha, A.L. and Ojha, J. (1976): Oxygen uptake capacity of gills and skin in relation to body weight of the air-breathing siluroid fish, *Clarias batrachus* (Linn.) Acta Physiol Acad Sci. Hungaricae 48:23-33.

Satchell, G.H. (1971) : *Circulation in Fishes.* Cambridge University Press Cambridge, 131 pp.

Singh, B.N. and Hughes, G.M. (1971): Respiration of an air-breathing catfish. *Clarias batrachus* (Linn). J. Exp. Biol. 55: 421-434.

Singh, B.N. and Hughes, G.M. (1973): Cardiac and respiratory responses in the climbing perch, Anabas testudineus, J. Comp. Physiol, 84: 205-226.

Singh, B.N. and Hughes, G.M. (1995a): Oxygen dissociation curve and respiratory properties of the blood of an air-breathing catfish. *Saccobranchus fossilis* (Bloch). J. Aqua. Trop. 10 : 355-360.

Singh, B.N. and Hughes, G.M. (1995b) : Oxygen dissociation curve and respiratory properties of the blood of an air-breathing catfish. *Clarias batrachus* (Linn.). J. Aqua. Trop. 10: 361-368.

Singh, B.N. and Sahoo, C. (1995): Growth tissue composition and certain blood parameters of Indian catfish, *Clarias batrachus* fed on diets with varying levels of animal protein. Abstract proceedings of National Seminar on current and emerging trends in Aquaculture and its impact on Rural Development, Berhampur.

Sinha, Y.K.P. and Kumar, K. (1991): Haematology of *Anabas testudineus* Bloch. J. Appl. Zool. Res. 2(1): 13-16.

Srivastava, A.K. (1968a) : Studies on the haematology of freshwater teleosts. I Erythrocyte, Anat. Anz. 123: 233-249.

Srivastava, A.K. (1968b) : Studies on the haematology of freshwater teleosts IV Haemoglobin. Folia Haematologia. Leipzig. 90: 411-418.

Willmer, E.N. (1934) : Some observations on respiration of certain tropical freshwater fishes. J. Exp. Biol. 11 : 283-306.

References

Alexander, R. McN, (1962): The Structure of the Weberian apparatus in the Ctprini, Proc. Zool, Soc. London, Vol. 139, pt. 3, pp 451-473.

Alexander, R. McN, (1964a): The structure of the Weberian apparatus in the Siliri, Proc. Zool. Soc. London, vol. 142, pt. 3, pp. 419-440.

Anderson, J. (1970) : Metabolic rates of spiders: Comp. Biochem, Physiol. 33 : 1-72.

Augustinsson, K.B., R. Fange, A. Johnels and E. Ostlund (1956) : Histological, physiological and biochemical studies on the heart of two cyclostomes, hagfish (Myxine) and lamprey (Lampetra). J. Physiol., Lond. 131 : 257-276.

Awarti, P.R. and D.W. Bal (1934): Studies in Indian puffer or Globe fish part II. The blood vascular system of Tetradom oblongus, J. Univ. Bombay 2 :58-74.

Bamford, T.W. (1948): The cranial development of *Goleichthys felis*. Proc. London, vol. 118, no. 2, pp. 364-391.

Bettex-Galland, M and Hughes, G.M. (1972): Demonstration of Contractile filamentous material in the pillar cells of fish gills. *J. Cell Sci.* 13 : 359-370 actomyosin-like protein in the pillar cells of fish gills. *Experientia* 28: 744.

Beamish, F.W.H. (1964): Seasonal change in the standard rate of oxygen consumption of fishes. *Can J. Zool*, 42 : 189-194.

Beamish, F.W.H. and Mookerji, P.S. (1964): *Respiration of fishes with special emphasis on standard oxygen consumption*. I. Influence of weight and temperature on respiration of gold fish, *Carassius auratus* (L.) can J. Zool, 42 : 161-175.

Beamish, F.W.H. (1970): Oxygen consumption of large mouth bass, *Micropterus salmoides* in relation to swimming speed and temperature, Can. J. Zool, 48: 1221-1228.

Bertmar, G. (1959) : On the ontogeny of the chondral skull in Characidae, with a discussion on the chondrocranial base and the visceral chondrocranium in fishes. Act. Zool. Stockholm, vol. 40, pp. 203-364.

Bertmar, G. (1962) : On the ontogeny and evolution of the arterial vascular system in the head of the African characidean fish *Hepsetus adoe*. Ibid. vol. 43, pp. 255-295.

Bertalanffy, L. von (1957): Wachstum; Kukenthal's Handb, d. Zoologie Bd 8 4(6) (Berlin: De Gruyter and Co.) 68 pp.

Boake, B (1865-1866): *On the air-breathing fishes of Ceylon*, J. Ceylon Brch R. Asiat. Soc. 4 : 128-142.

Brett, J.R. (1965): The relation of size to rate of oxygen consumption and sustained swimming speed of sockeye salmon (*Oncorhynchus nerka*). J. Fish, Res. Bd. Can. 22 (191-1501).

Brett, J.R. (1972): The metabolic demand for oxygen in fish, particularly salmonids and comparison with other vertebrates; *Respir. Physiol.* 14 : 151-170.

Bridge, T.W. and A.C. Haddon (1893) : Contribution to the anatomy of fishes, II. The air-bladder and Weberian ossicles in the siluroid fishes. Phil. Trans. Roy. Soc. London ser. B, vol. 84, pp. 65-353.

Bertalanffy, L. Von (1957) : *Wachstum kukenthal's Hand b.d. Zoologie Bd.* 8: 4 (6) (Berlin: De Gruyter and Co.) 68 pp.

Boland, E.J. and K.R. Olson (1979): Vascular organisation of the catfish gill filament. Cell Tissues Res. 198-487-500.

Benninghoff, A, Das Herz (1933): In: Bulk *et al.,* Handbuch der Vergleichenden Anatomie der Wirbeltiere, 6, Berlin and Wien.

Bevelander G.A (1935) : A comparative study of the branchial epithelium in fishes with special reference to external excretion. *J. Morph.* 57: 335-351.

Bevan, D.J., and D.L. Kramer (1987) : The respiratory behaviour of an air-breathing catfish. *Clarias macrocephalus* (Clariidae). Can. J. Zool, 65: 348-353.

Boland, E.J., and K.R. Olson (1979) : Vascular organization of the catfish gill filament. Cell Tissue. Res. 198: 487-500.

Bruton, M.N. (1979) : The survival of habitat desiccation by air breathing clariid catfishes. Environ. Biol. Fishes, 4 : 273-280

Bevelander G.A,(1936) : Branchial glands in fishes. *J. Morph.* 59 :215-222.

Bevelander G.A,(1946) : Secretory cells in the branchial epithelium of fishes. *Biol. Bull.* 91: 230-231.

Bertin, I. (1958) : Organes de la respiration aquatique. In: *Traite de Zoologie (vol 13, Grasse, (Ed) Masson Paris* 1303-154.

Biswas N, Ojha J and Munshi J.S.D. (1981) : Morphometrics of the respiratory organs of an estuarine goby, *Boleophthalmus boddaerti; Japan J. Ichthyology,* 27 : 310-326.

Boake B (1866) : On the air-breathing fish of Ceylon. J. Ceylon Brch. R. Asiat. Soc – 4: 138-142.

Berrill, N.J. (1955): The origin of vertebrates, *Oxford University Press*, London.

Bridges, C.R. (1988): Respiratory adaptations in intertidal fish. Am. Zool. 28 : 79-96.

Burns J. and D.E. Copeland (1950): Chloride excretion in the head region of *Fundulus heteroclitus*. Biol. Bull.99 : 381-385.

Casselmann, W.G.B. (1959): Gistochemical technique, John Wiley and Sons Inc. New York.

Chan, D.K.O. and Woo, N.Y.S. (1978): The respiratory metabolism of the Japanese eel, *Auguilla japonica*. Effect of ambient oxygen, temperature, season, body weight and hypophysectomy. *Gen. Comp. Endocrinol.* 35 (2): 160-168.

Copeland D.E. (1948a) : The cytological basis of chloride transfer in the gills of *Fundulus heteroclitus*, *J. Morph.* 82 : 201-218.

Copeland D.E. (1948b) : Adaptive behavious of the Chloride cell in *Fundulus heteroclitus*, *Anat. Rec.* Vol 100, pp 652.

Choudhary D.P. (1979) : *Some Aspects of Respiratory Physiology of Channa striata (Bloch) (Channidae Channiformes); Ph.D. Thesis*, Bhagalpur University, Bhagalpur.

Choudhury, D.P. (1992): Morphometrics of the respiratory organs of the Indian snake headed fish, *Channa striata* (Bloch) (Ophicephaliformes, Channidae). J. Freshwater Biol. 4 (2):81-98.

Carter, G.S. (1957); Air-breathing. In *The physiology of fishes* 1: 65-79. Brown, M.E. (Ed.) New York: Academic Press.

Carter, G.S. and Beadle, L.C. (1931: The fauna of the swamps of the *Paraguayan chaco* in relation to its environment. II. *Respiratory adaptation in the fishes, J. Linn, Soc. Lond. (Zool)* 37: 327-368.

Das, B.K. (1927) : The bionomics of certain air-breathing fishes of India together with an amount of the development of their air-breathing organs. Philos. Trans R. Soc. Lond. B. 216 : 183-219.

Dornesco, P.G.T. and W. Santa (1963): La structure des aortas et des vaiseaux sanguins de la carpe (*Cyprinus carpio* L.) Anat. Anz. 113 : 136-145.

Dandotia, O.P. (1978): Studies on the functional capacity of the respiratory organs of a freshwater amphibious fish. *Channa* (=Ophiocephalus) *gachua*. Ph.D. thesis Bhagalpur University, India.

Day, F. (1868) : Observation on some of the freshwater fishes of India. Proc. Soul Soc. Lond. 1368 : 274-288.

Day, F. (1876) : On the respiration of some species of Indian freshwater fishes. Proc. Zool., Soc. Lond 1874 : 312-321.

Dejours, P (1973): Problem of control of breathing in Fishes. *In: Comparative physiology* Locomotion, Respiration, Transport and Blood, L. Bolis, K. Schmidt-Nielsen and S.H.P., Maddrell (eds). North Hilland/American Elsavier, Amsterdam and New York, pp. 117-133.

Dejours, P. (1976): *Principles of comparative respiratory physiology (Amsterdam, New York, North Holland, American Elsevier)*, pp. 262.

Dejours, P (1976): Water versus air as the respiratory media. In: Respiration of Amphibious Vertebrates, G.M. Highes (ed). Academic Press, New York, London, pp.1-15.

Dejours, P. (1975): *Principles of comparative respiratory physiology. North-Holland Publishing Co., Amsterdam*, 253 pp.

Dutta, H.M. (1996): A composite approach for evaluation of the effects of pesticides on Fish. *In: Fish Morphology* – Horizon of New Research, J.S. Datta Munshi and H.M. Dutta (eds). Science Publishers, Inc., New Hampshire, U.S.A., pp. 249-277.

Dutta, H.M., Munshi, J.S.D. Roy, P.K. Singh, N.K., Adhikari, S. and Killus, J. (1996): Ultrastructural changes in the respiratory lamellae of the catfish, *Heteropneustes fossilis* after sublethal exposure to Malathion. Environmental Pollution, 92(3): 329-341.

Datta Munshi, J.S. and Srivastava, M.P. (1988) : Natural History of Fishes and systematics of Freshwater Fishes of India, Narendra Publishing House, Delhi, pp 1-403.

De Beer, G.R. (1958): Embryos and ancestors, 3rd (ed), *Oxford University Press*, London and New York.

Dehadrai, P.V. and Tripathi, S.D. (1976): Environment and ecology of freshwater air-breathing teleosts. In : *Respiration of Amphibious Vertebrates*. (Ed. Hughes, G.M.),Academic Press, London, pp. 39-72.

Dejours, P. (1988) : Respiration in Water and Air Adaptations. Regulation, Evolution pp1-179, New York, Oxford.

Danforth, C.H (1912) : The heart and arteries of polyodon. J. Morphol. 23, 400-454.

Day, F. (1868): Observations on some of the freshwater fishes of India; Proc. Zool, Soc. Londoni 1868: 277-288.

Das, B.K (1927): The bionomics of certain air-breathing fishes of India, together with an account of the development of their air-breathing organs. Phil. Trans. R.Soc., London B 216: 183-219.

Dubale M.S. (1951) : A comparative study of the extent of gill surface in some representative Indian fishes and its bearing on the origin of airbreathing habit; J. *University. Bombay* (N.S.) B 18 6-13.

Dubale, M.S. (1971): A comparative study of the extent of gill-surface in some representative Indian fishes and its bearing on the origin of air-breathing fishes. J. Univ. Bombay (N.S.) 198 : 90-191.

Dehadrai, P.V. (1962): Respiratory function of the swimbladder of *Notopterus* (Lacapede) Proc. Zool. Soc. London, 139: 341-357.

Dehadrai, P.V. (1962): Observation on certain physiological relations in *Ophiocephalus striatus* exposed to air; Life Sci. II 655-657.

Dube, S.C. and Munshi, J.S.Datta (1974b) : Diffusing capacity of gills of climbing perch *Anabas testudineus* (Bl.) in relation to body weight. Indian J. Exp. Biol. 12: 207-208.

Eaton, T.H. (1948): Form and function in the head of the channel catfish (*Ictalurus lacustris punctatus*). Jour, Morph. Vol. 83, pp. 181-194.

Ellis, A. E. (1977): The leucocytes of fish: A review. Journal of Fish Biology 11, 453-491

Ellis, A.E., Munro, A.L.S. and Roberts, R.J. (1976): Defence mechanism in fish. I. A study of the phagocytic system and the fate of intraperitoneally injected particulate material in the plaice (*Pleuronectes platesa*), Journal of Fish Biologt 8, 67-78.

Fange, R. (1976): Gas exchange in the swimbladder. In *Respiration of Amphibious Vertebrates*, pp. 189- 211 (Ed. G.M. Hughes). Academic Press, London-New York-San Francisco.

Fry, F.E.J. (1957): *The aquatic respiration of fish*; In *The Physiology of Fishes Vol, I pp 1-63 ed M.E. Brown (New York: Academic Press).*

Ferguson, H.W. (1975): Phagocytosis by the endocardial lining cells of the strium of place (Pleuronectes platessa). Journal of Comparative Pathology 85, 561-569.

Faussek, V. (1902): *Beitrage zur Histologie der Kiemenbei Fishen und Amphibien, Arch. Mikrosk. Anat. Ents. Mech.* 60 :157-174.

Fawcett, D.W. (1966): An atlas of fine structure : *The cell, its organelles and inclusion. Philadelphia : W.B. Saunders.*

Fullmer, H.M. and R.D. Lillie (1950) : A selective stain for elastic tissue (Orcinol new fuchsin). Stain technol. 31 : 27-29.

Gannon, B.J., G. Campbell and D.J. Randall (1973) : Scanning electron microscopy of vascular casts for the study of vessel connections in a complex vascular bed – the trout gill *31st Ann. Proc. Electron Microscopy. Soc. Am* 31: 442-443.

Getman, H.C. (1950): Adaptive changes in the Chloride cells of *Anguilla rostrata. Biol. Bull.* Vol. 99, pp. 439-445.

Ghosh, E. (1934) : An experimental study of the asphyxiation of some air-breathing fishes of Bengal. J. Asiat. Soc. Bengal (N.S.) 29: 327-332.

Ghosh, T.K. and Biswas, N. (1980): Rhythmic behaviour in the bomodal oxygen uptake of *Anabas testudineus* (Bloch) Biol. Bull. India 2 : 8-11.

Ghosh, T.K., Biswas, N. and Munshi J.S.D. (1980) : Oxygen uptake in relation to body weight in swamp-eel, *Monopterus (=Amphipnous) cuchia* (Ham.) (Symbranchiformes, Amphipnoidae). India Biol. Jour. 2 (2) : 49-54.

Ghosh, T.K. and Munshi, J.S.D. (1987): Bimodal oxygen uptake in relation to body weight and seasonal temperature of an air-breathing climbing perch. *Anabas testudineus* (Bloch). *Zool. Beitr.* 33(3): 357-364.

Ghosh, E. (1934) : An experimental study of the asphysiation of some air-breathing fishes of Bengal; *J. Asiat. Soc. Bengal* (N.S.) 29: 327-332.

Ghosh, T.K, Kunwar, G.K and Munshi, J.S.D. (1990): Diurunal variation in the bimodal oxygen uptake in an air-breathing cat fish, *Clarias batrachus,* Japan. J. Ichthyol. 37 (1): 56-59.

Gosline, W.A. (1959): Mode of life, functional morphology, and the classification of modern teleostean fishes. Syst. Zool., vol. 8 no. 3, pp 160-164.

Gosline, W.A. (1961): Some osteological features of modern lower teleostean fishes. Smithsonian Misc. Coll. Vol. 142, no. 3, pp. 1-42.

Goniakowska, Witalinska, L. (1978) : Ultrastructural and Morphometric study of the lung of European, Salamdar, L Cell tissue res. – 191: 343-356.

Goniakowska, Witalinska, L. (1980) : Scanning and transmission Electron Microscpy study of the lung of the newt.

Gregory, W.K. (1933): Fish skulls: a study of the evolution of natural mechanisms. Trans. Amer, Phil. Soc. New ser., vol. 23 art 2, pp. 75-481.

Granel F. (1923): *Etude histologique et embryologique sur is pseudobranchie des teleosteans; Arch d' Anat. Histol. Et. Embr.* Tome 2 153-195.

Gray, I.E. (1954): Comparative study of the gill area of marine fishes. *Biol. Bull (Woods Hole, Mass)* 107: 219-225.

Grarstang, W. (1894): Preliminary note on a new theory of the phylogeny of the chordate, *zoologischer Anzeiiger* 17: pp 122-125.

Graham, J.B. (1976) : Respiratory adaptations of marine air-breathing fishes. In *Respiration of Amphibious Vertebrates.* pp. 165-187 (Ed. G.M. Hughes). Academic Press, London-New York-San Francisco.

Gunther, A.C, L.G. (1880) : As introduction to the study of fishes. Edinburgh : A, and C, Black.

Goodrich, E.S. (1930): Studies on the Structure and Development of Vertebrates *Macmillan Co., London.*

Ghosh, T.K., G.K. Kunwar and J.S.D. Munshi (1990) : Diurnal variation in the bimodal oxygen uptake in an air-breathing catfish, *Clarias batrachus. Jpn. J. Ichthyol.* 37: 56-59

Graham, J.B. (1995): Air-breathing fishes: Evolution, diversity and adaptation. In press.

Graham, J.B. (1997) : *Air-breathing Fishes – Evolution, Diversity and Adaptation, Academic Press. New York, Sydney, Tokyo, Toronto.*

Gagenbaur, C. (1801) : Conus arteriosus der Fische.Morph. Jaharb, 17, 596-610.

Goodrich, E.S. (1930) : Studies on the structure and development of vertebrate, London, Macmillan and Co.

Grodizinski, Z (1954) : Contractions of the isolated heart of the European Glass Ed. Anguilla anguilla L. Bull. Acas. Pol. Sci. et II, Ser. Sci. Biol. 2, 19-22.

Graham, J.B., T.A. Baird and W. Stockmann (1987): The transition to air breathing in fishes IV. Impact of branchial specializations for air-breathing on the aquatic respiratory mechanisms and ventilatory costs of the swamp eel *Synbranchus marmoratus*. J. Exp. Biol. 129-83-106.

Hakim, A., Munshi, J.S.D. and Hughes, G.M. (1983): Oxygen uptake from water through the respiratory organs in *Channa punctata* (Bloch) in relation to body weight, Proc. Indian natn. Sci. Acad, B. 49 : 73-85.

Hossain, A. and Dutta, H.M. (1986): Acid phosphatase activity in the intestine and caeca of bluegill, exposed to methyl mercuric chloride. Bulletin of Environmental Contamination and Toxicology 36: 460-467.

Hakim, A., Munshi, J.S. Datta and Hughes, G.M. (1978): Morphometrics of the respiratory organs of the Indian green snake-headed fish, *Channa punctata*, J. Zool. Lond. 184, 519-543.

Hakim, A., Munshi, J.S. Datta and Hughes, G.M. (1978): Morphometrics of the respiratory organs of the Indian green snake-headed fish, *Channa punctata*, J. Zool. Lond. 184, 519-543.

Harrington, R.W. (1955): The osteocranium of the American cyprinid fish, Notropis bifrenatus, with an annotated synonymy of teleost skull bones. Copeia, no. 4, pp. 267-296.

Henninger, G. (1907): Die Labyrinthorgans bei Labyrinthfischen. *Zool. Jb. (Anat.)* 25: 251-304.

Hemingsen, A. (1960): *Energy metabolism as related to body size and respiratory surfaces and its evolution; Rept. Steno. Meml. Hosp. Nord. Insulin Lab.* 9 1-110.

Heisler, N. (1982) : Intercellular and extracellular acid-base regulation in the tropical freshwater teleost fish, *Symbranchus marmoratus* in response to the transition from water breathing to air-breathing, J. Exp. Biol., 99: 9-28.

Heisler, N. (1984): Acid-base regulation in fishes. *In: Fish Physiology,* W.S. Hoar and D.J. Randall (eds.). Academic Press, New York, pp. 315-392.

Hora, S.L. (1934) : Trade in live fish (Jiol Machh), In Calcutta. J. As. Soc. Bengal (N.S.) 30,1-15.

Hora, S.L. (1935) : Physiology bionomics and evolution of the air-breathing fishes of India. Trans. Natn. Inst. Sci. India 1: 1-16.

Hora, S.L. (1935) : *Physiology, bionomics and evolution of the air-breathing fishes of India. Trans. Natn. Inst. Sci. India* 1: 1-16.

Howell, B.J. (1970) : Acid-base balance in transition from water breathing to air-breathing Fed. Proc., 29: 1120-1134.

Hughes, G.M. (1966b) : The dimensions of fish gills in relation to their function, J. Exp. Biol. 45 : 177-195.

Hughes, G.M. (1966): The dimensions of fish gills in relation to their function. *Academic Press, London, New York*, 1-299.

Hughes, G.M. (1966) : Evolution between air and water. In: *Ciba Foundation Symposium on Development of the Lung, de Reuck and Porter* (Eds.) *Churchill Ltd*. Place 64-80.

Hughes, G.M. (1970b) : Gill dimensions in relation to other respiratory parameter, Proc. 9th Int. Cong. Anat., Leningard.

Hughes, G.M (1972): Morphometrics of fish gills. Resp. Physiol. 48: 1-12.

Hughes, G.M. (1976): On the respiration of *Latimeria chalumnae*. J. Linn. Soc. Lond. (Zool.) 59 (2): 195-208.

Hughes, G.M. (1977): Dimensions and the respiration of lower vertebrates. In: *Scale effect in Animal Locomotion* (ed) T.J. Pedley. Academic Press. London, pp. 57-81.

Hughes, G.M. (1984) : General anatomy of the gill; in Fish Physiology Vol. xA pp 1-72 eds W S Hoar and D.J. Randall (London; Academic Press).

Hughes, G.M. (1984) : Scaling of respiratory areas in relation to oxygen consumption of vertebrates. Experientia. 40: 519-524.

Hughes, G.M. and B.N. Singh (1970) : Respiration in an air-breathing fish, the climbing perch, *Anabas testudineus*. I. O$_2$ uptake and CO$_2$ release into air and water. J. Exp. Biol. 53: 265-280, Figures 1-7.

Hughes, G.M. and B.N. Singh (1971): Gas exchange with air and water in an air-breathing catfish, *Saccobranchus fossilis*. J. Exp. Biol., 55: 667-682.

Hughes, G.M. and Gray *I.E.* (1972): Diemnsions and ultrastructure of toadfish gills; Biol. Bull. Mar. boil. Lab: Woods Hole 143 : 150-161.

Hughes, G.M., Dube S.C. and Munshi J.S.D. (1973): Surface area of the respiratory organs of the climbing perch, *Anabas testudineus, J. Zool, Lond* 170: 227-243.

Hughes, G.M. and Morgan M (1973) : The structure of fish gill in relation to their respiratory function, Biol. Rev. 48: 419-475.

Hughes, G.M. and Ojha J. (1986): Critical study of the gill area of a a free swimming river carp. *Catla catla* and an airbreathing perch, *Anabas testudineus*; Proc. Indian nain. Sci. Acad. B51 391-404.

Hughes, G.M., Singh B.R., Guha G., Dube S.C. and Munshi J.S.D. (1974): Respiratory surface area of an air-breathing siluroid fish, *Saccobranchus, (Heteropneustes)* fossilis in relation to body size. J. Zool. Lond. 172 : 215-235.

Hughes, G.M., Roy, P.K. and Munshi, J.S.D. (1992): Morphometric estimation of oxygen diffusing capacity for the air-sac in the catfish *Heteropneustes fossilis*. J. Zool. (London) 227: 193-209.

Hughes, G.M., Singh, B.R., Thakur, R.N. and Munshi, J.S.D. (1974a): Areas of the air-breathing surfaces of *Amphipnous cuchia* (Ham.), Proc. Nat. Acad. Sci. India 408: 379-392.

Hughes, G.M., Singh, B.R., Guha, G., Dube, S.C. and Munshi, J.S.D. (1974b): Respiratory surface area of an air-breathing siluroid fish, *Saccobranchus. (=Heteropneustes) fossilis* in relation to body size. J. Zool. Lond. 192: 215-232.

Hughes, G.M. (1984): General anatomy of gills. *In: Fish Physiology, W.S. Hoar* and *D.J. Randall* (Eds.), *Academic Press*, London XA 1-72.

Hughes, G.M. (1984): Measurement of gill area in fishes. *Practice and problems. J. Mar. Biol. Ass.* U.K. 64 : 637-655.

Hughes, G.M. (1998) : The Gills of *Latimeria chalumnae Latimeriidac.* What can they teach us? *Ital. J. Zool,* 65 Suppl. 425-429.

Hughes, G.M. (1999): *Fish Gills: Old Fourlegs and Air-breathing Fishes.* In *Waterl Air Transition in Biology* (Eds. Mittal, A.K, Eddy, F.B. and Munshi, J.S.D.) Science Publishers, Inc. USA.

Hughes, G.M. and Grimstone, A.V. (1965): The fine structure of the secondary lamellae of the gills of *Gadus pollachius. Q.J. Microsc. Sci.*106: 343-353.

Hughes, G.M. and Munshi, J.S.D. (1968): Fine structure of the respiratory surfaces of an air-breathing fish, the climbing *perch Anabas testudineus* (Bloch.) *Nature,* 219 (5/6): 1382-1384.

Hughes, G.M. and Munshi, J.S.D. (1973a): Fine structure of the respiratory organs of the climbing perch. *Anabas testudineus. J. Zool.* (Lond.) 170:201-225.

Hughes, G.M. and Munshi, J.S.D.(1973b): Nature of the air-breathing organs of the Indian fishes, *Channa, Amphipnous, Clarias* and *Saccobranchus* as shown by electron microscopy. J. Zool. (Lond.) 170: 245- 270.

Hughes, G.M., Singh, B.R., Guha, G., Dube, S.C. and Munshi, J.S.D. (1974): Respiratory surface area of an air-breathing siluroid fish, *Saccobranchus. (=Heteropneustes) fossilis* in relation to body size. J. Zool. Lond. 192: 215-232.

Hughes, G.M. and Munshi, J.S.D. (1978): Scanning electron microscopy of the respiratory surface of *Saccobranchus (=Heteropneustes) fossilis* (Bloch.) *Cell Tissue Res.* 195 : 99-109.

Hughes, G.M. and Munshi, J.S.D. (1979): Fine structure of the gills of some Indian air-breathing fishes. *J. Morphol.*160: 169-194.

Hughes, G.M. and A.K. Mittal (1980): Structure of the gills of *Barbus sophor,* Ham, a cryprinid with tertiary lamellae. *J. Fish Biol.* 16: 461-467.

Hughes, G.M. and Munshi, J. S.D. (1986): Scanning electron microscopy of the accessory respiratory organs of the snake-headed fish, *Channa striata* (Bloch.) (Channidae, Channiformes). *J. Zool.Lond.* 209:305- 317.

Hughes, G.M., J. Ojha and Munshi, J. S.D. (1986): Post-embryonic development of water and air-breathing organs *Anabas testudineus* (Bloch), *J. Fish Biol.* 29 : 443-450.

Hughes, G.M., Roy P.K. and Munshi, J,S.D. (1992): Morphometric estimation of oxygen diffusing capacity for the air-sac in the catfish, *Heteropneustes fossilis.* J. Zool. Lond., 227: 193-209.

Hughes, G.M. (1964): Fish respiratory homeosrasis. Soc. Exp. Biol. 18: 81-107

Hughes, G.M., Singh, B.R., Thakur, R.N. and Munshi, J.S.D. (1974b). Areas of the air-breathing surface of *Amphipnous cuchia* (Ham). Proc Indian natl. Sci. Acad. 40b: 379-392.

Hutchison, V.H., Whitford, W.G. and Kohl. M. (1968) : Relation of body size and surface area to gas exchange in anurans. Physiol. Zool. 41 : 63-85.

Hughes, G.M. (1966) : The dimension of fish gills in relation to their function. J. exp. Biol. 45 : 177-195.

Hughes, G.M. (1970a) : Morphological measurements on the gills of fishes in relation to their respiratory function. Folia morph. Praha 18 : 78-95.

Hughes, G.M. (1970b) : Gill dimensions in relation to other respiratory parameters. In Congr. Anat. 9.

Hughes, G.M. (1972) : Morphometrics of fish gills. Resp. Physiol. 14 : 1-26.

Hughes, G.M., Dube, S.C. and Munshi, J.S.D. (1973) : Surface areas of the respiratory organs of the climbing perch, *Anabas testudineus*. J. Zool, Lond. 170 : 227-243.

Hughes, G.M. (1970) : Ultrastructure of the air-breathing organs of some lower vertebrates. 7th Intern Cong. Electron Microscopic, pp 599-600.

Hughes, G.M. (1973) : Ultrastructure of the lung of Neoceratodus and Lepidosiron in relation to the lung of other vertebrates. Folia Morphol. Prague, 21 : 155-161.

Hughes, G.M. (1978) : A morphological and ultrastructural comparison of some vertebrate lungs pp 393-405. In Elika, E. (ed): XIX Congressrs Morphological Symposium. Charles Univ. Press Progue.

Hughes, G.M. and Weibel, E.R. (1976) : Morphometry of fish lungs pp 212-232. In Hughes G.M. (ed): Respiration of Amphibious Vertebrates. Acad. Pr. London and New York.

Hughes, G.M. (1966): Evolution between air and water. Ciba Foundation Symposium on Development of the Lung, pp. 64-80 (Eds. de Reuck and Porter). Churchill Press, London.

Hughes, G.M. and Weibel, E.R. (1978): Visualisation of layers lining the lung of the South American lungfish (*Lepidosiren paradoxa*) and a comparison with frog and rat. Tissue and Cell 10: 343-353.

Hughes, G.M. and Grimstone, A.V. (1965): The fine structure of the secondary lamellae of the gills of *Gadus pollachius*, Q.J. Microsc. Sci. 106: 343-353.

Hughes, G.M. and Singh, B.N. (1970): *Respiration in an air-breathing fish*, the climbing perch *Anabas testudeneous* Bloch, I. *Oxygen uptake and carbon dioxide release into air and water, J. exp. Biol.* 53: 265-280.

Hughes, G.M. and Wright, D.E. (1970): A comparative study of the ultrastructure of the water-blood pathway in the secondary lamellae of teleost and elasmobranch fishes, Benthic forms. Z. *Zellforsch Mikrosk. Anat.* 104: 478-493.

Hughes, G.M. and Weibel, E.R. (1972): Similarity of suppuring tissue in fish gills and the mammalian reticulo endothelium. *J. Ultrasir. Ibructuce Res.* 39: 106-114.

Ishimatsu, A, and Y. Itazawa (1983): Difference in the blood oxygen levels in the out-flow vessels of the heart of an air-breathing fish *Channa argus*: Do separate blood streams exist in a teleostean heart? J. Comp. Physiol. 149-435-440.

Ishimatsu, A. and Y. Itazawa (1981): Ventilation of the air-breathing organ in the snake head *Channa argus*; Japan J. Ichthyol 28: 276-282.

Inque, T., and A. Osatake (1988) : A new drying method of biological specimens for scanning electron microscopy. The –butyl alcohol freeze-drying method. Arch. Histol. Cytol. 51 : 53-59.

Johansen, K (1970): Air-breathing in fish. In *Fish Physiology* Vol. IV (Eds. W.S. Hoar and Randall, D.J.). Academic Press, New York, pp. 361-411.

Joseph, N.I. (1960): Osteology of *Wallago attu* Bloch and Schneider. Pt. 1. Osteology of head. Proc. Natl. Inst. Sci. India, ser B. vol. 26, pp. 205-233.

Job, S.V. (1955) : The oxygen consumption of *Salvelinus fontinalis, Univ. Toronto Stud. Biol. Ser., No. 61, Publ. Ontario Fisheries Res. Lab.,* 73 : 1-39.

Jones, R.M. (1950): Mc Clung's hand book of microscopical technique. Paul B. Hoeber Inc., New York. Medical Book Dept. of Harper and Brothers.

Johansen, K and Lentant, C. (1967): Hepatic vein sphincters in elasmobranches and their significance in controlling hepatic blood flow. J. exp. Biol. 46: 195-203.

Johansen, K. (1970): Air-breathing in fishes. In: Fish Physiology, W.S. Hoar and D.J. Randall, eds Academic Press. New York, 99 : 361-411.

Jordan, J. (1976) : The influence of body weight on gas exchange in the air-breathing fish, *Clarias batrachus. Comp. Biochem. Physiol.,* 53A : 305-310

Munshi, J.S.D. (1961) : The accessory respiratory organs of *Clarias batrachus* (Linn.), J. Morphol. 109 : 115-140

Julian Huxley (1953): Evolution in action based on the *Patten Foundation Lectures* delivered at Indiana University, U.S.A., Chatto and Windus, London, pp 1-159.

Jensen, D. (1965) : The neural heart of the lungfish. Ann. New York Acad.Sci 127: 443-458.

Johansen, K. (1966):. Air breathing in the teleost *Symbranchus marmoratus*. Comp. Biochem, Physiol.18: 383-395.

Johansen, K.and D. Hanson (1968) : Functional anatomy of the heart of lungfish and amphibians., Amer. Zoologist, 8 : 191-200.

Johansen, K and Lentant, C. (1967): Respiratory function in the South American lung fish. *Lepidosiren paradoxa* (Fitz.). J. Exp. Biol. 46: 205-218.

Kamler, E. (1972): Respiration in carps in relation to body size and temperature, *Pol. Arch. Hydrobiol.* 19: 325-331.

Kindered, J.E. (1919): The skull of *Ameiurus*. Illinois Biol. Monogr., vol. 5, pp. 1-120.

Kramer, D.L. (1978): Ventilation of the respiratory gas bladder in *Hoplerythrinus unitaeiatus* (Pisces, Characoidei, Erythrinidae). *Can. J.Zool.* 56: 931-938.

Kunwar, G.K., Pandey, A. and Munshi, J.S.D. (1989): Oxygen uptake in relation to body weight of two freshwater major carps *Catla catla* (Ham.) and *Labeo rohita* (Ham.). *Indian J. of Animal. Sci.* 59 (5) : 621-624.

Kingsley,J.S. (1926): Outlines of comparative anatomy of vertebrate.London: John Murray.

Karandikar, K.R. and S.S. Thakur (1954): Anatomy with notes on distribution and bionomics of Sciaenoides. Brunneus, Day. Univ. Bombay Zool memoirs No. 3 : 54-63.

Kharakter Svyazimezhdu (1972) : The nature of relationship between the body weight and heart weight in the perch, Ecologiya 1 : 58-65.

Keys A.B.(1931) : Chloride and water secretion and absorption by the gills of the eel. *Z vergleich, Physiol.* 15: 354.

Keys A.B. and E.N. Willmer (1932) : Chloride secreting cells in the gills fishes with special reference to the common eel, *J. Physiol* 79 :368-378.

Keys, A.B. and Willmer, E.N. (1932): "Chloride secreting cells" in the gills of fishes, with special reference to the eel. *J. Physiol.* 76: 368-378.

Krogh, A (1919) : The rate of diffusion of gases through animal tissue. J. Physiol. Lond. 52 : 391-408.

Krogh, A. (1939): Osmotic regulation in aquatic animals, *Cambridge Univ. Press*, New York and London.

Kunwar, G.K. (1984): The Structure and Function of the Respiratory Organs of a Major carp. *Catla catla* (Ham); *Ph.D. Thesis*, Bhagalpur University, Bhagalpur.

Landolt, J.C. and Hill L.G. (1975): Observations of the gross structure and dimensions of the gills at three species of Gars (Lepisosteidae); Copeia 3 470-475.

Leiner, M. (1938): Die Augenkiemendruse (Pseudobranchie) der knochenfische Experimentelle untersuchungen uberihre physiologieBedeutung; Z. vergl. Physiol. 26 : 416-466

Laurent, P. and Dunel, S. (1984) : The pseudobranch: Morphology and function. In Fish Physiology (eds. W.S. Hoar and D.J. Randall). Vol. 13. pp. 285-323. Academic Press Inc., New York.

Liu, C.K. (1942) : Osmolic regulation and chloride secreting cells in the paradise fish, *Macropodus opercularis, Sinensia* 13: pp 15-20.

Liem, KF. (1961): Tetrapod parallelism and other features in the functional morphology of the blood vascular system of *Fluta alba* Zuiew (Pisces: Teleostei), J, Morph. 108: 131-143.

Liem, K.F. (1963) : The comparative osteology and phylogeny of the Anabantoidei (teleostei, Pisces). Illinois Biol. Monogr., no. 30, 149 pp.

Lomholt, J.P. and Johansen, K. (1974): Control of breathing in *Amphipnous cuchia*, an amphibious fish. Resp. Physiol. 21:325-340.

Lomholt, J.P. and Johansen, K.(1976): Gas exchange in the amphibious fish *Amphipnous cuchia*, J, Comp. Physiol. 107: 141-157.

Laurent, P. and S. Dunel 91980) : Morphology of gill epithelia in fish. Am. J. Physiol., 238 : R147-R159.

Lewis, S.V. (1979a) : The morphology of the accessory air-breathing organs of the catfish. *Clarias batrachus* : A SEM study. J. Fish Biol., 14 : 187-191.

Lewis, S.V. (1979b) : A scanning electron microscope study of the gills of the air-breathing catfish, *Clarias batrachus* L. J. Fish Biol. 15 : 381-384.

Low, W.P., D.J.W. Lane and Y.K. Ip (1988) : A comparative study of terrestrial adaptations of the gills in three mudskipper. *Periophthalmus chrysospilos, Boleophthalmus boddaerti* and *Periophthalmus schlosseri*. Biol. Bull. 175 : 434-438

Laurent, P. and S. Dunel (1976): Functional organization of the teleost gill. Blood pathways. Acta Zool (Stockholm) 57: 189-209.

Lele, S.H. (1932) : The circulation of blood in the air-breathing chamber of *Ophicephalus punctatus* Jl. Linn Soc. London (Zool) 38: 49-54.

Liem, K.F. (1980) : Air-ventilation in advanced teleosts. Biochemical and Evolutionary aspects. *In: Environmental physiology*, M.A. Ali (Ed). Plenum Press, New York 57-91.

Lander, J. (1964) : The shark circulation B.Sc. med. Dissertation. University of Sydney.

Light, J.H. and W.S. harris (1973): The structure, composition and elastic properties of the teleost bulbus arteriosus in the carp. Comp. Biochem. Physiol. 46A : 699-708.

Lillie, R.D. (1954): Histopathologic techniques and practical histochemistry. The Blakiston Division, McGraw Hill Book Co. New York.

Lomholt, J.P. and Johansen, K. (1974): Control of breathing in *Amphipnous cuchia*, an amphibious fish. Resp. Physiol. 21:325-340.

Lomholt, J.P. and Johansen, K.(1976): Gas exchange in the amphibious fish *Amphipnous cuchia*, J, Comp. Physiol. 107: 141-157.

Liem, K.F. (1963): The comparative Osteology and Phylogeny of the Anabantoidei (Teleostei, Pisces). The University of Illinois Press, Urbana, pp 1-149.

Lenfant, C. and Johansen, K. (1968): Respiration in the African lungfish, *Propterus aethiopicus*. J. Exp. Biol. 49: 453-468.

Moussa, T.A. (1957): Physiology of the accessory respiratory organs of the teleost *Clarias lazera* (C and V). J. exp. Zool, 136: 419-454.

McMurrich, J.P. (1984) : The osteology of *Ameiurus catus* (L.) Gill. Proc. Canadian Inst. Toronto, new ser., vol. 2, pp. 270-310.

Masurekar, V.B. (1962): Weberian apparatus in *Mystus gulio* (Ham-Buch) and *Mystus bleeberi* (Day). Proc. Indian Acad. Sci. ser. B. vol. 55, no. 1, pp 24-37.

Myers, G.S. (1938): Freshwater fishes and West Indian zoogeography. Ann. Rept. Smithsonian Inst. For 1937, pp. 339-364.

Myers, G.S. (1958): Trends in the evolution of teleostean fishes, Stanford Ichthyol. Bull. Vol. 7, no. 3, pp. 27-30.

Marlier, G. (1938): Consideration sur les organs accessories servant a la respiration aeriienne chezles Teleosteens _Annls Soc. R. zool. Beig_ 69: 163-185.

Millonig, G. (1961): Advantages of a phosphate buffer for O_2O_4 solutions in fixation (Abstract). _Program 19th Annual Electron Microscope Society of America_, Pittsburg. 1961.

Milton P (1971): Oxygen consumption and the osmoregulation in the shanny _Blennius pholis; J. mar boil. Assoc._ UK 51 : 247-265.

Muir B. S. and Hughes, G.M. (1969) : Gill dimension for three species of tunny; J. exp. Boil. 51: 271-285.

Muller, J. (1839): _Vergleichende Anatomy der Myxinoidem III iiber das gefassystem – Abhandb, Akad, Wiss, (Berlin),_ 175- 303.

Munshi, J.S.D. (1960) : _Studies the structure of the gills of certain freshwater teleosts. Ind. J. Zool_ 4: 1-40.

Munshi, J.S.D. (1962b) : On the accessory respiratory organs of _Ophiocephalus punctatus_ (Bloch) and _O. striatus_ (Bloch.) J. Linn. Soc. London Zool. 44: 616-626.

Munshi, J.S.D. (1976): _Gross and fine structure of the respiratory organs of air-breathing fishes._ In: _Respiration of Amphibious Vertebrates,_ pp 73-104. (Ed. G.M. Hughes). _Academic Press, London-New York._

Munshi, J.S.D., Ojha, J. and Sinha, A.L. (1980): Morphometrics of the respiratory organs of an air-breathing cat fish, _Clarias batrachus_ (Linn.) in relation to body weight. Proc Indian. Natn. Acad., B 46 (5): 621-635.

Munshi, J.S. Datta(1968): The accessory respiratory organs of _Anabas testudineus._ Proc. Linn. Soc. Land (Zool) No. 179: 107-127.

Munshi, J.S. Datta and Singh, B.N. (1968): On the micro-circulatory system of the gills of certain freshwater teleostean fishes. J. Zool. Lond. 154: 365-376.

Munshi, J.S.D. and N. Mishra (1974): Structure of the heart of _Amphipnous cuchia_ (ham), Zool, Anz. 193: 228-239.

Munshi, J.S.D. (1985) : The structure, function and evolution of the accessory respiratory organs of air- breathing fishes of India. In: _Fortschrittle der Zoologie,_ Band 30, pp. 353-366,(Eds. Duncker/Fleischer). Gustav Fischer Verlag, Stuttgart-New York.

Munshi, J.S.D. and Singh, B.N. (1968) : On the microcirculatory system of the gills of certain freshwater teleostean fishes J. Zool. Lond., 164 : 365-376.

Munshi, J.S. Datta, Olson, K.R., Ojha, J. and Ghosh, T.K. (1986): Morphology and vascular anatomy of the accessory respiratory organs of the air-breathing climbing perch _Anabas testudineus_ (Bloch). Am. J. Anant. 176: 321-331

Munshi, J.S.D. Hughes, G.M., Gehr, P. and Weibel, E.R. (1989): Structure of the air-breathing organs of a swamp mud eel, *Monopterus cuchia* (Ham). Japan. J. Ichthyol. 35 (4) : 453-465.

Menon, A.G.K. (1974) : A check list of fishes of the Himalayan and the Indo-gangetic Plains. Spec. Publ. Inland Fish. Soc. India No. 1 : 90-91.

Milton, P. (1971) : Oxygen consumption and the osmoregulation in the shanny, *Blennius pholis*. J. mor. Boil.Ass. U.K. 51 : 247-265.

Muir, B.S. and Hughes, G.M. (1969) : Gill dimensions for three species of tunny., J. exp. Biol. 51 : 271-285

Munshi, J.S.D. (1962): On the accessory respiratory organs of *Ophicephalus punctatus* (Bloch) and *Ophicephalus striatus* (Bloch). J. Linn. Soc. Lond. (Zool). 44 :616-626.

Munshi, J.S.D. (1976) Gross and fine structure of the respiratory organs of air-breathing fishes. In Respiration of amphibious vertebrates 73-104. Hughes G.M. (Ed.). London : Academic Press.

Munshi, J.S.D. and Dube, S.C. (1973): Oxygen uptake capacity of gills in relation to body size of the air-breathing fish, *Anabas testudineous* (Bloch). Acta Physiol. 44: 113-129.

Munshi, J.S.D., Sinha, A.L. and Ojha, J. (1976): Oxygen uptake capacity of gills and skin in relation to body weight of the air-breathing siluroid fish, *Clarias batrachus* (Linn): Acta Physiol, 48 : 23-33.

Munshi, J.S.D., Pandey, B.N. Pandey, P.K. and Ojha, J. (1978): Oxygen uptake in *Saccobranchus fossilis (Heteropneustes fossilis)*. J. Zool, Lond. 184: 171-180.

Munshi, J.S.D., Patra, A.K. and Hughes, G.M. (1982): Oxygen consumption from air and water in *Heteropneustes (=Saccobranchus) fossilis* (Bloch) in relation to body weight at three different seasons: Proc. Indian Natn. Sci. Acad. B 48: 715-729.

Milonig, G. (1961) : Advantages of a phosphate buffer for O_sO_4 solution's in fixation (Abstract). Program 19th Annual Electron Microscope Society of America, Pittsburg.

Mishra, A.B. and J.S.D. Munshi : (1958) On the accessory respiratory organs of *Anabas scandens* XVth Inter Cong. Zoology. No. 32, London.

Mishra, N, P.K. Pandey, J.S.D. Munshi and B.R. Singh (1977) : Haematological parameters of an air-breathing mud-eel. *Amphipnous cuchia* (Ham.) (Amphipnoidae Pisces). J. Fish Biol. 10 : 567-573.

Munshi, J.S.D. (1958): *Studies of some of Indian Freshwater Fishes*, Ph.D. Thesis, Banaras Hindu University, India, 1-328.

Munshi, J.S.D. (1960): The structure of the gills of certain freshwater teleosts. Ind. J. Zoot. 4:1-40 +Plates I – XVII.

Munshi, J.S.D. (1961): The accessory respiratory organs of *Clarias batrachus* (Linn.) J. Morphol.109(2): 115-139.

Munshi, J.S.D. (1962a) : On the accessory respiratory organs of *Heteropneustes fossilis* (Bloch). Proc. R. Soc. (Biol.) Edin. 68 : 128-146.

Munshi, J.S.D. (1962b) : On the accessory respiratory organs of *Ophiocephalus punctatus* (Bloch) and *O. striatus* (Bloch.) J. Linn. Soc. London Zool. 44: 616-626.

Munshi, J.S.D. and Singh, B.N. (1968) : A study of the gill-epithelium of certain freshwater teleost fishes with special reference to the air-breathing fishes. Ind. J. Zoot. 9 : 91-107.

Munshi, J.S.D. (1976): Gross and fine structure of the respiratory organs of air-breathing fishes. In: *Respiration of Amphibious Vertebrates*, pp. 73-104. (Ed Hughes, G.M.). Academic Press, London, New York.

Munshi, J.S.D. (1980) : *The structure and function of the respiratory organs of air-breathing fishes of India*, Presidential address, *section of zoology, Entomology and Fishes, 67th Session. Indian Science Congress Association, Calcutta*, pp. 32-70.

Munshi, J.S.D. (1985): The structure, function and evolution of the accessory respiratory organs of air- breathing fishes of India. In: *Fortschrittle der Zoologie, Band* 30, pp. 353-366, (Eds. Duncker/Fleischer). *Gustav Fischer Verlag, Stuttgart-New York.*

Munshi, J.S.D. (1996): Microcirculation in gills and accessory respiratory organs and the problem of Homeostasis in air-breathing teleostean fishes. *Proc. Indian National Science Academy.* B 63: No. 5 pp. 303-336.

Munshi, J.S.D. and Singh, B.N. (1968a): On the micro-circulatory system of the gills of certain freshwater teleostean fishes. *J. Zool. Lond.* 154: 365-376.

Munshi, J.S.D. and Singh, B.N. (1968b): On the respiratory organs of *Amphipnous cuchia* (Ham.). *J. Morph,* 124:423-444.

Munshi, J.S.D. and Hughes, G.M. (1986): Scanning electron microscopy of the respiratory organs of juveniles and adult climbing perch, *Anabas testudineus. Japan J. Ichthyol.* 33: 39-45.

Munshi, J.S.D. and Hughes, G.M. (1991): Structure of the respiratory islets of accessory respiratory organs and their relationship with the gills in the climbing fresh, *Anabas testudineus. J. Morph,* 209:241-256.

Munshi, J.S.D. and Hughes, G. M. (1992) : *Air-breathing Fishes of India Oxford and IBH,* New Delhi, pp 338.

Munshi, J.S.D. Olson, KR., Ojha, J. and Ghosh, T.K. (1986): Morphology and vascular anatomy of the accessory respiratory organs of the air-breathing climbing perch *Anabas testudineus* (Bloch.) *Am. J. Anat.*176: 321-331.

Munshi, J.S.D., Weibel, E.R., Gehr, P. and Hughes, G.M. (1986): Structure of the respiratory air-sac of *Heteropneustes fossilis* (Bloch) (Heteropneustidae, Pisces) An electron microscope study. *Proc. Indian Natn. Sci. Acad.* B 52(6):703-713.

Munshi, J.S.D., Hughes, G.M., Gehr, P. and Weibel, E.R. (1989): Structure of the air-breathing organs of a swamp *mud-eel, Monopterus cuchia* (Ham.) *Japan. J.Ichthyol,* 35 (4):453-465.

Munshi, J.S.D. and Ghosh, T.K (1993): Metabolic wheel hypothesis as applied to air-breathing of India. pp 70-78. In *Advances in Fish Biology*, (ed. H.R Singh) *Hindustan Publishing Corp. (India) Delhi*.

Munshi, J.S.D., Olson, KR., Ghosh, T.K and Ojha, J. (1990): Vasculature of the head and respiratory organs in an obligate air-breathing fish the swamp eel, *Monopterus* (= *Amphipnous*) *cuchia*. *J.Morph.* 203:181-201.

Munshi, J.S.D., Roy, P.K. Ghosh, T.K. and Olson, K.R. (1994): Cephalic circulation in the air-breathing snakehead fish, *Channa punctata*, C. *gachua* and C. *marulius* (Ophiocephalidae, Ophiocephaliformes). *The Anatomical Record.* 238: 77-91.

Munshi, J.S.D. (1960): The cranial muscles of some freshwater teleosts. *Ind. J. Zoot* 3, 1-76

Munshi, J.S.D. (1962a) : On the accessory respiratory organs of *Heteropneustes fossilis* (Bloch).*Proc. Roy. Soc. (Biol.) Edinburgh*, 68: 128-146.

Munshi, J.S.D. (1962b) : On the accessory respiratory organs of *Ophiocephalus punctatus* and *O. striatus* (Bloch). *J. Linn. Soc. Zool*, 44, 616-628.

Munshi, J.S.D. (1964): Chloride cells in the gills of freshwater teleosts. *Quart. J. Microscope Sci.* 105: pp 79-89.

Munshi, J.S.D. and Ghosh T.K. (1993) : Metabolic wheel hypothesis as applied to air-breathing fishes of India, pp 70-78. In A*dvances in Fish Biology* (Ed. H.R. Singh) Hindustan Publishing Corp. (India), Delhi.

Munshi, J.S.D., Patra, A.K., Biswas Niva and Ojha J. (1979): Interspecific variations in the circadian rhythm of bimodal oxygen uptake in four species of murrels, Japan J. Ichthyol, 26 : 69-74.

Munshi, J.S.D., Ojha, J. and Sinha, A.L. (1980): Morphometrics of the respiratory organs of an air-breathing cat fish, *Clarias batrachus* (Linn.) in relation to body weight. Proc Indian. Natn. Acad., B 46 (5): 621-635.

Munshi, J.S.D., Patra, A.K. and Hughes, G.M. (1982) : Oxygen consumption from air and water in *Heteropneustes* (= *Saccobranchus fossilis*) (Bloch) in relation to body weight at three different seasons. Proc. Indian Natn. Sci. Acad. B 48 (6): 715-729.

Munshi, J.S.D. Hughes, G.M., Gehr, P. and Weibel, E.R. (1989): Structure of the air-breathing organs of a swamp mud eel, *Monopterus cuchia* (Ham). Japan. J. Ichthyol. 35 (4) : 453-465.

Munshi, J.S.D. and Dube, S.C. (1973): Oxygen uptake capacity of gills in relation to body size of the air-breathing fish, *Anabas testudineous* (Bloch). Acta Physiol. 44: 113-129.

Munshi, J.S.D., Sinha, A.L. and Ojha, J. (1976): Oxygen uptake capacity of gills and skin in relation to body weight of the air-breathing siluroid fish, *Clarias batrachus* (Linn.): Acta Physiol, 48 : 23-33.

Munshi, J.S.D., Pandey, B.N. Pandey, P.K. and Ojha, J. (1978): Oxygen uptake in *Saccobranchus fossilis (Heteropneustes fossilis)*. J. Zool, Lond. 184: 171-180.

Munshi, J.S.D., Patra, A.K. and Hughes, G.M. (1982): Oxygen consumption from air and water in *Heteropneustes (=Saccobranchus) fossilis* (Bloch) in relation to body weight at three different seasons: proc. Indian natn. Sci. Acad. B 48: 715-729.

Moitra A. and Munshi, J.S.D. (1997) : A note on the problem of buoyancy in air-breathing fishes. J. Freshwater Biol. 9(1) : 57-61.

Mc. Clelland J (1845) : Apodal fishes of Bengal Calcutta, J. Nat. Hist. 5 : 192-195.

Mitra, B. and E. Ghosh (1932): On the hypobranchial artery of *Cirrhina mrigala* and *Catla catla* with short notes on their heart and afferent and efferent branchial system. Zool. Anz. 100 : 67-73.

Morr. J.C.: Cardiovascular system (1957) : In Physiology of fish. Ed. M.E. Brown, 81-108.

Munshi, J.S.D. and Singh, B.N. (1968) : On the respiratory organs of *A. cuchia*, J. Morph. 124 : 423-444.

Munshi, J.S.D. (1980) : *The structure and function of the respiratory organs of air-breathing fishes of India*, Presidential address, section of zoology, Entomology and Fishes, 67th Session. Indian Science Congress Association, Calcutta, pp. 32-70.

Munshi, J.S.D. (1990) : The brearing of ecological factors on the evolution of air-breathing fishes pp. 15-18. In *Impacts of Environment on Animals and Aquaculture* (Eds. G.K. Manna and B.B. Jana) Publ. Saraswaty Press Ltd., Calcutta.

Munshi, J.S.D. and Hughes, G.M. (1992): *Air-breathing Fishes of India*. Oxford and IBH Publishing Co. Pvt. Ltd., New Delhi, Bombay, Calcutta.

Mnshi, J.S.D. and Singh, R.K (1971) : Investigation on the effects of insecticides and other chemical substances on the respiratory epithelium of the predatory and weed fishes. Indian J. Zool. XII (2): 127-134.

Munshi, J.S.D. and Singh, A. (1992): Scanning electron microscopic evaluation of effects of low pH on gills of *Channa* (Bloch.), J. Fish Biol. 41 : 83-89.

Munshi, J.S.D., Roy, P.K., Ghosh, T.K. and Olson, K.R. (1994): Cephalic circulation in the air-breathing snakehead fish, *Channa punctata, C. gachua and C. marulius* (Ophiocephalidae, Ophiocephaliformes) Anatomical Record.238: 77-91.

Munshi, J.S.D., Singh, N.K. Mishra, N. and Ojha, J. (1990): Cytology of macrophages in normal and mercury treated air-breathing fish, *Channa punctata* (Bloch.). J. Fish Biol. 37 : 651-653.

Munshi, J.S.D., A.L. Sinha and J. Ojha (1976) : Oxygen uptake capacity of gills and skin in relation to body weight of the air-breathing siluroid fish, *Clarias batrachus* (Linn.) Aeta Physiol. Acad. Sci Hungaricae, Tomus, 48 : 23-33.

Munshi, J.S.D., K.R. Olson, J. Ojha and T.K. Ghosh (1986) : Morphology and vascular anatomy of the accessory respiratory organs of air-breathing climbing perch, *Anabas testudineus* (Bloch). Am. J. Anat. 176 : 321-331.

Munshi, J.S.D., K.R. Olson, T.K. Ghosh and J. Ojha (1990) : Vasculature of the head and respiratory organs in an obligate air-breathing fish, the swamp eel *Monopterus (=Amphipnous cuchia)*, J. Morphol, 203 : 181-201.

Murakami, J. (1971) : Application of the Scanning Electron Microscope to study the fine distribution of the blood vessels *Arch. Histol. Jpn.* 32 : 445-454.

Munshi, J.S.D. and Dube, S.C. (1973) : Oxygen uptake capacity of gills in relation to body size of the air-breathing Fish, *Anabas testudineus* (Bloch) Acta Physiol. 44 : 113-123

Munshi, J.S.D., Sinha, A.L. and Ojha, J. (1976): Oxygen uptake capacity of gills and skin in relation to body weight of the air-breathing siluroid fish, *Clarias batrachus* (Linn.) Acta Physiol Acad Sci. Hungaricae 48:23-33.

Munshi, J.S.D., Pandey, B.N., Pandey, P.K. and Ojha, J.(1978) : Oxygen uptake through gills and skin in relation to body weight of an air-breathing siluroid fish, *Saccobranchus* (= *Heteropneustes*) *fossilis*. J. Zool. (London) 184: 171-180.

Nekvasil, N.P. and Olson, K.R. (1986b): Extraction and metabolism of circulatiug catecholamines by the trout gill. Am. J. Physiol. 250 R-526-R531.

Nawar, G. (1955) : On the anatomy of *Clarias lazera*. III. The vascular system. J. Morphol. 97 : 179-214.

Newstead, D. (1967): Fine structure of respiratory lamellae of teleostean gills. Z. *Zellforsch mikrosk. Anat* 79: 396-428.

Norman, J.R. (1963): *A history of fishes* (Second Edition by P.H. Greenwood). London: *Ernest Benn Limited.*

Nachilas, M.M., K.C. Tsou, R. Desouza, C.H. Cheng and A.M. Seligman (1957) : Cytochemical demostrantion of succinic dehydrogenase by the use of a new p-nitrophenyl substituted ditetrazole. J. Histochem. CYtochem 5: 420-436.

Munshi, J.S.D. (1961) : The accessory respiratory organs of *Clarias batrachus* (Linn.). J. Morph. 109: 115-139.

Munshi, J.S.D. and Dube, S.C. (1973) : Oxygen uptake capacity of gills in relation to body size of the air-breathing Fish, *Anabas testudineus* (Bloch) Acta Physiol. 44 : 113-123

Munshi. J.S.D., Pandey, B.N., Pandey, P.K., Ojha, J. (1977) : Oxygen uptake trough gills and skin in relation to body size of juvenile and mature air-breathing catfish, *Heteropneustes fossilis* (Bloch), J. Zool. London.

Nawar, G. (1954): On the anatomy of *Clarias lazera.* I. Osteology. Jour. Morph. Vol. 94 pp. 551-585.

Nelsons, E.M. (1948): The comparative morphology of the Weberian apparatus of the Catostomidae and its significance in systematics Jour, Morph. Vol. 83, pp. 225-251.

Ojha, J. and Munshi, J.S.D. (1975): Oxygen consumption in relation to body size and respiratory surface area of a freshwater eel, *Macrognathus aculeatum* (Bloch). Indian J. Expt. Biol. 13: 353-357.

Ojha,J. and Munshi, J.S.D. 1974: Morphometric studies on the gill and skin dimensions in relation to body weight in a freshwater mud eel, *Macrognathus aculeatum* (Bloch.). Zool. Anz. 193: 364-381.

Oliva, O (1960) : The respiratory surface of gills is teleosts. The respiratory surface of the gills in the viviparous blenmy (*Zoarces viviparous*); *Acta. Biol. Cracov* 3: 71-81.

Olson, K.R., Kullman, D. Narkates, A.J. and Oparil, S. (1986): Angiotensin extraction by trout tissue in vivo and metabolism by the perfused gill, *Am. J. Physiol*, 250 (Regul. Intgr. Comp. Physiol. 19): R532-R538.

Olson, K.R., Munshi, J.S.D., Ghosh, T.K and Ojha, J. (1986): Gill microcirculation of the air-breathing climbing *perch,Anabas testudineus* (Bloch.) Relationships with the accessory respiratory organs and systemic circulation. *Am. J. Anat.* 176:,305-320.

Olson, K.R., Lipke, D., Munshi, J.S.D., Moitra, A. Ghosh, T.K., Kunwar, G., Ahmad, M., Roy, P.K., Singh, O.N., Nasar, S.S.T., Pandey, A., Oduleye, S.O. and Kullman, D. (1987) : Angiotensin converting enzymes in organs of air-breathing fish. *Gen.Comp. Endocrinol.* 68 : 486-491.

Olson, KR., Munshi, J.S.D., Ghosh, T.K and Ojha, J. (1990): Vascular organization of the head and respiratory organs of the air-breathing catfish, *Heteropneustes fossilis*, *J. Morph.* 203:165-179.

Olson, K.R. (1983) : Effects of perfusion pressure on the morphology of the Cental sinus in the trout gill filament. *Cell Tiss. Res.* 232 : 312-325.

Olson, K.R. (1985) : Preparation of fish tissues for electron microscopy, *J. Electron Microsc. Technique* 2 : 217-228.

Olson, K.R. (1991) : Vasculature of the fish gill. Anatomical correlates of physiological function. *J. Electron Microse. Technique*, 19 : 389-405.

Olson, K.R. (1995) : Scanning electron microscopy of the fish gill. In Horizon of New Research on Fish Morphology in the 20th Century, J.S.D. Munshi and H.M. Dutta, eds. Oxford and IBH Publishing Co. Pvt. Ltd., New Delhi.

Olson, K.R. and B. Kent (1980) : The microvasculature of the elasmobranch gill. *Cell Tiss. Res.* 209 : 49-63.

Olson, K.R., J.S.D. Munshi, T.K. Ghosh and J. Ojha (1990) : Vascular organization of the head and respiratory organs of the air-breathing catfish. *Heteropneustes fossilis*. *J. Morphol.* 203 : 165-179.

Olson, K.R. (1992): Blood and extracellular fluid volume regulation: Role of the Renin-Angiotensin system, Kalikrein-kenin system and atrial natriuretic peptides. In: *Fish Physiology*. Vol. XII Part B. The cardiovascular System, W.S. Hoar, D.J. Randall and A.P., Farell (eds) Academic Press, Sandiego, New York, Boston. London, Sydney, Tokyo, Toronto, pp. 135-254.

Olson, KR., Roy, P.K Ghosh, T.K and Munshi, J.S.D. (1994): Microcirculation of gills and accessory respiratory organs from the air-breathing snakehead fish, *Channa punctata*, *C. gachua* and *C. marulius*. *The Anatomical Record*. 238: 92-107.

Ojha, J and Munshi, J.S.D. (1974) : Morphometric studies on the gill and skin dimensions in relation to weight in a freshwater mud-eel, *Macrophages aculeatum* (Blch). *Zool. Anz.* 193:364-381.

Olson, K.R.(1980) : Application of corrosion casting procedures in identification of perfusion distribution in a complex microvasculature. In: Scanning Electron Microscopy, Vol. 3, O. Jhari (Ed.), SEM Inc. Chicago II, 357-364.

Olson, K.R.(1984) : Distribution of flow and plasma skimming in isolated perfused gills of three teleosts. J. Exp. Biol. 109, 97-108.

Olson, K.R.(1994) : Circulatory Anatomy in Bimodally breathing Fish. Amer. Zool, 34: 280-288.

Olson, K.R., Munshi, J.S.D., Ghosh, T.K. and Ojha, J. (1986) : Gill micro-circulation of the air-breathing climbing perch, *Anabas testudineus* (Bloch) : Relationships with the accessory respiratory organs and systemic circulation. Am. J. Anat. 176: 305-320.

Olson, K.R., Munshi, J.S.D., Ghosh, T.K. and Ojha, J. (1990): Vascular Organisation of the head and respiratory organs of the air-breathing catfish, *Heteropneustes fossilis*. J. Morphol. 203: 165-179.

Olson, K.R., Roy, P.K., Ghosh, T.K. and Munshi, J.S.D. (1994): Micro-circulation of gills and accessory respiratory organs from the air-breathing snakehead fish, *Channa punctata, C. gachua* and *C. marulius*. The Anatomical Record, 238 : 92-107.

Olson, K.R., Ghosh, T.K., Roy, P.K. and Munshi J.S.D (1995): Microcirculation of gills and accessory Respiratory organs of the walking cat fish, *Clarias batrachus*. The *Anatomical Record* 242: 383-399.

Olson, K.R., Kullman, D. Narkates, A.J. and Oparil, S. (1986): Angiotensin extraction by trout tissue in vivo and metabolism by the perfused gill, Am. J. Physiol, 250 (Regul. Intgr. Comp. Physiol. 19): R532-R538.

Olson, K.R., Munshi, J.S.D., Ghosh, T.K and Ojha, J. (1986): Gill microcirculation of the air-breathing climbing *perch,Anabas testudineus* (Bloch.) Relationships with the accessory respiratory organs and systemic circulation. Am. J. Anat. 176:,305-320.

Olson, K.R., Lipke, D., Munshi, J.S.D., Moitra, A. Ghosh, T.K., Kunwar, G., Ahmad, M., Roy, P.K., Singh, O.N., Nasar, S.S.T., Pandey, A., Oduleye, S.O. and Kullman, D. (1987) : Angiotensin converting enzymes in organs of air-breathing fish. Gen.Comp. Endocrinol. 68 : 486-491.

Olson, KR., Munshi, J.S.D., Ghosh, T.K and Ojha, J. (1990): Vascular organization of the head and respiratory organs of the air-breathing catfish, *Heteropneustes fossilis*, J. Morph. 203:165-179.

Ojha, J., O.P. Dandotia, A.K. partra and J.S.D. Munshi (1978): Bimodal oxygen uptake in relation to body weight in a freshwater murrel, *Channa (=Ophiocephalus) gachua*. Z. Tier, Physiol. Tierernabrg. U. Futtermittelkde. 40: 57-66, Figures 1-2.

Ojha, J.and Munshi, J.S.D. (1975): Cytochemical differentiation of muscle fibres by succinic dehydrogenase activity in the respiratory muscle of air-breathing fish, *Channa punctata*. Anat. Anaz.138: 62-68.

Ojha, J., Patra, A.K. and Munshi, J.S.D. (1978): Bimodal oxygen uptake in relation to body weight in a freshwater murrel, *Channa* (=*Ophiocephalus*) *gachua*, Z. Tiemphysiol. 40, 57-66.

Omarkhan, M. (1949a) : The morphology of the chondrocranium of *Gymnarchus niloticus*. Jour. Linnean Soc. London, Zool, vol. 61, pp. 452-481.

Omarkhan, M. (1949b) : the lateral sensory canals of larval *Notopterus*. Proc. Zool. Soc. London vol. 118, pt. 4, pp. 938-970.

Omarkhan, M. (1950) : The development of the chondrocranium of *Notopterus*. Jour. Linnean Soc. London, Zool, vol. 61, pp. 608-624.

Palmer, J.D. (1976): An Introduction to Biological Rhythms. Academic Press, New York, 1-30.

Patra, A.K., Biswas, Niva, Ojha, J. and Munshi J.S.D. (1978) : Circadian rhythm in bimodal oxygen uptake in an obligatory air-breathing swamp eel, *Amphipnous* (=*Monopterus*) *cuchia* (*Ham.*). Indian J. Exp. Biol. 16: 808-809.

Parvatheswararao, V. (1960): Studies on the Oxygen consumption in feshwater fish *Puntius sophore* (Ham.) in relation to size and temperature, *Proc. Natn. Inst. Sci.* India 26 : 64-72.

Palotheimo, J.E. and Dickie, L.M. (1965): Food and growth of fishes. II. Effect of food and temperature on the relation between metabolism and body weight. J. Fish. Res. Bd. Can 23: 864-908.

Prosser, C.L. and Brown F.A. Jr. (1961): Comparative animal Physiology: 164 pp. 2nd edn, (Philadelphia: W.B. Saunders).

Pearse, A.G.E. (1968): Histochemistry : Theoritical and Applied,London J.and A, Churchill

Priede, I.G. (1976): Functional morphology of the bulbus arteriosus of rainbow trout (*Salmo gairdneri* Richardson). J. Fish Biol. 9 : 209-216.

Payan, P., J.P. Girad, and N. Mayer-Gostan (1984) : Branchial ion movements in teleosts. The roles of respiratory and chloride cells. In: Fish Physiology, W.S. Hoar and D.J. Randell, eds. Academic Press, Inc. Orlando, pp. 39-63.

Pettersson, K. and K. Johansen (1982) : Hypoxic vasoconstriction and the effects of adrenaline on gas exchange efficiency I fish gills. J. Exp. Biol. 97 : 263-272.

Patra, A.K., Biswas, Niva, Ojha, J. and Munshi J.S.D. (1978) : Circadian rhythm in bimodal oxygen uptake in an obligatory air-breathing swamp eel, *Amphipnous* (=*Monopterus*) *cuchia* (*Ham.*). Indian J. Exp. Biol. 16: 808-809.

Patra, A.K., Munshi, J.S.D. and Hughes, G.M. (1983): Oxygen consumption of the freshwater air-breathing Indian siluroid fish, *Clarias batrachus* (Linn.) in relation to body size and seasons. Proc. Indian Natn. Sci. Acad. B. 49: 808-809.

Pattle, R.E. (1976) : The lung surfactant in the evolutionary tree. In: *Respiration of Amphibious Vertebrates*, pp. 233-255 (Ed. G.M. Hughes), Academic Press, London-New York-San Francisco.

Pattle, R.E., Hopkinson, D.A.W. (1963) : Lung lining in bird, reptile, and amphibian, *Nat. Lond*. 200, 894.

Pattle, R.E., Gandy, G., Schock, C., Creasey, J.M. (1974) : Lung inclusion bodies: different ultrastructure in simian and non-simian mammals. *Experientia* 30: 797-798.

Pattle, R., Schock, C. Creasey, V., Hughes, G.M. (1977): Surpellic films, lung surfactant and their cellular origin in newt, caecilan and frog. *J. Zool, Lond*. 182 : 125-136.

Price, J.W. (1931) : Growth and gill development in the small mouthed black bass. *Micropterus dolomieu* Lacepede. St. Univ. Stud. 41 : 1-46.

Parsons, C.W. (1929): The conus arteriosus in fishes. Quart. J. Micr. Sci., London 73: 145-176.

Prakash, R. (1953) : The heart of common Indian Catfish : *Heteropneustes fossilis* with a special reference to the conducting system, P.Z.S. Bengal 6 : 113-118.

Pease, D.C. (1962): Microscopic and submicroscopic anatomy of arterial and arteriolar systems. In Blood vessels and lymphatics. Abrahamson, D. I. (Ed.) New York : Academic Press.

Philpot, C.W. and Copeland, D.E. (1963): Fine structure of chloride cells from three species of *Fundulus. J. cell Biol*. 18: 389-404.

Plehn, M. (1901): Zum feineren Bau der Fischkieme, Zool, Anz. 24: 439-443.

Pearse, A.G.E. (1968): Histochemistry : Theoritical and Applied, London J.and A, Churchill.

Parvatheswararao, V. (1960): Studies on the Oxygen consumption in feshwater fish *Puntius sophore* (Ham.) in relation to size and temperature, *Proc. Natn. Inst. Sci.* India 26 : 64-72.

Palotheimo, J.E. and Dickie, L.M. (1965): *Food and growth of fishes. II. Effect of food and temperature on the relation between metabolism and body weight. J. Fish. Res. Bd. Can* 23: 864-908.

Prosser, C.L. and Brown F.A. Jr. (1961): Comparative animal physiology: 164 pp. 2nd edn, (Philadelphia: W.B. Saunders).

Quast, J.C. (1965) : Osteological characteristics and affinities of the hexagrammid fishes, with a synopsis. Proc. California Acad. Sci., ser. 4, vol. 31, no. 21, pp, 563-600.

Rai, D.N., and Datta-Munshi, J. (1979): The influence of thick floating vegetation (water hyacinth, *Eichhornia crassipes*) on the physico-chemical environment of a freshwater wetland. Hydrobiologia (Hague) 62: 65-69.

Rai, D.N. and Munshi, J.S.D. (1979): Observations on diurnal of some. physico-chemical factors of three tropical swamps of Darbhanga (North Bihar). Com. Physiol. Ecol. 4:52-55.

Roy, P.K. and Munshi, J.S.D. (1984): Oxygen uptake in relation to body weight and respiratory surface area in *Cirrhinus mrigala* (Ham.) at two different seasonal temperatures. *Proc. Indian Natn. Sci. Acad*., B 50 (4): 387-394.

Rauther, M. (1910): Die akzessorischenAtmungsorgane der Knochenfische- Ergeb; Fortchr. Zool 2: 517- 585.

Reynolds, E.S. (1963): The use of lead citrate of high pH as an electron opaque stain in electron microscopy. J. cell Biol. 17 : 208-212.

Rajbanshi, V.K. (1977) : The architecture of the gill surface of the catfish. *Heteropneustes fossilis* (Bloch) : SEM study. J. Fish Biol. 10 : 325-329.

Ristori, M.T. and P. Laurent (1977): Action de Phypoxie sur le systeme vasculars branchial de is lete perfuse de truite. C.R. Seancess Soc. Biol. 171 : 809-813.

Rahn, H. Rahn, K.B. Howell, B.J. Gans, C. and Tenney, S.M. (1971) : Air- breathing of the garfish

Rahn, H, Rahn, KB., Howell, B.J., Gans, C. and Tenney, S.M. (1971): Air-breathing of the garfish, *Lepisosteus osseus* Resp. Physiol. 11:285-307.

Romer, A.S. (1946): The Early Evolution of Fishes, *Quart, Rev. Biology* pp. 21-33.

Romer, A.S. (1971): The vertebrate story (Rev. Ed), University of Chicago Press. An elementary account of vertebrate evolution.

Roy, P.K. and Munshi, J.S.D. (1984): Oxygen uptake in relation to body weight and respiratory surface area in *Cirrihinus mrigala* (Ham.) at two different seasonal temperatures. Proc. Indian. Natn. Sci. Acad. B 50(4) : 387-394.

Roy, P.K. and Munshi, J.S.D. (1995): Morphometrics of the respiratory system of air-breathing fishes of India. In: Fish Morphology: Horizon of new Research, J.S.D. Munshi and H.M. Dutta (eds). Oxford and IBH Publishing Co. Pvt. Ltd., New Delhi, pp. 203-234.

Rahn, H, and Howell, B.J. (1976): Bimodal gas exchange. In *Respiration* of *Amphibious Vertebrates*, pp. 271-285. (Ed. G.M. Hughes). Academic Press, London-New York.

Roy, P.K. and Munshi, J.S.D. (1986): Morphometrics of the respiratory organs of a freshwater major carp. *Cirhinus mrigala* in relation to body weight. Japan. J. Ichthyol. 33 (3): 269-279.

Randall, D.J. (1968): Functional morphology of the heart in fishes. Am. Zool. (Amer.) 8 : 179-189.

Randall, D.J. (1970): The circular system. In Fish physiology. Vol. IV (W.S. Hoar and D.J. randall, ed). Academic Press, New York, 133-172.

Singh, B.N. (1976): Balance between aquatic and aerial respiration. In: Respiration of Amphibious Vertebrates pp. 125-164. (Ed. G.M. Hughes) Academic Press, London, New York.

Ramaswami, L.S. (1948): The homalopterid skull. Proc. Zool. Soc. London, vol. 118 pt. 2, pp. 515-538.

Ramaswami, L.S. (1952a): Skeleton of cyprinoid fishes in relation to phylogenetic studies. 1. The systematic position of the genus *Gyrinocheilus vaillant*. Proc. Natl. Inst. Sci. India. Vol. 18, no. 2, pp. 125-140.

Ramaswami, L.S. (1952b):Skeleton of cyprinoid fishes in relation to phylogenetic studies. 2. The systematic position of *Psilorhynchus McClelland*. Ibid., vol. 18, no. 2, pp. 141-150.

Ramaswami, L.S. (1952c): Skeleton of cyprinoid fishes in relation to phylogenetic studies. 4. The skull and other skeletal structures of gastromyzonid fishes. Ibid, vol. 18, no. 6, pp. 519-538.

Ramaswami, L.S. (1953): Skeleton of cyprinoid fishes in relation to phylogenetic studies. 5. The skull and gasbladder capsule of the Cobitidae. Ibid. vol. 19, no. 3, pp. 323-347.

Ramaswami, L.S. (1955a): Skeleton of cyprinoid fishes in relation to phylogenetic studies. 6. The skull and Weberian apparatus in the subfamily Gobioninae, vol. 36, pp. 127-158.

Ramaswami, L.S. (1955b):Skeleton of cyprinoid fishes in relation to phylogenetic studies. 7. The skull and Weberian apparatus of Cyprininae (Cyprinidae). Ibid, vol. 36. pp. 199-242.

Ramaswami, L.S. (1957): Skeleton of cyprinoid fishes in relation to phylogenetic studies. 8. The skull and Weberian ossicles of Catostomidae, Proc., Zool, vol. pp. 293-303

Ridewood, W.G. (1904a): On the cranial osteology of the fishes of the families Elopidae and Albulidae, with remarks on the morphology of the skull in the lower teleostean fishes generally, Proc. Zool. Soc. London, vol. 2 pp. 35-81.

Ridewood, W.G. (1904b) : On the cranial osteology of the fishes of the families Mormyridae, Notopteridae, and Hyodontidae. Jour. Linnean Soc. London, Zool, vol. 29, pp. 188-217.

Ridewood, W.G. (1905a) : On the cranial osteology of the fishes of the families Osteoglossidae, Pantodontidae and Pharactolaemidae. Ibid, vol. 29, pp 252-282.

Ridewood, W.G. (1905b) : On the cranial osteology of the fishes of the clupeoid fishes. Proc. Zool. Soc. London, vol. 2, pp. 448-493.

Ridewood, W.G. (1905c) : On the skull of *Gonorhynchus greyi*. Ann. Mag. Nat. Hist. ser. 7, vol. 15, no. 88, art. 45, pp. 361-372.

Singh, B.N. and Hughes, G.M. (1971): Respiration of an air-breathing catfish. *Clarias batrachus* (Linn). J. Exp. Biol. 55: 421-434.

Singh, B.N. and Hughes, G.M. (1973): Cardiac and respiratory responses in the climbing perch, *Anabas testudineus*, J. Comp. Physiol, 84: 205-226.

Satchell, G.H. (1971) : *Circulation in Fishes*. Cambridge University Press Cambridge, 131 pp.

Satchell, G.H. 1976: The circulatory system of air-breathing fish. In: *Respiration of Amphibious Vertebrates* (Ed.) Hughes, G.M.Academic Press, London, pp. 105-123.

Singh, B.R., A.P., Mishra, M. Sheel and I. Singh (1982) : Development of the air-breathing organ in the catfish, *Clarias batrachus* (Linn) Zool. Anx. Jena 208: 100-111.

Smith, D.G. (1977) : Sites of cholinergic vasoconstriction in trout gills. Am. J. Physiol. 233 : R222-R229.

Schmidt-Nielsen, K. (1984): *Scalling why is animal size so important ?* Cambridge University Press, Cambridge.

Singh, B.N. (1976): Balance between aquatic and aerial respiration. In: *Respiration of Amphiibious Vertebrates*.pp. 125-164 (Ed G.M. Hughes) Academic Press, London – New York

Singh, B.R. and Thakur, R.N. (1979): Oxygen uptake capacity of the amphibious fish, *Amphipnous cuchia* (Ham.) (Symbranchiformes, Amphipnoidae). Acta Physiol. Acad. Sci. Hung. 54 : 13-21.

Singh, B.R., Prasad, M.S., Mishra, A.P. and Manju Sheel (1997) : Oxygen uptake in the early life of air-breathing teleosts. In: Advances in Fish Research. Vol. 2 B.R. Singh (ed). Narendra Publishing House, Delhi, pp. 163-186.

Scarpelli, E. (1995): R.E. Pattle and the discovery of lung surfactant. *Am. J. Perimatol,* 12 : 377-378.

Shelbourne, J.E. (1957a) : Site of Chloride regulation in marine fish larvae. *Nature, London* pp 920-922.

Shelbourne, J.E. (1957b) : The effect of water conservation on the structure of marine fish embryos and larvae. *J. Mar. Biol. Ass. U.K. Vol.35*, pp. 275-286.

Smith, H.W. (1930): The absorption and excretion of water and salts by marine teleosts. *Amer J. Physiol*. 93: 480.

Smith, H.W. (1931): The absorption and excretion of water and salts by the Elasmobranch fishes. *Amer. J. Physiol*. 98: 279-310.

Smith, H.W. (1945): The freshwater fishes of Siam or Thailand. *Bull. U.S. Nat. Mus*, No. 188 pp, 1-107.

Shelbourne, J.E. (1957a) : Site of chloride regulation in marine fish larvae. *Nature, London*, pp 920-922.

Shelbourne, J.E. (1957b): The effect of water conservation on the structure of marine fish embryos and larvae, *J. Mas. Biol. Ass. UK* Vol 35, pp 275-286.

Sinha, N.D.P. and Dejours, P. (1980): Ventilation and blood acid base balance of the crayfish as function of water oxygenation (40 – 1500 Torr.), *Comp. Biochem, Physiol*, Vol. 65 : 427-432.

Sinha, N.D.P. and Dejours, P. (1981): Intravascular components of an air-breathing teleost *Channa punctatus* (Bloch) – Their dimensions and responses to osmotic stress. India. J. Exp. Biol. 19 : 763-765.

Satchell, G.H. (1970) : A functional appraisal of the fish heart. Fed. Proc. 29 1120-1123.

Satchell, G.H. (1971) : *Circulation in Fishes*. Cambridge University Press Cambridge, 131 pp.

Saxena, D.B. and F.L. Bakshi (1965) :Cardiovascular system of some fishes of the torrential streams in India. Part I. Heart of *Orienus plagiostomus* and *Botia birdi*. Jap. J. Ichthyol. 12 : 70-81.

Seyama, I. and H. Irisawa (1967) : The effect of sodium ion concentration on the action potential of the skate heart. J. gen. Physiol. 50 (1967) 505-517.

Senior, H.D. (1907) : The conus arteriosus in *Tarpan allenticus*, Biol. Bull. 12 : 146-151.

Senior, H.D. (1907) : Note on the conus arteriosus of *Megalops cyprinoids*, Biol. Bull 12 : 378-379.

Singh, G.P. (1960) : The structure of the heart of some freshwater teleosts. Indian J. Zootomy 1: 1-26.

Skramlik, E. Von (1935) : Uber den Kreislauf bei den Fischen. Ergehn Biol. 2 : 1-130.

Singh, S.P. (1973) : Pharmacological study of the heart of catfish. Abstr 60th Ind. Congr. (Sec. Zoo and Ent.) 1973.

Singh, B.N. and Hughes, G.M. (1973): Cardiac and respiratory responses in the climbing perch, *Anabas testudineus*, J. Comp. Physiol, 84: 205-226.

Saxena D.B. (1962): Studies on the physiology of respiration in fishes V. Comparative study of the gill area in the freshwater fishes *Labeo rohita, Ophicephalus* (=*Channa*) *Striatus* and *Anabas testudineus*; Ichthyologica 1: 59-70.

Singh R (1979): Interspecific variation in the functional capacity of the respiratory organs of certain siluroid fishes of the genus Mystus. Ph.D. thesis, Bhagalpur University, Bhagalpur, India.

Sharma S.N., Guha G and Singh B.R. (1982) : Gill dimensions of a hill stream fish. *Botia lohchata* (Pisces: C obitidae). Proc. Indian Natn. Sci. Acad B48: 81-91.

Singh, B.N. and Hughes, G.M. (1973): Cardiac and respiratory responses in the climbing perch, *Anabas testudineus*, J. Comp. Physiol, 84: 205-226.

Singh, B.N. (1976): Balance between aquatic and aerial respiration, In: *Respiration of Amphibious Vertrbrates* pp.125-164. (Ed. G.M. Hughes.) Academic Press, London-New York.

Smith H.W. (1935): The metabolism of the lung fish. II. Effect of feeding meat on the metabolic rate; J. Cell Comp. Physiol. 6 : 335-349.

Shelden, E.E. (1937): Osteology, mycology and probable evolution of the nematognath pelvic girdle. Ann. New York Acad, Sci vol. 37, no. 1 pp. 1-96.

Srivaachar, H.R. (1957): Development of the skull in catfishes, II. Development of chondrocranium in *Mystus* and *Rita* (Bagaridae). Morph. Jahrb. Vol. 98. no. 2, pp. 224-262.

Srivaachar, H.R. (1958): Development of the skull in catfishes, 5. Development of skull in *Heteropneustes fossilis* Bloch. Proc. Natl. Inst. Sci. India, ser. B. vol. 24, pp. 165-190.

Srivaachar, H.R. (1959): Development of the skull in catfishes, IV. The Development of chondrocranium in *Arius jella day* (Ariidae) and *Plotosus canius* (Plotosidae) with account of their inter-relationship. Morph. Jahrb, vol 99, no. 4, pp. 986-1016.

Starks, E.C. (1905): The osteology of *Caularchus maeandricus* (Girad). Biol. Bull. Vol. 9 no. 5, pp. 292-303.

Starks, E.C. (1923): The osteology and relationship of the uranoscopoid fishes, with notes on other fishes with jugular ventrals. Stanford Univ. Publ., Univ. Ser. Biol. Sci. vol. 3 No. 3, pp. 259-290.

Starks, E.C. (1930): The primary shoulder girdle of the bony fishes. Ibid. Univ. Serv. Biolo. Sci. vol. 6, pp. 147-239.

Swinnerton, H.H. (1903): The osteology of *Cromeria nilotica* and *Galaxias attenuatus*. Zool. Jahrb, Jena vol. 18 no. 1, pp. 58-70.

Tovell, P.W., Morgan, M and Hughes, G.M. (1970): *Ultrastructure of trout gills*. Congr. Int. Microsc. Electron No. 17, 3 : 601.

Tchernavin, V.V. (1947) : Six specimens of *Lyomeri* in the British Museum (with notes on the skeleton of Lyomeri). Jour. Linnean Soc. London, Zool. Vol. 41, pp 287-350.

Tilar, R. (1963a) : The osteocranium and the Weberian apparatus of the fishes of the family Sisoridae (Siluriodea); a study in adaptation and taxonomy. Zeitschr. Wiss. Zool. Leipzig. Vol. 168. nos. 3-4, pp. 281-320.

Tilar, R. (1963b) : The osteocranium and the Weberian apparatus of a few representatives of the families Siluridae and Plotosidae (Siluroidea); a study of inter-relationship. Zool. Anz., Vol. 171, nos. 11-12, pp 424-439.

Tilar, R. (1964) : The osteocranium and the Weberian apparatus of the fishes of the family Schilbeidae (Pisces: Siluriodea); proc. Zool. Soc. London Vol. 143. pt. 1,pp 1-36.

Tilar, R. (1965) : The comparative morphology of the osteocranium and the Weberian apparatus of tachysuridae. (Pisces: Siluroidei), Jour. Zool. Vol. 146, 146, pt. 2, pp 150-174.

Ursin E. (1976): A Mathematical model of some aspects of fish growth, respiration and mortally; J. Fish, Res. Rd. Can 24 : 2355-2453.

Vogel, W.O. (1985) : Systemic vascular anastomoses, primary and secondary vessels in fish, and the phylogeny of lymphatics, A.B. Symp. 27 : 143-159.

Vergara, G.A., Hughes, G.M. (1980): Phospholipids in washings from the lungs of the frog, *Rana pipiens*. J. Comp. Physiol, B 139: 117-120.

Vickers, T. (1961): A study of the so called chloride secretary cells of the gills of Teleosts, *Quart, J. Micr. Sci.* 102: 507-518.

Vivekanandan, E. (1977): Ontogenetic development of surfacing behaviour in the obligatory air-breathing fish *Channa* (*=Ophiocephalus*) *striatus*; Physiology Behaviour 18 559-562.

Whitear, Mary (1965): Presumed sensory cells in fish epidermis, *Nature, Land* 208: 703-704.

Whitear, Mary (1971): Cell specialization and sensory function in fish epidermis. J. *Zool., Lond.* 163: 237-264.

Watson, D.M.S. (1925): The structure of that group with other bony fish *Proc. Zool.Soc.* London pp 815-870.

Watson, D.M.S. (1934): The interpretation of the Arthrodires. *Proc. Zool. Soc. London* pp 453-460.

Welch, P.S. (1948) : *Limnology*. Methods McGraw-Hill Book Co. New York, 538 pp.

Winberg, G.G. (1956) : Rate of metabolism and food requirement of fishes, Res. Board, Canada, Trans. Ser. No.194.

Wu, H.W. and C.K. Liu (1943): On the blood vascular system of *Monopterus javanensis*, an air-breathing fish. Sinensia 14 61-87.

Welch, P.S. (1952) : *Limnology*. McGraw-Hill Book Co. New York, 538 pp.

Willmer, E.N. (1970): *Cytology and Evolution, 2ⁿᵈ Ed. Academic Press, London*

Willen, V. and Boelaert, R. (1937): Les manoeuvres respiratory de *Periophthalmus*, Bull. Acad. *Roy, Sci. Belg.* (Ser. 5).23: 942-959.

Winberg, G.G. (1956) : Rate of metabolism and food requirement of fishes, *Res. Board, Canada, Trans. Ser. No.194.*

Welch, P.S. (1952) : *Limnology*. McGraw-Hill Book Co. New York, 538 pp.

Willmer, E.N. (1970): Cytology and Evolution, 2ⁿᵈ Ed. Academic Press, London.

Willen, V. and Boelaert, R. (1937): Les manoeuvres respiratory de *Periophthalmus*, Bull. Acad. Roy, Sci. Belg. (Ser. 5).23: 942-959.

Winberg, G.G. (1956) : Rate of metabolism and food requirement of fishes, Res. Board, Canada, Trans. Ser. No.194.

Weitzman, S.H. (1954): The osteology and the relationship of the South American characid fishes of the subfamily Gasteropelecinae. Stanford Ichthyol. Bull. Vol. 4 no. 4, pp 213-263.

Weitzman, S.H. (1962): The osteology of *Brycon meehi*, a generalized characid fish, with an osteological definition of the family. Bid, vol. 8, no. 1, pp. 1-77.

Weitzman, S.H. (1964): Osteology and relationship of South American characid fishes of sub families Lebiasininae and Erythrininae, with special reference to subtribe Nannostomina. Proc. U.S. Natl. Mus., Vol. 116, no. 3499, pp. 127-169.

Young, J.Z. (1981): *The life of vertebrates* third edition, Clarendon Press, Oxford, pp I-XV, 1-645.

Zoraff, N. (1888): On the construction and purpose of the so-called labyrinthine apparatus of the labyrinthic fishes, *Q. Jl. Microsc. Sci.* 28: 501-512.

Zeuthen, E. (1953) : Oxygen uptake in relation to body size in organization : Q. Rev. Biol. 28: 1-12.

Author Index

Subject Index